APPLICATIONS of BLOCKCHAIN and BIG IoT SYSTEMS

Digital Solutions for Diverse Industries

APPLICATIONS of
BLOCKCHAIN and BIG IoT SYSTEMS

Digital Solutions for Diverse Industries

Edited by
Arun Solanki, PhD
Vishal Jain, PhD
Loveleen Gaur, PhD

AAP | APPLE
ACADEMIC
PRESS

First edition published 2023

Apple Academic Press Inc.
1265 Goldenrod Circle, NE,
Palm Bay, FL 32905 USA

4164 Lakeshore Road, Burlington,
ON, L7L 1A4 Canada

CRC Press
6000 Broken Sound Parkway NW,
Suite 300, Boca Raton, FL 33487-2742 USA

4 Park Square, Milton Park,
Abingdon, Oxon, OX14 4RN UK

Apple Academic Press exclusively co-publishes with CRC Press, an imprint of Taylor & Francis Group, LLC

Library and Archives Canada Cataloguing in Publication

Title: Applications of blockchain and big IoT systems : digital solutions for diverse industries / edited by Arun Solanki, PhD, Vishal Jain, PhD, Loveleen Gaur, PhD.
Names: Solanki, Arun, 1985- editor. | Jain, Vishal, 1983- editor. | Gaur, Loveleen, editor.
Description: First edition. | Includes bibliographical references and index.
Identifiers: Canadiana (print) 20220171262 | Canadiana (ebook) 20220171297 | ISBN 9781774637456 (hardcover) | ISBN 9781774637463 (softcover) | ISBN 9781003231332 (ebook)
Subjects: LCSH: Blockchains (Databases)—Industrial applications. | LCSH: Internet of things—Industrial applications. | LCSH: Big data—Industrial applications.
Classification: LCC QA76.9.B56 A67 2023 | DDC 005.74—dc23

Library of Congress Cataloging-in-Publication Data

....

CIP data on file with US Library of Congress

....

ISBN: 978-1-77463-745-6 (hbk)
ISBN: 978-1-77463-746-3 (pbk)
ISBN: 978-1-00323-133-2 (ebk)

About the Editors

Arun Solanki, PhD
Assistant Professor, Department of Computer Science and Engineering, Gautam Buddha University, Greater Noida, India

Arun Solanki, PhD, is an Assistant Professor in the Department of Computer Science and Engineering, Gautam Buddha University, Greater Noida, India, where he has also held various additional roles over the years. He is the Co-Convener of the Center of Excellence in Artificial Intelligence. He has supervised more than 60 MTech dissertations. His research interests span expert systems, machine learning (ML), and search engines. He has published many research articles in SCI/Scopus-indexed international journals and has participated in many international conferences. He has been a technical and advisory committee member of many conferences and has chaired and organized many sessions at an international conferences, workshops, and seminars. Dr. Solanki is working as an Associate Editor for the International Journal of Web-Based Learning and Teaching Technologies (IJWLTT). He has also been working as a guest editor for special issues in *Recent Patents on Computer Science*. Dr. Solanki is the editor of many books and is working as the reviewer for journals published by Springer, IGI Global, Elsevier, and others. He received an MTech degree in computer engineering from YMCA University, Faridabad, Haryana, India. He has received his PhD in Computer Science and Engineering from Gautam Buddha University, Greater Noida, India.

Vishal Jain, PhD
Associate Professor, Department of Computer Science and Engineering, School of Engineering and Technology, Sharda University, Greater Noida, U.P. India

Vishal Jain, PhD, is presently working as an Associate Professor at the Department of Computer Science and Engineering, School of Engineering and Technology, Sharda University, Greater Noida, U.P. India. Before that, he has worked for several years as an Associate Professor at Bharati Vidyapeeth's Institute of Computer Applications and Management (BVICAM), New Delhi. He has more than 14 years of experience in academics. He obtained a PhD

(CSE), MTech (CSE), MBA (HR), MCA, MCP, and CCNA. He has more than 370 research citation indices with Google Scholar (h-index score 9 and i-10 index 9). He has authored more than 70 research papers in reputed conferences and journals, including Web of Science and Scopus. He has authored and edited more than 10 books with various reputed publishers, including Springer, Apple Academic Press, CRC, Taylor and Francis Group, Scrivener, Wiley, Emerald, and IGI-Global. His research areas include information retrieval, semantic web, ontology engineering, data mining, ad hoc networks, and sensor networks. He received a Young Active Member Award for the year 2012–2013 from the Computer Society of India, Best Faculty Award for the year 2017, and Best Researcher Award for the year 2019 from BVICAM, New Delhi.

Loveleen Gaur, PhD
Professor and Program Director (Artificial Intelligence and Business Intelligence and Data Analytics), Amity International Business School, Amity University, Noida, India

Loveleen Gaur, PhD, is a Professor cum Program Director (Artificial Intelligence and Business Intelligence and Data Analytics) at Amity International Business School, Amity University, Noida, India. She is also a senior IEEE member and series editor. An established author and researcher, she has filed three patents in the area of IoT. For over 18 years she served in India and abroad in different capacities. Prof. Gaur has significantly contributed to enhancing scientific understanding by participating in over 300 scientific conferences, symposia, and seminars; by chairing technical sessions; and by delivering plenary and invited talks. She has specialized in the fields of information sciences IoT, data analytics, e-commerce and e-business, data mining, and business intelligence. Prof. Gaur has authored and co-authored around 10 books with Elsevier, Springer, and Taylor and Francis. She is invited as a guest editor for Springer NASA journals and Emerald Q1 journals. She has chaired various committees for international conferences and is a reviewer with IEEE, SCI, and ABDC journals. She is actively involved in various projects of the Government of India and abroad. She has been honored with prestigious national and international awards, such as the Senior Women Educator & Scholar Award by the National Foundation for Entrepreneurship Development on Women's Day, the Sri Ram Award from the Delhi Management Association (DMA), and a Distinguished Research Award by Allied Academies, which was presented in Jacksonville, Florida, and an Outstanding Research Contributor award by Amity University.

Contents

Contributors

Purnima Ahirao
Faculty of Department of Information Technology, K.J. Somaiya College of Engineering, Mumbai, Maharashtra, India

Nada M. Alhakkak
Computer Science Department, Baghdad College for Economic Science University, Baghdad, Iraq, E-mails: nadahakkak@hotmail.com; dr.nada@baghdadcollege.edu.iq

Okuogume Anthony
Lapland University of Applied Sciences, Tornio, Finland, E-mail: Anthony.Okuogume@lapinamk.fi

Abhik Banerjee
Department of Computer Science and Engineering, Netaji Subhash Engineering College, Kolkata, West Bengal, India, E-mail: abhik.banerjee.1999@gmail.com

Arpit Bhardwaj
Associate Professor, Department of Computer Science and Engineering, BML Munjal University, Haryana, India, Email Id: arpit.bhardwaj@bmu.edu.in

Harshit Bhardwaj
Department of Computer Science and Engineering, University School of Information and Communication Technology, Gautam Buddha University, Greater Noida, Uttar Pradesh, India, E-mail: hb151191@gmail.com

Dinesh Bhatia
Department of Biomedical Engineering, North Eastern Hill University, Shillong – 793022, Meghalaya, India, E-mail: bhatiadinesh@rediffmail.com

Kunal Bohra
UG Student of Department of Information Technology, K.J. Somaiya College of Engineering, Mumbai, Maharashtra, India

Yogita Borse
Faculty of Department of Information Technology, K.J. Somaiya College of Engineering, Mumbai, Maharashtra, India, E-mail: yogitaborse@somaiya.edu

Rajdeep Chakraborty
Department of Computer Science and Engineering, Netaji Subhash Engineering College, Kolkata, West Bengal, India, E-mail: rajdeep_chak@rediffmail.com

Dhritiman Chanda
Assistant Professor, Faculty of Commerce and Management, Vishwakarma University, Pune, India, Email: operationsdchanda@gmail.com

Naveen Dahiya
Department of Computer Science and Engineering, MSIT, New Delhi, India

Shantashree Das
Software Research Analyst, SelectHub, Denver, United States, Email: dshantashree26@gmail.com

Nidhi Dedhia
UG Student of Department of Information Technology, K.J. Somaiya College of Engineering, Mumbai, Maharashtra, India

Divya
Department of Computer Science and Applications, CDLU, Sirsa, Haryana, India; Department of Computer Science and Engineering, MSIT, New Delhi, India, E-mail: divyajatain@msit.in

D. Ghose
Associate Professor, Department of Business Administration, Assam University, Silchar, India, E-mail: operationsdghosh@gmail.com

Sanjukta Ghosh
Srishti Manipal Institute of Art Design and Technology, Bangalore, India

Sounak Ghosh
Department of Computer Science and Engineering, Netaji Subhash Engineering College, Kolkata, West Bengal, India, E-mail: sounakghosh.official@gmail.com

Anuj Gupta
Department of Information Technology, Netaji Subhas University of Technology (Formerly known as Netaji Subhas Institute of Technology), New Delhi, India, E-mail: ganuj32@gmail.com

Meenu Gupta
Department of Computer Science and Engineering, Chandigarh University, Punjab, India, E-mail: gupta.meenu5@gmail.com

Tejas Gupta
Department of Information Technology, Netaji Subhas University of Technology (Formerly known as Netaji Subhas Institute of Technology), New Delhi, India

Yash Jain
UG Student of Department of Information Technology, K.J. Somaiya College of Engineering, Mumbai, Maharashtra, India

Rohit Kasale
UG Student of Department of Information Technology, K.J. Somaiya College of Engineering, Mumbai, Maharashtra, India

D. Kesavaraja
Associate Professor, Department of Computer Science and Engineering, Dr. Sivanthi Aditanar College of Engineering, Tiruchendur, Tamil Nadu, India, Tel.: 9865213214, E-mail: dkesavraj@gmail.com

Nimrita Koul
School of Computing and Information Technology, REVA University, Bangalore, Karnataka – 560064, India, E-mails: nimritakoul@reva.edu.in; emailnk1@gmail.com

G. H. Arun Kumar
Department of Computer Science and Engineering, Bapuji Institute of Engineering and Technology, Davangere, Karnataka, India (Affiliated to Visvesvaraya Technological University, Belagavi, Karnataka, India)

Ravinder Kumar
Department of Mechanical Engineering, Amity University, Noida, Uttar Pradesh, India, E-mail: rkumar19@amity.edu

Sumit Kumar
Gopal Narayan Singh University, Bihar, India, E-mail: sumit170787@gmail.com

Meet Kumari
Department of Electronics and Communication Engineering, Chandigarh University, Punjab, India,
E-mail: meetkumari08@yahoo.in

Unmesh Madke
UG Student of Department of Information Technology, K.J. Somaiya College of Engineering, Mumbai,
Maharashtra, India

Sunilkumar S. Manvi
School of Computing and Information Technology, REVA University, Bengaluru, Karnataka – 560064,
India, E-mail: ssmanvi@reva.edu.in

Nilanjan Mazumdar
Assistant Professor, Department of Business Administration, University of Science and Technology
Management, Guwahati, India, Email: nilanjanmazumdar@ustm.ac.in

Animesh Mishra
Department of Cardiology, North Eastern Indira Gandhi Regional Institute of Health and Medical
Sciences, Shillong, Meghalaya, India, E-mail: animesh.shillong@gmail.com

S. Porkodi
Scholar, Department of Computer Science and Engineering, Dr. Sivanthi Aditanar College of
Engineering, Tiruchendur, Tamil Nadu, India, Tel.: 7339464560,
E-mail: Ishwaryaporkodi6296@gmail.com

N. Pradeep
Department of Computer Science and Engineering, Bapuji Institute of Engineering and Technology,
Davangere, Karnataka, India (Affiliated to Visvesvaraya Technological University, Belagavi, Karnataka,
India)

Anoop Kumar Prasad
Royal School of Engineering and Technology, Assam Science and Technology University, Guwahati,
Assam, India, E-mail: anoopkprasad@rgi.edu.in

Sujata Pudale
Srishti Manipal Institute of Art Design and Technology, Bangalore, India

G. M. Roopa
Department of Computer Science and Engineering, Bapuji Institute of Engineering and Technology,
Davangere, Karnataka, India (Affiliated to Visvesvaraya Technological University, Belagavi, Karnataka,
India), E-mail: roopa.rgm@gmail.com

Aditi Sakalle
Department of Computer Science and Engineering, University School of Information and
Communication Technology, Gautam Buddha University, Greater Noida, Uttar Pradesh, India,
E-mail: aditi.sakalle@gmail.com

Deepak Kumar Sharma
Department of Information Technology, Netaji Subhas University of Technology (Formerly known as
Netaji Subhas Institute of Technology), New Delhi, India, E-mail: dk.sharma1982@yahoo.com

Uttam Sharma
Department of Computer Science and Engineering, University School of Information and
Communication Technology, Gautam Buddha University, Greater Noida, Uttar Pradesh, India,
E-mail: uttamsharma.usc@gmail.com

Priyanka Singh
Department of Civil Engineering, Amity School of Engineering and Technology, Amity University
Uttar Pradesh, Noida, Uttar Pradesh, India, E-mail: priyanka24978@gmail.com

Sunil Kumar Singh
Mahatma Gandhi Central University, Bihar, India,
E-mails: sksingh@mgcub.ac.in; sunilsingh.jnu@gmail.com

Tarana Singh
Gautam Buddha University, Greater Noida, Uttar Pradesh, India

Vikram Singh
Department of Computer Science and Applications, CDLU, Sirsa, Haryana, India

Arun Solanki
Gautam Buddha University, Greater Noida, Uttar Pradesh, India

Shrikant Tangade
School of Electronics and Communication Engineering, REVA University, Bengaluru, Karnataka, India

Pradeep Tomar
Department of Computer Science and Engineering, University School of Information and
Communication Technology, Gautam Buddha University, Greater Noida, Uttar Pradesh, India,
E-mail: parry.tomar@gmail.com

Shivani A. Trivedi
S.K. Patel Institute of Management and Computer Science-MCA, Kadi Sarva Vishwavidyalaya, Gujrat,
India, E-mail: satrivedi@gmail.com

Chetanya Ved
Department of Information Technology, Bharati Vidyapeeth's College of Engineering, Maharashtra,
India, E-mail: chetanyaved@gmail.com

Abbreviations

ADG	direct acyclic graph
AE	autoencoder
AI	artificial intelligence
AoI	age of information
APO	area post office
ARL	army research laboratory
ATA	Agent of Trusted Authority
BAT	basic attention token
BC	blockchain
BFT	Byzantine fault tolerance
BIM	building information modeling
BIoT	blockchain-based IoT
BTC	bitcoin
CAESAIR	collaborative analysis engine for situational awareness and incident response
CCC	contract-compliance-checker
CCMSNNB	cyberthreat classification and management system using neural network and blockchain
CE	circular economy
CRM	customer relationship management
CS	cloud server
CTI	cyber threat intelligence
CV	computer vision
CVE	common vulnerability and exposure
DApp	decentralized application
DBT	direct bank transfer
DDOS	distributed denial of service
DEC	design, engineering, and construction
DEMB	decentralize electronic-medical blockchain-based system
DL	deep learning
DLT	distributed ledger technology
DNS	domain name system
DoS	denial of services
DPoS	delegated proof-of-stack

DSRC	dedicated short-range communication
DTC	distributed time-based consensus algorithm
DTLS	datagram transport-level security
EAB	education consultative board
ECDSA	digital signature elliptical curve algorithm
EHRs	electronic healthcare records
ELIB	efficient, lightweight, integrated blockchain
ERP	enterprise resource planning
ETH	Ethereum
ETSI	European Telecommunications Standards Institute
FedBlock	federated blockchain
GDP	gross domestic product
GDPR	general data protection regulation
GUID	global unique identifier
HER	electronic health record
ICO	initial coin offering
ICS	indicator centric schema
ICTs	information and communication technologies
ID-MAP	identity-based message authentication using proxy vehicles
IERC	European research cluster on the Internet of Things
IIoT	industrial internet of things
IoBT	internet of battlefield things
IoC	indicators of corruption
IoMT	internet of medical things
IoTs	Internet of things
IoV	internet of vehicles
IPO	initial public offering
ITS	intelligent transportation system
KB	knowledge base
LPoS	leased proof-of-stack
LTE	long-term evolution
M2M	machine-to-machine
MCDM	multi-criteria decision making
MHR	medication health record
MIN	miner identifier number
MISP	malware information sharing platform
ML	machine learning
NB-IoT	narrowband internet of things
NPV	negative predictive value

OBU	on-board-unit
OSINT	open-source intelligence
OSU	open-source university
P2M	participant to machine
P2P	participant to participant
P2P	peer-to-peer
PBFT	practical Byzantine fault tolerance
PHR	personal health records
PII	personally identifiable information
PKI	public-key infrastructure
PoA	proof of authority
PoB	proof-of-burn
PoC	proof-of-capacity
PoET	proof-of-elapsed time
PoI	proof-of-importance
PoL	proof-of-luck
PoS	proof of stack
PoSp	proof-of-space
PoV	proof-of-vote
PoW	proof of work
PoX	proof-of-eXercise
PPV	positive predictive value
PRISMA-SGR	preferred reporting items for systematic reviews and meta-analysis
RBAC	role-based access management
RFQ	request for quotation
RIRN	Rencana Induk Riset Nasional
RPL	routing protocol for low-power
RSP	rock-scissor-paper
RSUs	roadside units
RTO	regional transport office
SCM	supply chain management
SCs	smart contracts
SMR	state machine replication
SMS	short message service
SOC	security operation center
SPF	single point of failure
SPV	simplified-payment-verification
SSCM	sustainable supply chain management

SSOT	single source of truth
SVM	support vector machine
TA	trusted authority
TCA	tournament consensus algorithm
TEE	trusted execution environment
UAI	unique address identifier
UIDs	unique identifiers
V2I	vehicle to infrastructure
V2P	vehicle to people
VANET	vehicular ad hoc network
WAVE	wireless access in vehicular environment
WHO	World Health Organization
WSN	wireless sensor network
XT	eXercise transaction

Preface

Blockchain (BC) and the Internet of Things (IoT) are two trendy and powerful technological names that have already proven their importance in various fields. The blockchain was born for the security of a magical cryptocurrency, "Bitcoin (BTC)," while the Internet of Things justifies its name. The Internet of Things is a fast-growing and easy-to-use technology that has also caught on in a concise period of time; and it covers almost all areas of life. The Internet of Things is now involved from everyday to high-level technical scenarios, so security is becoming a crucial issue.

The authors of this book come from research and academia, and their work demonstrates the power of knowledge. The chapters in this book are well written, easy to understand, and technically rich. They present knowledge about these two technologies, explaining them in different aspects.

It gives us immense pleasure to introduce to you the first edition of the book entitled *Applications of Blockchain and Big IoT Systems: Digital Solutions for Diverse Industries.* The primary intent of this book is to explore the various applications of blockchain and big IoT systems. It presents the rapid advancement in the existing business model by applying blockchain, big data, and IoT techniques. Several applications of blockchain, IoT, and big data in different industries are incorporated in the book. The wide variety of topics it presents offers readers multiple perspectives on various disciplines. This book will help the data scientists, blockchain engineers, big data engineers, and analytics managers.

Each chapter presents blockchain/big data and IoT use in application areas like agriculture, education, IoT, medical, smart city, and supply chain. The idea behind this book is to simplify the journey of aspiring engineers across the world. This book will provide a high-level understanding of various Blockchain algorithms, along with big data and IoT techniques in different application areas.

This book contains 19 chapters. Chapter 1 elaborates on these two categories as well. Further, it covers the consensus mechanism, and it is working along with an overview of the Ethereum (ETH) platform. Chapter 2 provides an in-depth analysis of IoT security issues and how federated learning along with BC technology can be used to solve them. Chapter 3 discusses the following consensus algorithms-PoW [2], proof of stake, delegated proof of stake, Byzantine fault tolerance (BFT), crash fault tolerance, hashgraph consensus algorithm, proof of elapsed time, and proof of authority (PoA).

Chapter 4 discusses blockchain technology in support of auto encoder deep neural networks, which is evaluated for managing and classifying the incidents and for validating its performance and accuracy. Chapter 5 proposes a hybrid algorithm that manages the decentralized network starting from joining the network as a new node until adding a new authorized block to the blockchain of network nodes. Chapter 6 proposes a blockchain-based security solution for IoV to authenticate vehicles, calculate reward points, and compute new trust value. Chapter 7 discusses the general architecture of smart cities using blockchain technology, applications, opportunities, and the future scope of blockchain technology in implementing smart cities. Chapter 8 talks about the recent implementation of a few major sectors in a city: healthcare, governance, energy, and social benefits. Chapter 9 discusses elaborate blockchain technology for biomedical engineering applications.

Chapter 10 discusses the decentralized and secured applications of blockchain in the biomedical domain. Chapter 11 discusses various use case studies of blockchain in the management of healthcare data. Chapter 12 discusses the future applications of blockchain in business and management. Chapter 13 discusses the development of blockchain-based cryptocurrency. Chapter 14 analyzes the enablers of blockchain technology by using DEMATEL techniques. Chapter 15 presents an overview of and motive for enabling blockchain technology in the construction industry and project management, for the smooth functioning without much duplicity in the system. Chapter 16 discusses the mitigation of various wastages generated across the supply chain. Chapter 17 shows system-level thinking pertaining to the current food supply chain; it than elaborates on multiple steps associated with service design, followed by integrated supply chain information and secured blockchain frameworks. Chapter 18 adopts distributed ledger technology (DLT) that allows the recorded data in the system to fan-out amongst the farmers, consumers, and all the actors involved in the system. Chapter 19 discusses the transformation of higher education system using blockchain technology.

We hope that readers make the most of this volume and enjoy reading this book. Suggestions and feedback are always welcome.

—*Arun Solanki, PhD*
Vishal Jain, PhD
Loveleen Gaur, PhD

PART I

Blockchain Mechanisms for IoT Security

CHAPTER 1

Blockchain Technology: Introduction, Integration, and Security Issues with IoT

SUNIL KUMAR SINGH[1] and SUMIT KUMAR[2]

[1]Mahatma Gandhi Central University, Bihar, India,
E-mails: sksingh@mgcub.ac.in; sunilsingh.jnu@gmail.com

[2]Gopal Narayan Singh University, Bihar, India,
E-mail: sumit170787@gmail.com

ABSTRACT

Blockchain (BC) was mainly introduced for secure transactions in connection with the mining of cryptocurrency bitcoin (BTC). This chapter discusses the fundamental concepts of BC technology and its components, such as block header, transaction, smart contracts (SCs), etc. BC uses the distributed databases, so this chapter also explains the advantages of distributed BC over a centrally located database. Depending on the application, BC is broadly categorized into two categories; permissionless and permissioned. This chapter elaborates on these two categories as well. Further, it covers the consensus mechanism, and it is working along with an overview of the Ethereum (ETH) platform. BC technology has been proved to be one of the remarkable techniques to provide security to IoT devices. An illustration of how BC will be useful for IoT devices has been given. A few applications are also illustrated to explain the working of BC with IoT.

1.1 INTRODUCTION

With the emergence of new communication and information technology, security always has been a major concern. In recent, many well-known organizations have faced security breaches. For example, a well popular search engine Yahoo experienced a major attack in the year 2016, resulting in the conciliation of billions of accounts [1]. After doing the security-related research on many companies, it observed that 65% of the data infringement has happened because of a weak or reeved password. Further, it is found that many times sensitive information stealing was done by phishing e-mails.

Blockchain (BC) technology was conceived mainly to address the security issue of cryptocurrency bitcoin (BTC). It has several benefits and is well suited to handle the security issue. In the BC system, there is no central database, and it is a kind of system that does not trust the people. This system assumes that anyone can attack on the system, whether part of the system or outsider, can attack the system; therefore, it is a system that is devoid of human consuetude. Moreover, it is enabled with cryptographic features, which can be like hashing and digital signature. BC is immutable [1] also, therefore, anyone can store the data. Finally, as many users are involved in the BC system, changing or adding new blocks in the system needs to be validated by the majority of the users.

BTC is one of the first digital currency [2], created in 2009, underlying BC technology. As BTC is known as the first cryptocurrency, it was marked as a spire performing currency in the year 2015 and considered a spanking commodity in 2016. Nowadays, besides BTC, BC is applied in many other areas like medicine, economics, the Internet of Things (IoT), software engineering, and many more.

BC technology is getting popular for offering better and foolproof security by removing intermediaries. It also results in reducing the cost of transactions. It is a shared data structure that is amenable for collecting all the transactional history. In BC technology, blocks are connected in the form of chains. The beginning block of the BC is recognized as the Genesis block [3]. All other blocks are simple blocks. The chain in the BC is the link or the pointers connecting the blocks. Blocks, in turn, keeps the transactions that take place in the system.

Many organizations have defined BC technology in different ways. The Coinbase, the bulkiest cryptocurrency exchange across the globe, has established the BC as "a distributed, public ledger that contains the history of every BTC transaction" [3]. Oxford dictionary bestows a familiar definition stated as "a digital ledger in which transactions made in BTC or another cryptocurrency are recorded chronologically and publicly" [4]. Another description is given by Sultan et al., which narrates a very general definition of BC technology as "a decentralized database containing sequential, cryptographically linked blocks of digitally signed asset transactions, governed by a consensus model" [4].

Fundamentals of BC technology are supposed to lie in between the 1980s and 1990s of the 20th century [1] though it gained popularity very recently. It is widely recognized in 2008 after the inquisition of cryptocurrency BTC. BC became widely prevalent after the legendary work of Nakamoto [5], though it is a fictitious name and still has not been explored who the actual person is. Nakamoto proposed a technique to replace the centralized architecture with a pear-to-pear network-based architecture. Initially, BC technology was named as two words, "block" and "chain;" however, at the end of the year 2016, these two words have been combined to make its BC.

BC uses the concept of a ledger which may be seen as a database to maintain the records or a list of transactions. This is similar to the ledger of a hotel. For example, when you check-in in a hotel, the receptionist asks your identity and enters the record in a hardbound register (called ledger). This entry is maintained date and time-wise. One cannot add or remove the entries in between and can only append in the ledger. Thus, the entries cannot be made in between the two entries as well as cannot be deleted in between. One can consider the entry as a transaction and pages of the ledger as a block. So, it becomes a chain of blocks in the ledger. In case of any eventuality, this hotel ledger is to be consulted for security purposes. Though, this type of ledger is a centralized database of the hotel.

Intermediation is one of the prominent solutions for screening the ownership of assets or transaction processing. Intermediaries' role is to check and validate the participating parties along with the chain of intermediaries. This validation process, apart from time taking, incurs a significant amount of cost. In case the validation fails, it has credit risk too. The BC technology promises a way to overcome, representing "a shift from trusting people to trusting math" [6], i.e., free from human intervention or minimum human involvement.

IoTs is an upcoming technology that indicates the billions of tangible devices across the globe agglutinated to the Internet which collects and shares the information. IoT is the term coined by Kevin Ashton of MIT in 1999 during his work at Procter and Gamble (company) [7]. It promises the world to make it perceptive and proactive by enabling the things to talk with each other [8, 9]. In the IoTs, the collected data from the sensors [42] are maintained in central servers, which may lead to many intricacies when the devices try to communicate with each other through the internet [10]. Centralized locations may also suffer from security issues resulting in their misuses. BC technology can provide a solution in the form of a decentralized model. A distributed model can execute billions of operations between different IoT devices. An IoT with BC has been depicted in Figure 1.1, wherein distributed BC replaces the concept of the central server and big data processing at a centralized location. This minimizes the building and maintenance costs associated with the centralized location server. It also reduces the single point failure in the absence of a third party. This chapter deliberates on the BC and its relevance concerning IoT.

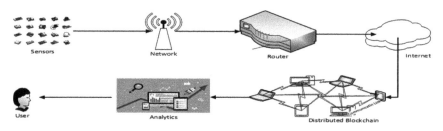

FIGURE 1.1   Data flow in the IoT-blockchain.

1.2   COMPONENTS OF BLOCKCHAIN (BC) TECHNOLOGY

BC is a network of blocks (nodes) that are connected with one another following some topology rather than being connected with a central server. It has the potential to store the transactions in the ledger effectively and confirming transparency, security, and auditability. Few crucial components of BC technology are as follows.

1.2.1   BLOCK

Block in the BC technology is the decentralized nodes/miners equipped with the databases, and it contains the digital piece of information. Blocks are linked together containing the hash value of the previous block into the current block. In general, block structure can be visualized into two parts: block header and a list of transactions.

Block header equipped with the following information:

- Version number indicates the version number of the block and uses 4 bytes for its representation.
- Previous block hash is a pointer between the previous and current block and uses 32 bytes.

- Timestamp uses 4 bytes and stores the time of the creation of the block.
- Merkle tree is represented by 32 bytes and is a hash of every transaction that takes place in a block.
- Difficulty target is indicated by 4 bytes and basically it is used to measure the intricacy target of the block.
- Nonce also uses 4 bytes and computes the different hashes.

Figure 1.2 shows a generic diagram of a block with its important components. It also shows the working Merkle root which is generated from the hash values of the transaction. In Figure 1.2, A, B, C, and D are the transactions and H(A), H(B), H(C), and H(D) are their respective hash values.

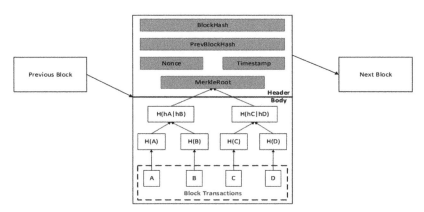

FIGURE 1.2　Diagram of a block.

1.2.2　GENESIS BLOCK

In a BC, genesis block is considered as a foundered block because it is the first block in the chain. The block height of the first block is always zero, and no block precedes the genesis block. Every block which is the part of the BC comprises of a block header along with transaction counter, and transactions.

1.2.3　NONCE

A nonce, an abridgment for "number only used once" is a one-time code in cryptography. It is a number appended to the hashed (encrypted) block in a

BC. When it is rehashed, it ensures the difficulty level of antagonism. The Nonce is the number for which BC miners solve a complex problem. It is also associated with the timestamp to limit its lifetime; that is why if one performs duplicate transactions, even then a different Nonce is required.

1.2.4   USER AND MINER

A computationally advanced node that tries to solve a complex problem (which requires high computation power) to retrace a new block which is recognized as a miner. The miners are capable of working alone or in a collective routine in order to find the solution to the given mathematical problem. The process of locating a novel block is opened by sharing new transaction information among every user in the BC network. It is the responsibility of each user to collect the new transactions into blocks and put their efforts to find the proof-of-work of the block. Proof-of-work is defined as a user is required to solve a computational complex puzzle for publishing a new block, and the solution of the puzzle will be its proof. This whole phenomenon is known as proof of work (PoW).

1.2.5   CHAIN AND HEIGHT

In BC technology, the chain is a virtual string that connects the miners in the accrescent set of blocks with hashes [11]. The chain keeps growing as and when a new block is appended. Blocks in the chain are generally indicated by their block height in the chain which is nothing but a sequence number starting from zero. The height of a block is defined as the number of blocks in the chain between the genesis block and the given block (for which height is to be calculated).

1.2.6   TRANSACTION

A BC transaction is represented in the form of a smaller unit of the tasks; and is warehoused in public records. After verification by more than 50% of the users of the BC network, records get implemented and executed. Its outcomes are stored in the BC. Previously stored records can be reviewed at any time, but the updation of the records are not permitted. The size of the transaction is a crucial parameter for the miners because the bigger size transaction requires larger storage space in the block. It also requires significantly more power, whereas the smaller size transaction requires less power. The structure of the BC [3] is shown in Table 1.1.

TABLE 1.1   Blockchain's Structure

Field	Size
Magic number	4 bytes
Block size	4 bytes
Header: Next 80 bytes	
Version	4 bytes
Previous block hash	32 bytes
Merkle root	32 bytes
Timestamp	4 bytes
Difficulty target	4 bytes
Nonce	4 bytes
Rest of Blockchain	
Transaction counter	Variable: 1 to 9
Transaction list	Transaction size-dependent: up to 1 MB

1.3   TYPES OF BLOCKCHAIN (BC)

Centered on the uses of BC technology for various applications in a different scenario, it is broadly categorized into two categories: permissioned and permissionless [1].

1.3.1   PERMISSIONED

In this, one is required to take some sort of permission from that particular organization or owner of the BC to access any or parts of the BC. For example, to read a BC would not allow us to perform any other operations in the block. One needs to take permission to access or transact the block. Permissioned BC is categorized into two categories, as follows.

1.3.1.1   PRIVATE BLOCKCHAIN (BC)

The private BC is fully permissioned, and if a node is willing to join, it has to be a member of that single organization. This new node needs to send an original transaction and required to take part in the consensus mechanism. The private BC is useful and is generally favored for individual enterprise solutions to record the track of data transfer between different departments [12]. Examples of private BC are Ripple and Hyperledger.

1.3.1.2   FEDERATED BLOCKCHAIN (FEDBLOCK)

Federated blockchain (FedBlock), also known as a consortium BC, shares a lot of similarity to a private BC. It is a 'semi-private' system that has a controlled user group. A FedBlock is taken as an auditable and credibly synchronized dispersed database that preserves the track of data exchange information between consortium members taking part in the system. Like a private BC, it does not annex the processing fee and incurs a low computational cost to publish new blocks. FedBlock ensures the auditability and contributes comparatively lower latency in transaction processing. Examples of FedBlock s are EWF, R3, Quorum, and Hyperledger, etc.

If we compare with the public BC (mentioned in Section 1.3.2.1), private is more comfortable because of less number of users. It requires less processing power and time for verifying a new block. It also provides better security because the nodes, which are within the organization, can read the transactions.

1.3.2   PERMISSIONLESS

A permissionless BC is simple, with no restriction for entry to use it. As the name indicates, anyone and anything can be a part of it without taking permission.

1.3.2.1   PUBLIC BLOCKCHAIN (BC)

A public BC is a permissionless BC in which the validation of transactions depends on consensus. Mostly it is distributed, in which all the members take part in publishing the new blocks and retrieving BC contents. Ina public BC, every block is allowed to keep a copy of the BC, which is used in endorsing the new blocks [12]. A few popular applications of public BC execution are cryptocurrency networks which are like BTC, ethereum (ETH), and many others. It has an open-source code maintained by a community and is open for everyone to take part in [1, 13].

A public BC is difficult to hack because for adding a new block, it involves either high computation-based puzzle-solving or staking one's cryptocurrency. In this, every transaction is attached to some processing fee. A comparison of various available technologies [11] is shown in Table 1.2.

TABLE 1.2   Comparison of Blockchain Technologies

	Public Blockchain	**Private Blockchain**	**Consortium/ Federated Blockchain**
Participation in Consensus	Every node	Solo organization	Some specified nodes in multiple organizations
Access	Read/write access allowed to all	High access restriction	Comparatively lower access restriction
Identity	Pseudo-anonymous	Accepted participants	Accepted participants
Immutability	Fully immutable	Partially immutable	Partially immutable
Transaction Processing Speed	Low	High	High
Permission Required	No	Yes	Yes

1.4   SMART CONTRACT

The smart contract is the term, introduced by Szabo in 1997 [11, 14], which combines computer protocols with users to run the terms of the contract. A smart contract is a self-enforcing agreement (an agreement enforced by the party itself) embedded in computer code managed by the BC. It is governed by the computer protocols under which the performance of a reliable transaction occurs without the participation of any third parties. The transactions performed under the smart contract can be tracked and is irreversible. A smart contract basically consists of the following components: lines of code, storage file, and account balance. It can be created by a node to initiate a transaction to the BC. The lines of code, i.e., program code is immutable and cannot be moderated once it is created.

Figure 1.3 shows the contract's storage file associated with the miner and stored in the public BC. The network of miners is responsible for executing the program logic and acquiring the consensus on the execution's output.

Only that particular node (miner) is enabled to hold, access, and modify the data in the BC. The contract's code follows a reactive approach, i.e., it is executed whenever it receives a message from the user or any other nodes in the chain. While during the execution of the code, the contract may access the storage file for performing the read/write operations.

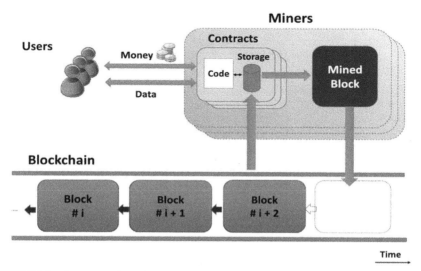

FIGURE 1.3   Structure of the distributed cryptocurrency system with smart contracts [14].

1.5   CONSENSUS MECHANISM

Consensus mechanisms [1] are the protocols that ensure the synchronization of all the nodes with each other in the BC. It validates the transaction if it is legitimate before adding it to the BC. This mechanism plays an essential character in the smooth and correct functioning of BC technology. It also ensures that all the nodes use the same BC and all the nodes must continuously check all performed transactions.

Many consensus mechanisms are available today. However, a few known prevalent BC consensus mechanisms are PoW, proof of stake (PoS), Delegated PoS, Ripple, and Tendermint [15]. The key difference among numerous consensus mechanisms can be identified, the way they depute and payoff the authentication of multiple transactions.

Even after the availability of the number of consensus mechanisms, many existing BC systems, including BTC and ETH uses PoW. PoW is the first and popular consensus mechanism. Its use is widely accepted in many of BC-based systems. It is mandatory for the users, participation in the BC network, to prove that the work is done for them to qualify and obtain the aptitude to add a new block to the ledger [1]. In the BC network, nodes are expected to receive the consensus and agree that the block hash provided by the miner is a valid PoW.

Figure 1.4 illustrates the working of the PoW mechanism in the BC. In this mechanism, every miner is first required to define and create a PoW puzzle in the BC. The created puzzle will be visible and accessible to every other node taking part in the system. However, the node, which can solve the PoW puzzle, is able to hold, access, and modify the data in the BC.

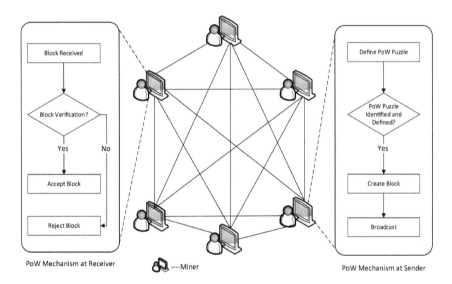

FIGURE 1.4   Working of consensus mechanism in blockchain.

1.6   ETHEREUM (ETH)

ETH is a distributed computing platform, used for public BC systems with an operating system featuring smart contract functionalities. It is an open-source platform proposed by VitalikButerin, a programmer and cryptocur-rency researcher in late 2013 [16].

ETH is also a validated platform used to deliver and execute smart contracts (SCs) reliably. It supports a modified form of the Nakamoto-consensus mechanism, which works on "Memory Hardness" despite fast computing power machines. ETH is a permissionless network, i.e., any node can join the network by creating an account on the ETH platform. Moreover, it uses its consensus model, which is identified as EthHash PoW. It is competent to run the scripts using a global network of public nodes. Miner nodes are ETH Virtual Machines provided by ETH BC. These nodes are adequate for providing cryptographic tamper-proof tenacious execution, and its implementation is called contracts. ETH reinforces its digital currency known as Ether [17]. ETH is one of the well-recognized platforms for executing SCs, though, it can execute other decentralized applications (DApps) and compatible to interact with many other BC s. It is also categorized as Turing-complete [18], a mathematical concept giving a hint that ETH programming language can be used as a platform to simulate other languages.

ETH platform may be used to regulate and configure various IoT devices [43–45]. Security keys are managed using the RSA algorithm, where private keys are stored on the devices, and BC controls public keys.

1.7   BLOCKCHAIN (BC) TECHNOLOGY IN IoT

BC technology can play a vital role for various privacy and security issues of the IoT. In IoT, sensing devices usually send the data at a centralized location for processing purposes. BC technology replaces the central server concept of IoT by introducing the concept of distributed ledger for every transaction with legitimate authentication [10]. It ensures that storing the transaction details with the intermediaries is no longer necessary because transaction records will be available on many computers of the chain. This system rejects the updation and breaching of one computer. However, to make it successful, multi-signature protection is required to authorize a transaction. If a hacker tries to steal the information by penetrating the network, multiple duplicate copies are available on many computers worldwide. For hacking the BC network successfully, the consensus of more than 50% of systems in the network is required [19].

1.7.1   BLOCKCHAIN (BC) INTEGRATION

Integration of BC with IoT opens a new door and wider domain of research and development in the area of IoT applications [16, 20, 21]. Over the last few years, unprecedented growth in the field of IoT has been observed, which enables wide opportunities like access and share of the information. Many times, accessing, and sharing information can induce challenges like security, privacy, and trust among the communicating parties. BC can solve various issues of IoT like privacy, security, and reliability. The distributed nature of BC technology can eliminate single point failure and makes it reliable.

We are all aware that BC has already proven its importance in financial transactions with the help of cryptocurrencies, such as BTC and ETH. It removed the third-party requirement between P2P payment services [18]. A few IoT enablers have chosen the BC technology and formed a consortium for standardization and reliable integration of BIoT (Blockchain-IoT). It is a group of 17 companies aimed to enable security, scalability, heterogeneity, privacy, and trust in distributed structure with the help of BC technology [16].

IoT devices can communicate with one another either directly, device to device, or through BC technology. There are three types of communication models in an integrated BC and IoT environment, which are as follows.

1.7.1.1   IoT-IoT COMMUNICATION

In IoT-IoT communication, IoT devices communicate directly without the involvement of the BC. This type of communication is also known as inter-IoT devices communication. It is the fastest communication model that does not associate high computation and time-consuming BC algorithms.

Figure 1.5 shows that BC is not involved in inter IoT communications that is why the system is not able to ensure data integrity, privacy, and security mechanisms. In this model, BC stores the communication/transaction history of the IoT devices. This is one of the fast communication models between IoT devices.

1.7.1.2   IoT-BLOCKCHAIN (BC) COMMUNICATION

In this model, all transactions among the IoT devices go through the BC. This model is enriched with the capability to ensure the data privacy, reliability, and safety of both data and transactions.

Figure 1.6 shows the IoT device communication model through BC, which ensures that stored records of each transaction will be immutable, and transaction details are traceable as its features can be verified in the BC. Although, BC upsurges the autonomy of IoT devices but it may suffer from BC overhead which causes latency.

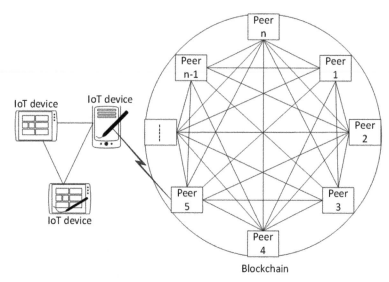

FIGURE 1.5   IoT-IoT communication model.

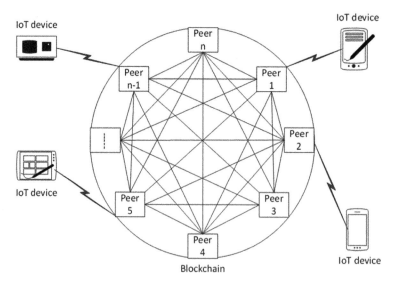

FIGURE 1.6   IoT-blockchain communication model.

1.7.1.3   HYBRID COMMUNICATION

The last communication model is a hybrid communication model, in which IoT communication involves the CLOUD/FOG networks. This model shifts partially or most of the computation load, such as encryption, hashing, and compression, from IoT devices to Fog nodes.

Figure 1.7 shows an IoT integrated with BC technology, which can transfer high computation load and time-consuming algorithm to Fog node. In this way, fog, and cloud computing comes into play and complement the shortcoming of BC and IoT [21].

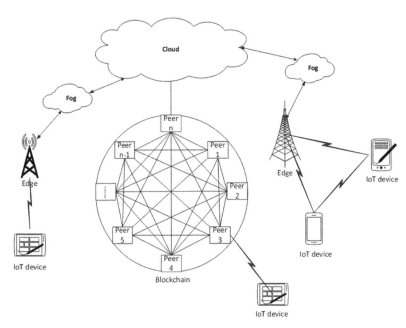

FIGURE 1.7   Hybrid communication model.

1.7.2   *SECURITY IN BLOCKCHAIN (BC) AND IoT*

The IoTs is a structure of machine-to-machine (M2M) associations, with no human involvement at all. Hence establishing faith with the participation of machines is a formidable challenge that IoT equipment still has not met broadly. The BC can take steps as a medium in this process, for improved scalability, protection of data, dependability, and privacy. This process can be done by BC technology to follow all devices which are connected to the IoT environment, and after that, it is used to make possible and/or synchronize all transaction processing. By using the BC function, we can fully remove a single point of failure (SPF) in IoT structure. In BC, data is encrypted using various algorithms like cryptographic algorithms and hashing techniques.

Therefore, the function of BC provides improved security services in an IoT. The function of BC technology is to repair the digital market. It has a guarantee and retaining both main and preliminary concerns of the function of the BC. The BC keeps the record of a group of sequential and sequence of information transactions since it can be read as a massive networked time-stamping system. The controllers are too concerned in BC's capability to recommend protected, confidential, immediately perceptible monitoring of transactions. Therefore, the BC can facilitate us to avoid the tampering and spoofing of data by the organization and securing the industrial IoT devices [22].

The BC records every transaction and provides a cross-border overall distributed confidence. Many times, it is possible that Trusted Third Party systems or central location-based services can be vitiated or hacked. In BC, when transactions are confirmed by consensus, then the block data are acceptable to all. The BC can be constructed as: (1) permissioned network, which is generally a private network; and (2) permissionless, a public network. Permissioned BC offers new privacy and improved access functionality. The BC can resolve these types of challenges effortlessly, strongly, and competently. It has been generally used for providing reliable and certified uniqueness registration, possession track, and monitor of products, supplies, and resources. IoT devices are not exempted, BC is able to identity all the connected IoT devices [17]. For security purposes, the BC supports the IoT as mentioned below.

1.7.2.1 DATA AUTHENTICATION AND INTEGRITY

The data transmission through IoT devices is linked to the BC network and it will be cryptographically proofed and signed via the correct correspondent to hold an exclusive public key and GUID (global unique identifier) which do not require any verification for its uniqueness, and thus it guarantees the verification and truthfulness of transmitting data. Additionally, all transactions complete toward or through an IoT device. Its transaction details are recorded on the BC ledger, which enables it to be tracked easily [17].

1.7.2.2 AUTHENTICATION, AUTHORIZATION, AND PRIVACY

In BC, SCs can offer a decentralized verification policy and sense to be capable of providing a particular and combined verification to an IoT device. The SCs are able to provide another effective permission access policy to link

IoT devices, employing a smaller amount of complexity while one compares among fixed approval protocols such as role-based access management (RBAC) [23], OpenID, etc. Nowadays, these protocols are generally used for managing, authorization, and verification of IoT devices [17].

1.7.2.3 SAFE AND PROTECTED CONNECTIONS

In general, communication protocols that are used by IoT applications; HTTP, MQTT, XMPP, and many other routing protocols that are not protected in design. These protocols are required to be wrapped with the new security protocols for providing secure communications. The new security protocols can be enriched with BC technology; which are DTLS (datagram transport-level security) or TLS [24] with the BC. Key management and identity allocation are completely removed from all IoT devices because it would contain its own single and distinctive GUID and the asymmetric key pair values once mounted and associated to the BC system [17].

Although the BC provides a strong approach for protected IoT, the consensus method depending on the miner's hashing control can be conceded, thus permit the hacker/attacker to host the BC. Also, the private keys among restricted uncertainty can be dried up to compromise the BC accounts. Efficient methods up till now need to be distinct to make sure the privacy of transactions and keep away from race attack, which can affect inside dual spending throughout the transactions [17].

The IoT device has very limited storage and computing power; it can still produce safe and protected keys. Once a key is created, the public key is attached to the Public Key data field in addition to the elected IoT receiver and mined with the BC. While protected data communication via BC is not suggested because access to all nature of a broadcast BC on a server, a BC-based public key swap permitted for IoT to set-up non-interactive key managing protocols [25]. With a Non-Interactive Protocol, session key series utilizing a mixture of BC data fields as 'salt' may provide an effective solution for updating the IoT session keys for safe and secure data transfer. Still, this research field is required to be explored further for better outcomes [26].

A few other aspects of BC technology are; it can resolve the IoT security issues considering its limited storage and low computation power. Because efficient, lightweight, integrated blockchain (ELIB) [27] with IoT devices, protects it from security breaches. ELIB easily copes with the computational complexity and several other issues like low bandwidth, delay, and overhead, etc.

The BC structure is offering a trustful background used for data storage and access. This structure has two characters. One is data integrity, and another is role-based data access characters. In data integrity, the structure avoids data stored within if it is being altered. In role-based data access, it is a guarantee that the structure recommends special data access permissions toward different users and IoT devices [28, 29].

Compared to the cloud-based centralized system, the BC system is a decentralized system that has a benefit in protecting certain specific attacks (e.g., distributed denial of service (DDOS) attacks). The BC system does concern with the particular point of failure problem, which can occur in the cloud-based centralized system. The centralized system is typically controlled by a manager. If the hacker/attacker pinch the manager's account, they can randomly change the system data. While we were well-known, the data or conversion in the BC system is altered conflict [29].

Privacy and security are most essential in the IoT environment. Within the cloud-based centralized system, user's data are stored randomly, which can simply be hacked by the attackers/hackers. The BC system can offer the independence service by the public-key cryptography method. Furthermore, communication in the IoT environment accepts the AES encryption algorithm, which is extremely flexible to the resource-constrained IoT mechanism. Access control is also an essential mechanism in the IoT system; the smart contract of the BC system be able to offer this type of security service [28].

Researchers have observed that associating BC with IoT is beneficial to handle security and privacy issues, which can probably transform many industries. It is pertinent to mention here that IoT security has always been a pressing concern. To explain this, let us take an example of six IoT devices; the Chamberlain MyQ Garage, the Chamberlain MyQ Internet Gateway, the SmartThings Hub, the Ubi from Unified Computer Intelligence Corporation, the Wink Hub, and the Wink Relay; that are tested by a US-based application security company "Veracode" in 2015. The Veracode team found five devices, out of six, had serious security issues. The team was responsible for observing the implementation and various security issues of the communication protocol used in IoT systems. The front-end (services between user and cloud) and back-end (between IoT devices and cloud) were examined, and it is found that except SmartThings Hub, the devices even unsuccessful to have a robust password. Besides, Ubi is deficient in encryption for user connection. These security breaches can cause to a *man-in-the-middle* attack. When the team examined the back-end connection, results were even worst. They also lacked the protection from replay attacks.

In the era of technological automation, hacking of IoT devices has severe consequences. The incorporation of BC technology in IoT [8] is being well adopted through a broad perspective of measures purported to reinforce security. Narrowband Internet of Things (NB-IoT) is one of the novel types of IoT which is built on cellular networks. It can directly be deployed on long-term evolution (LTE) architecture. BC technology is applied [30] to ensure reliable data integrity and authentication.

Several mission-critical [31] applications, moving towards automation, are getting popular. Ocado, an online supermarket in Britain, is fully equipped with IoT to stringently improve the warehouse. Installed RFID chips into the Ocado warehouse can sense when the new stock requirements are to be ordered or the status of the remaining number of items in the warehouse. BC technology is used to ensure data integrity, and its decentralized replication technique alienates the requirement to have entire IoT data collected at a central location. This is possible because SCs, stored on the BC, would not allow any modification to the contracts.

Although BC technology can protect from vulnerabilities, it still suffers from some issues. SCs in BC are visible to all the users that can cause bugs and vulnerabilities; these are the bugs that cannot be fixed in the stipulated time duration. Some other drawback includes its complexity, high computation, and sometimes resource wastage.

1.8   A COMPARATIVE STUDY

This section includes a comparative study on the previously developed system with the BC-IoT-based system. BC technology and IoT can be considered as emerging realities in the current epoch, and these two technologies can transform civilization at a rapid pace [32]. From Table 1.3, one can see that wireless sensor networks (WSN) [33] and the IoTs based on systems are not immutable, IoT-Cloud is partially immutable. At the same time, IoT-BC is a completely immutable system.

IoT-Cloud allows participant to participant (P2P) sharing while IoT-BC supports P2P as well as a participant to machine (P2M) and M2M sharing also. All other systems support limited sharing only. Table 1.3 lists the BC-enabled IoT system with respect to certain properties.

TABLE 1.3   A Comparison of Blockchain-IoT based System with Traditional SHM Systems

	Simple [34]	WSN [33, 35]	IoT [36, 37]	IoT-Cloud [30, 38]	IoT-Blockchain [32]
1. De-centralize	Completely centralized	Completely centralized	Completely centralized	Mostly decentralized	Entirely decentralized
2. Reliability	Highly not reliable	High data tempering	Data tampering is possible easily	Data tampering is possible easily	Tempering is not possible
3. Storage, Privacy, Security, and Confidentiality	Low	Low	Intermediate	Intermediate	Considerably higher
4. Immutable Behavior	Not immutable	Not immutable	Not immutable	Partially immutable	Fully immutable
5. Real-time	Nearly-real time	Real-time	Real-time	Real-time	Nearly-real time
6. Communication and Transparent information Sharing	Confined monitoring	Confined monitoring	Data processing and monitoring	Data processing, monitoring, and P2P information sharing	P2M and M2M communications, autonomous decision making using Smart contract-based analysis
7. Interoperability	Lower	Lower	Intermediate level	Intermediate level	High
8. Re-Active Maintenance	Lower	Low	Medium	Medium	Effectively high

Observations from Table 1.3 show that the BC-enabled IoT system is the most suitable system to ensure reliability, immutability, interoperability, and security, etc., as indicated in the table. Therefore, one can conclude that BC technology is the most suited technology for IoT-enabled systems. A BC-based decentralized system is most suitable for IoT networks; which is validated by a study of Rathore et al. [39]. In this, a review is done on centralized, distributed, and decentralized systems using various measures like accuracy, F-score, detection rate, etc. [40, 41]. Therefore, we can conclude that decentralized BC system is the most suitable system of IoT networks.

1.9   CONCLUSION

This chapter defines the fundamentals of BC technology, along with its components. A comparative study of various BC technology is also highlighted. Various application areas are mentioned in this chapter. A BC technology, ETH, is described that can be used to implement the public BC. It ensures the transparency of the information. The importance of BC is also explained with the help of the relevant examples.

IoT is an upcoming technology that is being introduced for a smart environment. With such a prevalent environment, security is a measure of concern. This chapter also introduces how BC can be used for security in IoT. BC, being a distributive technology, plays a good role in IoT security. The comparative study section of the chapter infers the same.

KEYWORDS

- **blockchain**
- **central database**
- **cryptography**
- **distributed denial of service**
- **Ethereum**
- **Internet of Things**

REFERENCES

1. Atlam, H. F., & Wills, G. B., (2019). Technical aspects of blockchain and IoT. In: *Advances in Computers* (Vol. 115, pp. 1–39). Elsevier.

2. Li, X., Jiang, P., Chen, T., Luo, X., & Wen, Q., (2017). A survey on the security of blockchain systems. *Future Generation Computer Systems.*

3. Dwivedi, A. D., Srivastava, G., Dhar, S., & Singh, R., (2019). A decentralized privacy-preserving healthcare blockchain for IoT. *Sensors, 19*(2), 326.

4. Sultan, K., Ruhi, U., & Lakhani, R., (2018). *Conceptualizing Blockchains: Characteristics & Applications.* arXiv preprint arXiv:1806.03693.

5. Nakamoto, S., (2019). *Bitcoin: A Peer-to-Peer Electronic Cash System.* Mangubat.

6. Nofer, M., Gomber, P., Hinz, O., & Schiereck, D., (2017). Blockchain. *Business & Information Systems Engineering, 59*(3), 183–187.

7. Ramakrishnan, R., & Gaur, L., (2016). Application of Internet of Things (IoT) for smart process manufacturing in Indian packaging industry. *Information Systems Design and Intelligent Applications* (pp. 339–346). Springer.

8. Kumar, S., & Raza, Z., (2018). Internet of things: Possibilities and challenges. In: *Fog Computing: Breakthroughs in Research and Practice* (pp. 1–24). IGI Global.

9. Ramakrishnan, R., & Gaur, L., (2019). *Internet of Things: Approach and Applicability in Manufacturing.* CRC Press,

10. Kumar, N. M., & Mallick, P. K., (2018). Blockchain technology for security issues and challenges in IoT. *Procedia Computer Science, 132*, 1815–1823.

11. Zhu, L., Gai, K., & Li, M., (2019). Blockchain and Internet of Things. In: *Blockchain Technology in Internet of Things* (pp. 9–28). Cham: Springer International Publishing.

12. Ali, M. S., Vecchio, M., Pincheira, M., Dolui, K., Antonelli, F., & Rehmani, M. H., (2018). Applications of blockchains in the Internet of Things: A comprehensive survey. *IEEE Communications Surveys & Tutorials, 21*(2), 1676–1717.

13. Ghimire, A., (2020). *Brief Survey and Testbed Development for Blockchain-Based Internet of Things.* The University of Mississippi.

14. Delmolino, K., Arnett, M., Kosba, A., Miller, A., & Shi, E., (2016). Step by step towards creating a safe smart contract: Lessons and insights from a cryptocurrency lab. In: *International Conference on Financial Cryptography and Data Security* (pp. 79–94). Springer.

15. Ahmad, F., Ahmad, Z., Kerrache, C. A., Kurugollu, F., Adnane, A., & Barka, E., (2019). Blockchain in internet-of-things: Architecture, applications and research directions. In: *2019 International Conference on Computer and Information Sciences (ICCIS)* (pp. 1–6). IEEE.

16. Syed, T. A., Alzahrani, A., Jan, S., Siddiqui, M. S., Nadeem, A., & Alghamdi, T., (2019). A Comparative analysis of blockchain architecture and its applications: Problems and recommendations. *IEEE Access, 7*, 176838–176869.

17. Khan, M. A., & Salah, K., (2018). IoT security: Review, blockchain solutions, and open challenges. *Future Generation Computer Systems, 82*, 395–411.

18. Fernández-Caramés, T. M., & Fraga-Lamas, P., (2018). A review on the use of blockchain for the Internet of Things. *IEEE Access, 6*, 32979–33001.

19. Kshetri, N., (2017). Blockchain's roles in strengthening cybersecurity and protecting privacy. *Telecommunications Policy, 41*(10), 1027–1038.

20. Panarello, A., Tapas, N., Merlino, G., Longo, F., & Puliafito, A., (2018). Blockchain and IoT integration: A systematic survey. *Sensors, 18*(8), 2575.

21. Reyna, A., Martín, C., Chen, J., Soler, E., & Díaz, M., (2018). On blockchain and its integration with IoT. Challenges and opportunities. *Future Generation Computer Systems, 88*, 173–190.

22. Miraz, M. H., & Ali, M., (2018). Blockchain-enabled enhanced IoT ecosystem security. In: *International Conference for Emerging Technologies in Computing* (pp. 38–46). Springer.
23. Ihle, C., & Sanchez, O., (2018). Smart contract-based role management on the blockchain. In *International Conference on Business Information Systems* (pp. 335–343). Springer.
24. Mahalle, P. N., Anggorojati, B., Prasad, N., R., & Prasad, R., (2013). Identity authentication and capability based access control (iacac) for the Internet of Things. *Journal of Cyber Security and Mobility, 1*(4), 309–348.
25. Singh, G., Gaur, L., & Ramakrishnan, R., (2017). *Internet of Things-Technology Adoption Model in India, 25,* 835–846.
26. Gagneja, K., & Kiefer, R., (2020). Security protocol for Internet of Things (IoT): Blockchain-based implementation and analysis. In *2020 Sixth International Conference on Mobile and Secure Services (MobiSecServ)* (pp. 1–6). IEEE.
27. Mohanty, S. N., et al., (2020). An efficient lightweight integrated blockchain (ELIB) model for IoT security and privacy. *Future Generation Computer Systems, 102,* 1027–1037.
28. Lemieux, V. L., (2016). Trusting records: Is blockchain technology the answer? *Records Management Journal.*
29. Xu, R., Lin, X., Dong, Q., & Chen, Y., (2018). Constructing trustworthy and safe communities on a blockchain-enabled social credits system. In: *Proceedings of the 15th EAI International Conference on Mobile and Ubiquitous Systems: Computing, Networking and Services* (pp. 449–453).
30. Hong, H., Hu, B., & Sun, Z., (2019). Toward secure and accountable data transmission in narrowband Internet of Things based on blockchain. *International Journal of Distributed Sensor Networks, 15*(4), 1550147719842725.
31. Hammoudeh, M., Ghafir, I., Bounceur, A., & Rawlinson, T., (2019). Continuous monitoring in mission-critical applications using the Internet of Things and blockchain. In*: Proceedings of the 3rd International Conference on Future Networks and Distributed Systems* (pp. 1–5).
32. Jo, B. W., Khan, M., A. r., & Lee, Y. S., (2018). Hybrid blockchain and internet-of-things network for underground structure health monitoring. *Sensors, 18*(12), 4268.
33. Stajano, F., Hoult, N., Wassell, I., Bennett, P., Middleton, C., & Soga, K., (2010). Smart bridges, smart tunnels: Transforming wireless sensor networks from research prototypes into robust engineering infrastructure. *Ad. Hoc. Networks, 8*(8), 872–888.
34. Yuan, Y., Jiang, X., & Liu, X., (2013). Predictive maintenance of shield tunnels. *Tunnelling and Underground Space Technology, 38,* 69–86.
35. Bennett, P. J., et al., (2010). Wireless sensor networks for underground railway applications: Case studies in Prague and London. *Smart Structures and Systems, 6*(5, 6), 619–639.
36. Zhou, C., & Ding, L., (2017). Safety barrier warning system for underground construction sites using internet-of-things technologies. *Automation in Construction, 83,* 372–389.
37. Ding, L., et al., (2013). Real-time safety early warning system for cross passage construction in Yangtze riverbed metro tunnel based on the Internet of Things. *Automation in Construction, 36,* 25–37.
38. Mahmud, M. A., Bates, K., Wood, T., Abdelgawad, A., & Yelamarthi, K., (2018). A complete Internet of Things (IoT) platform for structural health monitoring (shm). In: *2018 IEEE 4th World Forum on Internet of Things (WF-IoT)* (pp. 275–279). IEEE.

39. Rathore, S., Kwon, W. B., & Park, J. H., (2019). BlockSecIoTNet: Blockchain-based decentralized security architecture for IoT network. *Journal of Network and Computer Applications, 143,* 167–177.

40. Diro, A. A., & Chilamkurti, N., (2018). Distributed attack detection scheme using deep learning approach for Internet of Things. *Future Generation Computer Systems, 82,* 761–768.

41. Rathore, S., Sharma, K. P., & Park, J. H., (2017). XSSClassifier: An efficient XSS attack detection approach based on machine learning classifier on SNSs. *Journal of Information Processing Systems, 13*(4).

42. Pramanik, P. K. D., Solanki, A., Debnath, A., Nayyar, A., El-Sappagh, S., & Kwak, K. S., (2020). Advancing modern healthcare with nanotechnology, nanobiosensors, and internet of nano things: Taxonomies, applications, architecture, and challenges. In: *IEEE Access* (Vol. 8, pp. 65230–65266). doi: 10.1109/ACCESS.2020.2984269.

43. Rameshwar, R., Solanki, A., Nayyar, A., & Mahapatra, B., (2020). Green and smart buildings: A key to sustainable global solutions. In: *Green Building Management and Smart Automation* (). IGI Global: Hershey, PA, USA.

44. Krishnamurthi, R., Nayyar, A., & Solanki, A., (2019). Innovation opportunities through the Internet of Things (IoT) for smart cities. In: *Green and Smart Technologies for Smart Cities* (pp. 261–292). CRC Press: Boca Raton, FL, USA.

45. Solanki, A., & Nayyar, A., (2019). Green Internet of Things (G-IoT): ICT technologies, principles, applications, projects, and challenges. In: *Handbook of Research on Big Data and the IoT* (pp. 379–405). IGI Global: Hershey, PA, USA.

CHAPTER 2

Blockchain-Based Federated Machine Learning for Solving IoT Security Problems

DIVYA,[1,2] VIKRAM SINGH,[1] and NAVEEN DAHIYA[2]

[1]Department of Computer Science and Applications, CDLU, Sirsa, Haryana, India, E-mail: divyajatain@msit.in (Divya)

[2]Department of Computer Science and Engineering, MSIT, New Delhi, India

ABSTRACT

With the advent of technology, we are witnessing huge potential in devices enabled with sensors having advanced processing/computing capabilities. The internet, as a supporting technology has further helped the research community to gain momentum in the field of inter-sensor communication. Internet of things (IoT) has widely penetrated different aspects of our lives. As a result, many intelligent IoT services and applications are now emerging. However, due to insecure design, implementation, and configuration, these devices have potential vulnerabilities which can be potential problems. IoTs generate huge sets of data that need to be pre-processed, scaled, classified, and analyzed before putting to some use. Machine learning (ML) or artificial intelligence (AI) have proved to be very useful for this purpose, where we can use the enormous data to design and train a model for some analytics.

Traditional ML approaches were centralized and thus created issues related to the communication overhead, delay in processing, and privacy and security concerns owing to different computing capabilities and power of the connected devices. As a result, Google in 2016 has proposed a new method called federated ML, in which, we have numerous clients distributed over different environments that train on the data that is locally available to create a model. All such local models are then sent to the centralized server and merged to create a global model. Finally, this global model is sent as an update to the individual clients independently. Despite having advantages like being able to provide security and ensure privacy, and benefits of application to power constraint scenarios of sensor devices, there are still some areas that need proper attention like vulnerability of having a single centralized optimization at main server and scalability issues, etc. Moreover, the IoT devices are statistically heterogeneous and vulnerable due to insecure design, implementation, and configuration making it a challenging task to deploy Federated Learning directly. It is for this task blockchains (BCs) can be effectively used owing to their fault tolerance, transaction integrity and authentication, decentralization, etc.

In this chapter, the intent is to provide an in-depth analysis of IoT security issues and how Federated Learning along with BC technology can be used to solve them.

2.1 INTRODUCTION

In today's world, one can never forego the role of the Internet as an information provider and information disseminator. The growth of Web 3.0 and Web 4.0 at an exponential rate have witnessed an enormous growth in the number of internet users, where the data is no more the regular structured one, but is an unstructured Big Data. The term Big Data, coined in the 1990s, specifies the huge unstructured or semi-structured data sets that cannot be captured, stored, managed, processed, and analyzed by typical software tools [1]. These datasets have data in varying formats that span over text, sound, image, and/or video, and thus, it is a challenging task to process this data so as to have some useful outcomes. Essentially, nowadays, the rate of creation of this Big Data has captured the scenario in such a way that this accounts for almost 90% of all the data being actually created [2]. Talking about the Big Data, it can be characterized by seven Vs: Volume, Variety, Veracity, Velocity, Variability, Visualization, and Value. There are many enabling technologies that have contributed to the proliferation of Big Data, such as Internet of Things (IoT), Information, and Communication Technologies (ICTs), artificial intelligence (AI), etc.

With the advent of technology, we are witnessing huge potential in devices enabled with sensors having advanced processing/computing capabilities. The Internet, as a supporting technology has further helped the research community to gain momentum in the field of inter-sensor communication.

The IoTs basically consist of interrelated computing devices, which can be some mechanical or digital machines, objects, animals, or people that are having unique identifiers (UIDs). They have the ability to transfer data over a network without any human-to-human or human-to-computer interaction happening. Nowadays, IoTs has widely penetrated different aspects of our life. As a result, many intelligent IoT services and applications are now emerging. However, due to insecure design, implementation, and configuration, these devices have potential vulnerabilities which can be potential problems.

IoTs generates huge sets of data. In order to have some useful results or outcome from this data, it has to be pre-processed, scaled, classified, and analyzed. Machine learning (ML) or AI have proved to be very useful for this purpose, where we can use the enormous data to design and train a model for some analytics.

Traditionally, the ML approaches used to send the data to a central server where the data is processed and then the model is trained. But it created issues related to the communication overhead, delay in processing, and privacy and security concerns because each device may have different computing capabilities and power. As a result, Google in 2016 has proposed a new method called federated ML [3] in which, we have numerous clients distributed over different environments. Every client train itself on the data that is locally available to it and creates a model. All such local models are then sent to the centralized server, which merges them to create a global model. Finally, this global model is sent as an update to the individual clients independently.

There are many advantages of Federated Machine Learning like being able to ensure security and privacy, and ability to being deployed to power constraint scenarios of sensor devices, there are some areas that need proper attention like vulnerability of having a single centralized optimization at main server and scalability issues, etc. Moreover, the IoT devices are statistically heterogeneous and vulnerable due to insecure design, implementation, and configuration making it a challenging task to deploy Federated Learning directly.

Blockchain (BC), as the name suggests, is a chain of blocks wherein each block contains transaction information, hash of the previous block, and a timestamp. Although BC was initiated originally as a financial transaction protocol but due to benefits like fault tolerance, transaction integrity and authentication, decentralization, etc. It is seen as a promising candidate to ensure security and privacy in a variety of applications including IoT.

However, there are certain limitations related to scalability, latency in transactions, storage and energy constraints, intensive computations, etc., that need to be addressed before its applications to the IoT domain. In this chapter, the intent is to provide an in-depth analysis of IoT security issues along with the taxonomy, introduction to federated learning, federated block-chains (FedBlocks) and how federated learning along with BC technology can be used to solve them. The challenges faced by FedBlocks and the future direction of research trends is also discussed in further sections.

2.2   IoT AND ITS EMERGENCE

IoTs, or IoT was first used by Kevin Ashton [4]. It is basically a collection of autonomous sensors-enabled objects/things that communicate via internet [5, 6]. This is an interesting scenario wherein actual real physical and virtual objects both have their own roles to play over the internet infrastructure. As mentioned in Ref. [7], the European research cluster on the Internet of Things (IERC) states IoT to be a network of physical and virtual things that have identities and attributes, have self-configuring capabilities, and are seamlessly interconnected via the internet to an information network.

These smart devices have been used in a wide range of application areas viz., healthcare, transportation, industrial control, commerce, agriculture, energy, etc., to name a few [8]. All these and many more such interesting applications have mesmerized big tech giants like Amazon, Cisco, IBM, Apple, Google, Microsoft, etc., to jump into the field of IoT in the past few years. The European project Unify-IoT has identified that in the current market scenario, there are more than 300 IoT platforms, with an ever-increasing number. Having so many platforms, each with different infrastructures, protocols, interfaces, formats, and standards, the need for interoperability is growing day-by-day. McKinsey analysis [1] signifies the importance of having interoperability by highlighting the threat its absence creates to the economic values. The major application areas are described in Figure 2.1.

2.2.1   CONSUMER APPLICATIONS

Consumer applications include the application areas which are for simple consumers and include domains such as connected vehicles, home automation, wearable technology, etc., to name a few. Some of the detailed examples include:

1. Smart Home: This basically includes home automation, mood lighting, personalized heating, air conditioning, etc. These measures can benefit in the form of energy-saving, and certain psychological benefits like mood upliftment, etc. Smart home kits like Apple's HomeKit have certain dedicated applications in iOS devices like iPhone and Apple Watch. There are also dedicated smart home products such as the Amazon Echo, Google Home, Apple's HomePod, and Samsung's SmartThings Hub.

2. Elderly/Disabled Person Care: One of the most important and wonderful applications of IoT is to provide assistance to disabled and elderly people. These generally use owner specific information to accommodate the individualistic needs of the person. These may include voice control and cochlear implants for vocally disabled and hearing-impaired people, respectively. These might further be equipped with added security features also, thereby providing users a better quality of life experience.

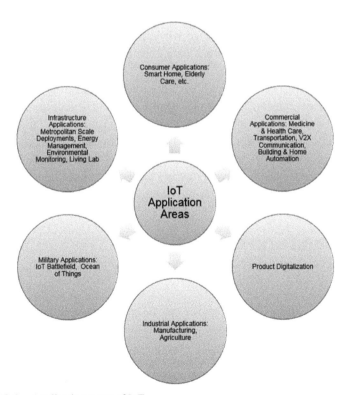

FIGURE 2.1   Application areas of IoT.

2.2.2 INDUSTRIAL APPLICATIONS

In industries, IoT monitoring systems can be used to regulate production scenario, automated updates of stock in storage area, monitoring of workforce, and other equipment so as to provide an a-priori information of critical nature, and saving lots of money and better asset utilization:

1. Manufacturing: The industrial IoT (IIoT) uses digital control systems for automation of process control, operator tools, service information systems for maintaining plant safety and security. This can be made feasible by having advanced sensors with abilities to sense, identify, process, communicate, actuate, and having networking capabilities. The industrial IoT systems enable fast manufacturing of new products, active response to demands of product, and optimization of manufacturing production. Asset management is also expected to have benefits from IIoT by having better network control, better management of manufacturing equipment, proper asset management, to name a few. All this could lead to Fourth Industrial Revolution, by generating business value while having predictive maintenance, maximization of reliability, energy optimization, better health, safety management, etc.

2. Agriculture: In the field of agriculture, collecting temperature, humidity, rainfall data or information about infection from pests or insects, soil mineral content information, etc., can be used for a variety of applications like taking better farming decisions, automation of farming techniques, minimization of crop wastage and better crop management. One interesting application may be to have automated fertilization or watering systems.

2.2.3 COMMERCIAL APPLICATION

Some of the commercial applications of IoT are listed below:

1. Medical and Healthcare: In the field of medical and healthcare internet of medical things (IoMT) is the buzzword. It collects data and analyzes it for research and monitoring of patients' records. Thus, smart healthcare aims for creating digitized system while connecting available resources and services. Remote health monitoring and health emergency notification systems is one such allied application domain.

Health monitoring devices can vary from simple blood pressure detection systems to advanced devices capable of monitoring specialized implants, such as pacemakers. Smart beds specialized to patient-specific needs, that help the patient without intervention of attendants and nurses are also underway. M-health, which uses mobile devices to support follow up of patients is also the latest intervention.

Homes can also be equipped with specialized sensors so that active monitoring of the health and well-being of persons living in them can be carried out, and before the happening of any emergency, actions can be taken well in advance. These sensors might be equipped with abilities for collection, processing, transfer, and analysis of information. Other devices such as FitBit or Apple Watch might promote healthy habits by maintaining calorie count, heart rate, number of steps, sleep hours, etc.

Thus, IoMT is empowering the doctors, patients, and other stakeholders to be a part of the monitoring system, by allowing dynamic access of information to them. This intervention has played a key role in managing chronic diseases and in disease prevention and control to a greater extent.

2. Transportation: Integration of communications, control, and information processing across various transportation systems has made smart traffic management, smart parking, e-toll collection, e-logistics, road assistance, vehicle control a reality now. Combined with ML, this field can prove to be a big hit amongst people.

3. V2X Communications: This basically signifies vehicle to everything communication. It basically has three components: intravehicular communication (V2V), vehicle to infrastructure (V2I) and vehicle to people (V2P) on-road communication (V2P); this is a major driving force behind autonomous driving infrastructure.

4. Building and Home Automation: IoT devices can be used to monitor and control the mechanical, electrical, and electronic systems used in public, private, industrial, and residential buildings. This might end up for having energy-efficient, smart buildings where a real time monitoring of occupant behavior can be done.

2.2.4   INFRASTRUCTURE APPLICATIONS

IoT devices can be used for controlling and monitoring critical urban and rural infrastructures infrastructure like bridges, railway tracks, etc. Using

IoT devices for these operations might improve management and response to critical emergency situations, saving time and costs, better quality workday, paperless workflow, reduction in risks and increased productivity. Fast decisions and corresponding reduction in money expenditure, improved quality of service is an added outcome of the implementation in infrastructure domain:

1.   Metropolitan Scale Deployments: There are several deployments of the IoT, to enable better management of cities in the form of smart cities. One such example is, Songdo, South Korea, the first fully equipped and wired smart city which has all the devices running on their own without any actual human intervention.

     Another project that is currently underway is in Santander, Spain. Here an app is developed which is connected to sensors that help the inhabitants to search parking space, environmental monitoring, etc.

2.   Energy Management: Deploying IoT enabled sensors might help in a wide range of applications like reducing energy spending, minimization of carbon emissions, integration of green energy, automation of processes, optimization of asset maintenance, cutting operational expenses, transparency to energy usage, efficient combating of power outages, and active prediction of consumption, to name a few. Today a large number of devices are already having internet connectivity, which allows them to communicate amongst themselves and even to send data to the cloud for active analysis. One such example is having a sensor enabled smart grid with advanced metering infrastructure to manage power distribution.

3.   Environmental Monitoring: Using IoT enabled sensor with having cloud to analyze and process data, one can achieve a number of goals including environmental protection, air quality monitoring, extreme weather monitoring, water, and land safety, monitoring movement of endangered species for protection, commercial farming, natural disaster early warning system, etc. Not all places are geographically conducive for active intervention by humans, and in those areas, sensor deployment can actually serve the purpose of data collection.

4.   Living Lab: This basically integrates research and innovation process within a public-private-people-partnership to collaborate and share knowledge between stakeholders for creating innovative and technological products. Living labs can be actively used for developing IoT services in many domains such as logistics in food supply chains,

development of smart cities, etc., to name a few. To obtain a win-win situation for all, the government may provide reliefs in tax procedures, cheap rate rents, improved transports, etc., and provide a good atmosphere in which the start-ups can flourish.

2.2.5   MILITARY APPLICATIONS

Internet of Military Things is an interesting application of IoT technologies in the domain of defense services for combat-related aims. It includes the use of sensors to enable defense personnel to have strategic advantages over their enemies. It includes the following initiatives:

1. Internet of Battlefield Things (IoBT): It is a project of the U.S. Army Research Laboratory (ARL) that focuses on the application of IoTs to improve the capabilities of the soldiers in the battlefield. This encompasses devices possessing intelligent sensing, learning, and actuation capabilities through virtual or cyber interfaces that are integrated into systems. These devices include sensors, smart vehicles, robots, Unmanned Ariel Vehicles, wearable devices, weapons, and other smart technology.
2. Ocean of Things: The Defense Advanced Research Projects Agency, USA undertook a special project called Ocean of Things which works over the large oceans to collect, monitor, and analyze data related to environment or vessel activity. The data collected from these sensors can be analyzed and used for a wide variety of applications, including oceanographic and meteorological models.

2.2.6   PRODUCT DIGITIZATION

Nowadays, one must have come across smart packing having QR Codes or NFC tags affixed over the packages. However, these are not strictly speaking IoT, but these QR codes basically are linked to URL, where one can find the digital information about the product in detail. To identify such products, a new term Internet of Packaging is actively being used, which basically describe applications using UIDs for automation of supply chains and/or by consumers to access digital information about the product.

The importance of having interoperability can be viewed from the perspective of providers as well as application developers. As the market is day-by-day becoming more mature, it is being costlier for the companies to support diversity and heterogeneity. Incompatibility between the IoT platforms leads to the applications becoming non-portable, thereby denying cross-domain and multiplatform application development. Due to these and many other factors, the need for interoperability is one of the key requirements by the industry, and to achieve this goal, a lot of research is currently being carried out. The research includes surveys dealing with enabling technologies, application areas and use cases [9–13] discussions about interoperability challenges and issues [14], semantic interoperability, security, and privacy, smart things and resilience and reliability [14], etc.

2.3   IoT SECURITY CHALLENGES

In a typical IoT environment, we have heterogeneous interconnected sensor-enabled devices that are uniquely identifiable. These devices generally have low power, less memory, and very constrained processing abilities. In the most common layered architecture of the IoT system, gateways connect them to the outside world, and there are some common protocols for key tasks like routing, authentication, messaging, key management, etc. Figure 2.2 shows some of the major issues faced by IoTs today.

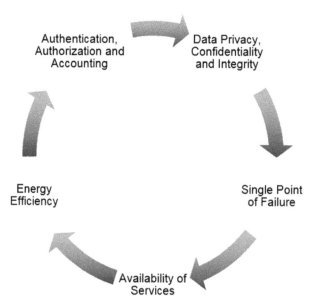

FIGURE 2.2   IoT issues.

1. **Authentication, Authorization, and Accounting:** In order to have secure communication, robust authentication is a need. Having a proper authorization and authentication mechanism ensures safety and security. However, owing to the diverse nature of IoT devices and their architectures, authentication gains even more importance. Variable environments pose even greater challenges, and so a global authentication mechanism is not a solution. Accounting for resource usage is also an issue.
2. **Availability of Services:** There are attacks like denial of service, sinkhole attack, adversarial jamming, replay attacks, etc., that can exploit the vulnerabilities in the devices and can deteriorate the quality of services provided to the end-users.
3. **Energy Efficiency:** In IoT devices the resources, especially power consumption and storage are major concerns. Keeping the devices functioning smoothly with limited resources to the maximum capabilities is a research area. There are various attacks that try to exploit these vulnerabilities and exhaust the resources for their advantage.
4. **Single Points of Failure:** IoT systems are particularly vulnerable to single point of failure (SPF), which can lead to a complete deterioration of services envisioned. Developing methods and mechanisms that make the system and devices fault-tolerant and attack-proof drives major research in the domain.
5. **Data Privacy, Confidentiality, and Integrity.** The security and privacy of data is central to any of the domains of research and application. Particularly in the case of IoT scenario, the data needs to be aggregated from multiple parties where each of the party may be having a different underlying technical architecture. In such a scenario maintaining the data privacy, confidentiality and integrity is a big challenge.

2.3.1 TAXONOMY OF IoT SECURITY ISSUES

IoT comprises of a wide variety of devices with different processing capabilities, and thus, there are a number of security challenges that researchers focus on. These can be identified as low threat level, medium threat level and high threat level challenges as mentioned in Figure 2.3. A detailed analysis of these challenges is described below.

The basic level security issues deal with the hardware and/or physical and data link layer of the communication network. These issues can be further classified as:

- As mentioned in Refs. [15, 16] jamming works by emitting radio frequency signals without protocol, which severely affects network operations by inhibiting the data sent or by malicious data sent to the legitimate nodes and thereby lead to serious deterioration or malfunctioning in the target network.
- To ensure proper functionality of the IoT system, a robust and secure mechanism needs to be maintained at the physical layer [17, 18]. Having this mechanism at place also makes the system secure to the unauthorized receivers.

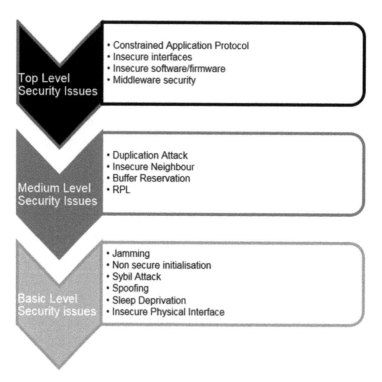

FIGURE 2.3   Taxonomy of IoT security issues.

- In Sybil and Spoofing the nodes assume fake identities to degrade the system's functionality, by having a node forge the MAC address, so as to deplete the precious network resources in a way that the actual, legitimate nodes may not have access to the resources [18, 19].

- Poor physical security [20] may wreck a havoc over the functioning of the devices in the network, and may even impact the testing/ debugging tools.
- In sleep deprivation attack [21], the sensor nodes, which are otherwise also having a limited battery, are made to stay awake, so that their energy rapidly depletes and result in a device being off the system soon.

Medium-level security issues are the issues concerning the network and transport layer, thus dealing mainly with communication, routing, and management of session. Some of these attacks are discussed as follows:

1. Duplication attack as mentioned by Kim et al. is basically a scenario where the reassembly of the legitimate packets (created by fragmentation of IPv6 packets) is hindered by duplicate, malicious nodes, resulting in depletion of resources, buffer overflows, and even device reboot [22, 23].
2. In IoT scenario, each device has a unique id, and the process of identification of the device uniquely needs to be end-to-end secure. The process of neighboring node discovery, if done without proper verification may have serious negative outcomes, with Denial-of-Service attack being most important.
3. As mentioned previously, since the packets need to be reassembled, the attacker might attack the system by sending incomplete packets [23]. This will lead to a wastage of the reserved buffer space, which a receiving node maintains for re-assembly of incoming fragments.
4. According to the authors in Ref. [24], eavesdropping, and resource wastage may be an outcome of having compromised nodes. The routing protocol for low-power and lossy networks (RPL) is particularly vulnerable to this sort of scenario.
5. There are certain other issues such as sinkhole and wormhole attack [25–28] which have implications in the form of eavesdropping, privacy issues and denial of service [29], phishing attacks [30, 31], issues related to authentication and secure communication [32, 33], transport layer security issues [32], and privacy violation on cloud-based IoT [34].

Top-level security issues are the issues related to applications that run over IoT. These are basically described as follows.

A constrained application protocol attack [35, 36] is a high-level attack in which the key management and authentication mechanisms are exploited by attackers. The interfaces which are used for accessing IoT services can have multifaceted nature, using either web services, or mobile or even cloud platform, and this may lead to potential pitfalls in data privacy context [20].

Serious vulnerabilities are found to exist due to the usage of insecure software/firmware [20]. The middleware used in IoT paradigm must be secure enough for communication in the heterogeneous environment [37, 38].

2.4 INTRODUCTION TO FEDERATED LEARNING

The Facebook data breach [39] exposed the vulnerabilities of the existing systems that are using the data extensively for their day-to-day operations. In the aftermath of this breach, people are now becoming more aware, concerned, and curious about how, when, and where their data is being used. Any sort of unauthorized use or exploitation of this data is a big threat to both the person and to the enterprise, and may have some really serious consequences. In keeping these things in mind, the countries all over the world are making new laws to protect data security and privacy. One such law is general data protection regulation (GDPR) [40] enacted by the European Union on May 25, 2018, which gives the users greater authority of their personal data [41]. Protection of user data privacy and security is the main aim of GDPR. The traditional models were transactional in the sense that the party responsible for the collection of data would transfer it to another party, which is, in turn, is responsible for data cleaning and fusing. But the Facebook breach and consequent GDPR enactment poses new challenges in the data-transaction procedures.

2.4.1 WHAT IT IS?

Federated learning, as introduced by Google [3, 42, 43] is a method to train ML models in such a manner that the data stays localized to the device, while the model is being trained in a distributed manner. The methodology is giving the clients permission to encrypt their models, and then uploading and aggregating these models at the cloud in a centralized manner. These encrypted models are then used for model building while keeping the client information and client data secure.

Let us have m data owners {F1,... Fm}, having their respective data {D1,... Dm}. In the conventional method, a ML model M' is trained by consolidating the data of individual owners as $D' = D1 \cup \cdots \cup Dm$

In the Federated Learning system, the data owners collaboratively train a Federated learning model Mfed, where some data owner Fk does not show its data Dk to others. Moreover, the system must be as accurate to the traditional method, i.e., if we specify the accuracy of the federated model Mfed, as Vfed, then it should be very close to the performance of M', V'.

Formally, δ-the accuracy loss of the Federated Learning Algorithm, be a non-negative real number represented as:

$$|Vfed - V'| < \delta \tag{1}$$

The traditional ML models were having issues in ensuring user data security and privacy. The primary advantage of Federated Learning in this context is solving issues related to scalability, improved accuracy, reduced training time, better throughput and greater privacy and security. Moreover, Federated Learning confirms to the GDPR as mandated by European Union law and thus, build more effective, secure, and private models.

2.4.2   TYPES OF FEDERATED LEARNING

In order to have a complete understanding of the taxonomy of Federated Learning, we have chosen to classify it on the basis of method of data partition, on the basis of modeling method used, on the basis of communication architecture, on the basis of level of privacy. These dimensions are then subdivided further to provide a detailed categorization of the systems as shown in Figure 2.4.

2.4.2.1   ON THE BASIS OF DATA PARTITIONS

On the basis of data partition, we further have three categorizations: horizontal federated learning, vertical federated learning, and federated transfer learning. This all depends upon the pattern of distribution of data among different parties in the feature and sample ID space.

Consider an example of two shopping websites having their markets spread over different geographical locations, thus they may have different set of users. They might also have a very small set of common users. However, the feature space in this context is the same because the nature of the business is the same. This is the case of *horizontal federated learning*; Shokri and Shmatikov [44] proposed a collaboratively deep-learning scheme within which the participants train independently and only subsets of updates of parameters is shared among them.

Google in 2017, proposed a horizontal federated-learning framework for Android phone model updates [43]. According to this model, the user having an Android mobile phone, locally updates the parameters and uploads these parameters to the Android cloud. So, in essence, every single user contributes and collaborates to train the centralized model along with other data owners. Secure aggregation scheme of Bonawitz et al., [45] ensures the privacy of aggregated user updates. In Ref. [46], the authors have used additively homomorphic encryption for aggregating the model parameters and ensuring security to the model.

Researchers in Ref. [47] have proposed a federated learning system of multi task style, where separate tasks can be performed in addition to knowledge sharing, security preservation and ensuring fault tolerance, all at once. Issues like high communication costs, and stragglers are also handled effectively. In the research Konečný et al. [3], facilitate the training of centralized models on the basis of data distributed over mobile clients, to improve communication costs. Deep gradient compression as proposed by Lin et al. [48] greatly reduces the communication bandwidth in large-scale distributed training.

Vertical federated learning is used in the cases where the two datasets share the same sample ID space but different feature space. Consider an example of two organizations, viz. a kindergarten school and a primary school in a small city. It is highly likely for their data set to contain most of the toddlers of the area, and so there is a large intersection of their userspace. However, their feature spaces are different owing to difference in the records. Thus, different features are aggregated in a privacy-preserving way wherein data is considered from both parties collaboratively.

When the data sets that are considered are different in both the sample space and the feature space, in such a scenario, we use *federated transfer learning* [49]. It is further classified as secure federated transfer learning [50] and federated transfer learning with secret sharing [51].

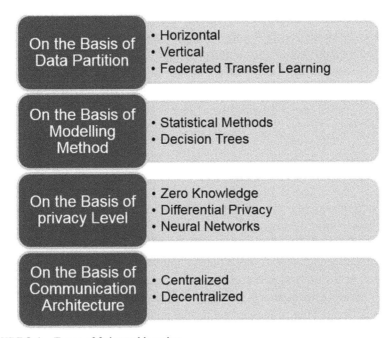

FIGURE 2.4   Types of federated learning.

2.4.2.2   ON THE BASIS OF MODELING METHOD

On the basis of the ML model used, the federated systems can be classified into three categories viz., *statistical method based federated learning* are the most commonly used methods due to ease of learning and have main representative methods as Linear regression and logistic regression [52–54]. However, the performance of *decision tree based federated learning* viz. gradient boosting decision trees [55], random forests [56] is better in many classification and regression tasks. Bonawitz et al. and Yang et al. [57, 58] in their research on neural network-based federated learning have also shown promising results.

2.4.2.3   ON THE BASIS OF PRIVACY LEVEL

On the manner in which the privacy levels are maintained in the federated learning, the system can be classified into three different privacy restriction levels. The first one is *complete zero-knowledge* systems, where the participants have knowledge only about the outputs, for instance, secure multi-party computation [59] and combination of cryptographic techniques such as secure aggregation, homomorphic encryption [45] and systems based on trusted processors such as IntelSGX [60]. The other methods are *differential privacy* [61], and *raw data protection* [50, 62]. All these methods have their own benefits and shortcomings, which make them usable in different scenarios.

2.4.2.4   ON THE BASIS OF COMMUNICATION ARCHITECTURE

On the basis of way of communication, the federated learning systems can be classified as having: *centralized architecture* [57] or *distributed architecture* [61, 62]. In the centralized method, as the name suggest, the communication is centralized in nature, with each participant sending the information to be aggregated at the central server. However, centralized design despite being used more widely, leads to a potential risk situation by being a SPF.

Using federated learning over traditional methods has a lot of advantages, some of which are scalability, improved accuracy, reduction in training time and training cost, enhanced privacy, and security.

2.5   BLOCKCHAINS (BCS) AND FEDERATED BLOCKCHAINS (FEDBLOCK)

Blockchains [63], as the name itself suggests, is a chain of blocks, where we have a group of interconnected systems that are fully open and transparent. The main use of BC technology is to make decentralized applications (DApps), which act as a collaborative network, wherein each system is having exactly the same copy of the database, and each modification to this is governed by a mathematically identified consensus mechanism. The database in this method is known as the public ledger and every transaction in the ledger is verified by consensus of a majority of the participants in the system. When any information is entered into the system, it essentially becomes permanent, verifiable, and secure. In contrast to the traditional systems, any central server or agent is not required to establish trust. The BC is the technological base for all those names like bitcoin (BTC), Hyperledger, Ethereum (ETH).

The BC exists as a peer-to-peer (P2P) network, stores data that is written by certain members, read by certain members, and has a very hard mechanism to modify or delete the historical records. Thus, it can be used as a way to develop an open, scalable, digital world with the convenience of being transparently distributed among various systems of the network, yet not being copied by anyone.

BCs have been shown to have both financial and non-financial applications. On the financial front, we have seen benefits of cryptocurrency like BTCs and ETH, while the non-financial applications also seem to be endless, such as putting proof of existence of all legal documents, health records, and loyalty payments in the music industry, notary, private securities, and marriage licenses in the BC. By storing the fingerprint of the digital asset instead of storing the digital asset itself, the anonymity or privacy objective can be achieved [64].

BC as a technology has shown its non-controversial nature and flawless working over a period of time, and this makes it a potential candidate to be integrated with the techniques like federated ML to solve the problems as faced by the IoT.

2.6   FEDERATED BLOCKCHAINS (FEDBLOCKS) AS A SOLUTION TO IoT SECURITY CHALLENGES

The rate with which the IoTs has grown might surprise a commoner, but the advent of internet enabled smart devices, mobile networks, and computing technology have provided a huge boost to the field. The potential that this domain actually has is enormous and has impacted almost all aspects of our modern life, including smart healthcare system, intelligent transportation infrastructure, etc., to name a few. The data gathered from these sensors deployed at varied locations need appropriate prediction and classification models. But the more important question remains-how to train ML model effectively and efficiently?

Having traditional ML methods may increase communication costs and transmission latency, as well as arising privacy issues by having sensitive data uploaded overcloud. Having the model trained on-site and updation being done to the remote cloud periodically is a solution, as done in Federated learning. However, this approach puts an additional pressure on IoT devices with limited computational, energy, and memory resources.

In the case of Federated Learning, the system is better than the traditional mechanisms, still having a central server may pose some added risks. In order to thwart these issues and to increase the security of the central Federated

Learning server, BC can be integrated with Federated Learning mechanism. In the previous sections, we have seen a detailed analysis of how conventional ML suffers from serious issues related to privacy and security of user data, and how Federated Learning has emerged as a potential candidate to solve these issues. In fact, BCs, as a decentralized, time-sequenced, origin preserving, immutable ledger technology [63] is found to provide an impressive way to integrate with federated learning and to solve the issues faced by IoT.

To this end, FedBlocks have emerged as an answer to the problem. Having integrated BC and Federated Learning, we have a robust method to train decentralized learning models that ensure privacy of the data. With the help of BCs, the trained model parameters along with the source of data origin can be securely maintained. As discussed by Ref. [64], the data that is collected from geographically separate sites for training of the centralized server are prone to issues like privacy attacks, network delay, etc. Rather, using a FedBlock enables to keep track of the model after every epoch, and even the performance of FedBlock is experimentally found to be comparable to normal federated learning method [65]. Some architecture Kim et al. in Ref. [66] have also simulated the on-device federated learning architecture over BC as a method for efficient model training.

In IoT scenario, updating models using the latest data as and when the sensors sensed it, has led to the development of many intelligent IoT applications adaptable to the changing environment [67]. For several applications like weather forecasting [68] this becomes of critical nature. To quantify the freshness of data in IoT [69] have proposed a metric called age of information (AoI). Using BC, a security layer can be added that ensures care for the presence of Byzantine devices. Moreover, the devices can be logged in a tamper proof manner, thus a proper record is there to look at the behavior of the deployed sensors.

2.7 CHALLENGES FACED BY FEDERATED BLOCKCHAINS (FEDBLOCKS)

FedBlocks is currently in a nascent stage of development, and thus there are a number of open ends that need to be dealt with before integrating the technology into some powerful mechanism to have strategic advantages. Some of the major challenges in front of research community working in this direction are given in Figure 2.5.

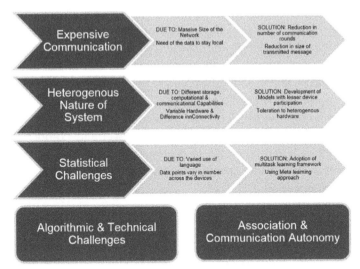

FIGURE 2.5   Challenges faced by federated blockchains.

The massive size of the IoT network makes communication quite expensive. This is further enhanced by the need of the data to stay local to the device, where each update is sent to the cloud, thereby making communication even more time consuming. One possible solution to the problem may be a reduction in the number of communication rounds taking place between the participating nodes. Thus, by deferring the updates for some time being and doing updations only when totally necessary. Even a reduction in the size of transmitted messages would provide the same outcome.

In traditional ML systems, the system can handle a less precise model at the cost of generalization as shown by Ref. [70]. But Federated Learning aims for a better and precise model. Huang et al. [71] in their research, have shown the superiority of computations of local updates over massive networks. One may also aim for reduction in communication rounds and reduction in size of a message.

The devices that collaborate amongst each other have heterogeneity in terms of storage, computational, and communicational capabilities. Moreover, the battery life and other such constraints result in having only a limited small number of devices out of the huge billions of devices being active at a time [57]. Also due to power issues, the dropout rate makes the system quite unreliable. So, keeping this in mind, the system must be robust to handle dropouts effectively.

In practical scenarios, the devices from which data is gathered are so varied in their nature that the data generation and collection process differ widely. This inherent variation adds to the complexity of modeling, analysis, and evaluation. The approaches specified by Stanford, Chiang, and Sanjabi [72]; Li, Khodak, Caldas, and Talwalkar [73] both allow for device specific modeling to handle statistical heterogeneity. There is a nice study of comparison done by Nilsson et al. [74] over stochastic gradient descent method, federated stochastic variance reduced gradient [42] and co-operative ML model [75], all of which are used to represent training being handled and controlled via a central server which hosts the shared global model for a round of communication.

In keeping up the autonomy of the communicating devices, the system addresses association autonomy and communication autonomy. These basically deal with the autonomy of the devices to associate or de-associate themselves from the communication and the discretion of a device to share the information fully, to some extent or not at all. For instance, Google's federated learning system [57] can tolerate the drop-outs of the devices. BC [76] is an ideal and transparent platform for multi-party learning.

Federated Learning solves privacy and security issues of the traditional methods to a large extent, but still, there are some challenges left in the system. Some of the recent works [77–79] have worked on preserving raw data on each device. Bonawitz et al. and Geyer, Klein, and Nabi [45, 80] have done their work on making the system more secure. Recently, a lot of work has also been done for ensuring privacy-preserving in the system. The major methods for achieving this are secure multiparty computation protocols [81, 82], three-party computation model [83, 84], differential privacy [85] or k-anonymity [86] and diversification [87]. Protecting client-side privacy protection, Geyer in 2017 [80] proposed a method that hides client's contributions during training.

Homomorphic encryption [88] has also been used as a way to protect the privacy of user data via the exchange of parameters [53, 89, 90]. A modified method, additively homomorphic encryption [91] uses homomorphic encryption for centralizing and training data on the cloud.

2.8 CONCLUSIONS AND FUTURE SCOPE

In this chapter, it has been discussed that how IoT has emerged as a game-changer in the current scenario, collecting, and fusing data from the sensors being a pivotal task. Moreover, discussion is done over how the traditional ML algorithms that were used to train models running over this big data are unsuitable for the task owing to a large number of issues, with privacy and security being the major ones. After the Facebook data breach, this issue gained much prominence, so a new computing paradigm, federated ML was introduced by Google. Integrating federated learning with BCs is a latest and an interesting paradigm that can open up enormous possibilities of safe, secure, private data transfer and/or manipulations. This can serve as a case study for various domain applications. In the context of IoT, this has significantly found to be quite interesting for the research community, addressing the challenges of IoT security in a very lucid manner. Stating all this, further a lot of improvements can still be done in the area, some of which are discussed as follows:

- Inter device communication between the heterogeneous devices is a domain which needs active research so as to maintain a tradeoff between the cost of communication and accuracy.
- Statistical heterogeneity, as in dealing with storage and computation in the devices need to get proper quantification, which in fact, is a very difficult task to achieve.
- Since the data generated by the sensors is either unlabeled or weakly labeled, appropriate handling of this data is of utmost importance.
- In the IoT scenario, the devices may not be available for sensing and communication always. Due to power constraints, the devices may drop out. Handling these scenarios while maintaining a robust network requires the development of new methods and techniques.
- According to Ref. [92], despite having a robust approach, BCs still have certain vulnerabilities, such as the dependence of consensus mechanism on the miner's hashing power. Moreover, the private keys can also be exploited to compromise the BC accounts.

Benchmarking is also one such area that is of particular interest to the research community. Since the concept of Federated Learning in itself is an active research area, integrating it with BCs needs some serious consideration of techniques, methods, and methodology for evolving into a higher, superior method of secure and private computations.

KEYWORDS

- **artificial intelligence**
- **Industrial Internet of Things**
- **information and communication technologies**
- **Internet of Medical Things**
- **Internet of Things**
- **unique identifiers**
- **vehicle to infrastructure**

REFERENCES

1. Manyika, J., (2011). In: Chui, M., Brown, B., Bughin, J., Dobbs, R., Roxburgh, C., & Byers, A. H., (eds.), *Big Data: The Next Frontier for Innovation, Competition, and Productivity*. Mc Kinsey and Company, Google Search.
2. Gantz, J., & Reinsel, D., (2011). *Extracting Value from Chaos*. IDC's Digital Universe Study, sponsored by EMC, Google Search.
3. Konečný, J., McMahan, H. B., Yu, F. X., Richtárik, P., Suresh, A. T., & Bacon, D., (2016). *Federated Learning: Strategies for Improving Communication Efficiency*.
4. Ashton, K., (2010). *Related Content RFID-Powered Handhelds Guide Visitors at Shanghai Expo Despite Sluggish Growth, Taiwan's RFID Industry Remains Committed Mobile RTLS Tracks Health-Care Efficiency RFID Journal LIVE! 2010 Report, Part 2 That "Internet of Things."*
5. Atzori, L., Iera, A., & Morabito, G., (2010). The Internet of Things: A survey. *Comput. Networks, 54*(15), 2787–2805. https://doi.org/https://doi.org/10.1016/j.comnet.2010.05.010.
6. Gubbi, J., Buyya, R., Marusic, S., & Palaniswami, M., (2013). Internet of things (IoT): A vision, architectural elements, and future directions. *Futur. Gener. Comput. Syst., 29*(7), 1645–1660. https://doi.org/10.1016/j.future.2013.01.010.
7. Rob van Kranenburg: The Internet of Things. A critique of ambient technology and the all-seeing network of RFID | The Mobile City http://themobilecity.nl/2008/10/16/rob-van-kranenburg-the-internet-of-things-a-critique-of-ambient-technology-and-the-all-seeing-network-of-rfid/2008 (accessed May 17, 2020).
8. Sundmaeker, H., Guillemin, P., Friess, P., & Woelfflé, S., (2010). *Vision and Challenges for Realizing the Internet of Things the Meaning of Things Lies Not in the Things Themselves, but in Our Attitude Towards Them*. Antoine de Saint-Exupéry.
9. Noura, M., Atiquzzaman, M., & Gaedke, M., (2019). Interoperability in Internet of Things: Taxonomies and open challenges. *Mob. Networks Appl., 24*(3), 796–809. https://doi.org/10.1007/s11036-018-1089-9.
10. Perera, C., Zaslavsky, A., Christen, P., & Georgakopoulos, D. (2014). Context aware computing for the Internet of Things: A survey. *IEEE Communications Surveys and*

Tutorials, 16(1), 414–454. https://doi.org/10.1109/SURV.2013.042313.00197 (accessed May 17, 2020).

11. Al-Fuqaha, A., Guizani, M., Mohammadi, M., Aledhari, M., & Ayyash, M., (2015). Internet of things: A survey on enabling technologies, protocols, and applications. *IEEE Commun. Surv. Tutorials, 17*(4), 2347–2376. https://doi.org/10.1109/COMST.2015.2444095.

12. Xu, L., He, W., & Li, S., (2014). Internet of things in industries: A survey. *IEEE Trans. Ind. Informatics, 10*, 2233–2243. https://doi.org/10.1109/TII.2014.2300753.

13. Bandyopadhyay, S., Sengupta, M., Maiti, S., & Dutta, S., (2011). Role of middleware for Internet of Things: A study. *Int. J. Comput. Sci. Eng. Surv., 2*. https://doi.org/10.5121/ijcses.2011.2307.

14. Gazis, V., Gortz, M., Huber, M., Leonardi, A., Mathioudakis, K., Wiesmaier, A., … Vasilomanolakis, E. (2015). A survey of technologies for the Internet of Things. *IWCMC 2015 – 11th International Wireless Communications and Mobile Computing Conference*, 1090–1095. https://doi.org/10.1109/IWCMC.2015.7289234.

15. Xu, W., Trappe, W., Zhang, Y., & Wood, T. (2005). The feasibility of launching and detecting jamming attacks in wireless networks. *Proceedings of the International Symposium on Mobile Ad Hoc Networking and Computing (MobiHoc)*, 46–57. https://doi.org/10.1145/1062689.1062697.

16. Noubir, G., & Lin, G., (2003). Low-power DoS attacks in data wireless LANs and countermeasures. *Mob. Comput. Commun. Rev., 7*, 29–30. https://doi.org/10.1145/961268.961277.

17. Hong, Y. P., Lan, P., & Kuo, C. J., (2013). Enhancing physical-layer secrecy in multiantenna wireless systems: An overview of signal processing approaches. *IEEE Signal Process. Mag., 30*(5), 29–40. https://doi.org/10.1109/MSP.2013.2256953.

18. Liang, X., Greenstein, L. J., Mandayam, N. B., & Trappe, W., (2009). Channel-based detection of Sybil attacks in wireless networks. *IEEE Trans. Inf. Forensics Secur., 4*(3). https://doi.org/10.1109/TIFS.2009.2026454.

19. Chen, Y., Trappe, W., & Martin, R. P. (2007). Detecting and localizing wireless spoofing attacks. *2007 4th Annual IEEE Communications Society Conference on Sensor, Mesh and Ad Hoc Communications and Networks, SECON*, 193–202. https://doi.org/10.1109/SAHCN.2007.4292831.

20. OWASP, Top IoT Vulnerabilities, (2016). URLhttps://owasp.org/www-project-internet-of-things/ – Google Search (accessed August 8, 2021).

21. Bhattasali, T., & Chaki, R., (2011). In: Wyld, D. C., Wozniak, M., Chaki, N., Meghanathan, N., & Nagamalai, D., (eds.), *A Survey of Recent Intrusion Detection Systems for Wireless Sensor Network BT – Advances in Network Security and Applications* (pp. 268–280). Springer Berlin Heidelberg: Berlin, Heidelberg.

22. Kim, H., (2008). *Protection Against Packet Fragmentation Attacks at 6LoWPAN Adaptation Layer*; 2008. https://doi.org/10.1109/ICHIT.2008.261.

23. Hummen, R., Hiller, J., Wirtz, H., Henze, M., Shafagh, H., & Wehrle, K., (2013). *6LoWPAN Fragmentation Attacks and Mitigation Mechanisms*.

24. Dvir, A., Holczer, T., & Buttyán, L., (2011). *VeRA – Version Number and Rank Authentication in RPL*. https://doi.org/10.1109/MASS.2011.76.

25. Weekly, K., & Pister, K. (2012). Evaluating sinkhole defense techniques in RPL networks. *Proceedings – International Conference on Network Protocols, ICNP*. https://doi.org/10.1109/ICNP.2012.6459948.

26. Ahmed, F., & Ko, Y. B., (2016). Mitigation of black hole attacks in routing protocol for low power and lossy networks. *Secur. Commun. Networks, 9*(18), 5143–5154. https://doi.org/10.1002/sec.1684.

27. Pirzada, A. A., & Mcdonald, C. (2005). Circumventing Sinkholes and Wormholes in Wireless Sensor Networks. *International Workshop on Wireless Ad-Hoc Net- Works.*

28. Wang, W., Kong, J., Bhargava, B., & Gerla, M., (2008). *Visualisation of Wormholes in Underwater Sensor Networks: A Distributed Approach, 3.*

29. Hamada, I. M., (2016). *Octopus: An Edge-Fog Mutual Authentication Scheme, 18.*

30. Zhang, K., Liang, X., Lu, R., & Shen, X., (2014). Sybil Attacks and their defenses in the Internet of Things. *Internet Things Journal, IEEE, 1*, 372–383. https://doi.org/10.1109/JIOT.2014.2344013.

31. Wang, G., Mohanlal, M., Wilson, C., Wang, X., Metzger, M., Zheng, H., & Zhao, B., (2012). *Social Turing Tests: Crowdsourcing Sybil Detection.*

32. Granjal, J., Monteiro, E., & Silva, J. S. (2013). End-to-end transport-layer security for Internet-integrated sensing applications with mutual and delegated ECC public-key authentication. *2013 IFIP Networking Conference, IFIP Networking 2013.*

33. Raza, S., Duquennoy, S., Chung, T., Yazar, D., Voigt, T., & Roedig, U., (2011). *Securing Communication in 6LoWPAN with Compressed IPsec.* https://doi.org/10.1109/DCOSS.2011.5982177.

34. Henze, M., Wolters, B., Matzutt, R., Zimmermann, T., & Wehrle, K. (2017). Distributed configuration, authorization and management in the cloud-based Internet of Things. *Proceedings – 16th IEEE International Conference on Trust, Security and Privacy in Computing and Communications, 11th IEEE International Conference on Big Data Science and Engineering and 14th IEEE International Conference on Embedded Software and Systems*, 185–192. https://doi.org/10.1109/Trustcom/BigDataSE/ICESS.2017.236.

35. Brachmann, M., Keoh, S. L., Morchon, O., & Kumar, S., (2012). *End-to-End Transport Security in the IP-Based Internet of Things.* https://doi.org/10.1109/ICCCN.2012.6289292.

36. Sethi, M., Arkko, J., & Keränen, A., (2017). *End-to-End Security for Sleepy Smart Object Networks Year: 2012.* Version: Post Print.

37. Conzon, D., Bolognesi, T., Brizzi, P., Lotito, A., Tomasi, R., & Spirito, M. A., (2012). The VIRTUS middleware: An XMPP based architecture for secure IoT communications. In *21st International Conference on Computer Communications and Networks (ICCCN)* (pp. 1–6). https://doi.org/10.1109/ICCCN.2012.6289309.

38. Liu, C. H., Yang, B., & Liu, T., (2014). Efficient naming, addressing and profile services in internet-of-things sensory environments. *Ad Hoc Networks, 18*, 85–101. https://doi.org/10.1016/j.adhoc.2013.02.008.

39. *Facebook Dataleak*, (2018). Wikipedia, Google Search.

40. Hoofnagle, C. J., Sloot, B. V. D., & Borgesius, F. Z., (2019). The European Union General Data Protection Regulation: What it is and what it means. *Inf. Commun. Technol. Law, 28*(1), 65–98. https://doi.org/10.1080/13600834.2019.1573501.

41. Goodman, B., & Flaxman, S., (2016). *European Union Regulations on Algorithmic Decision-Making and a 'Right to Explanation.'* https://doi.org/10.1609/aimag.v38i3.2741.

42. Konečný, J., McMahan, H. B., Ramage, D., & Richtárik, P. (2016). *Federated Optimization: Distributed Machine Learning for On-Device Intelligence.* 1–38. Retrieved from http://arxiv.org/abs/1610.02527.

43. Mcmahan, H. B., Moore, E., Ramage, D., & Com, B., (2012). *Federated Learning of Deep Networks Using Model Averaging* Blaise Agüera y Arcas.

44. Shokri, R., & Shmatikov, V., (2015). Privacy-preserving deep learning. In: *Proceedings of the 22nd ACM SIGSAC Conference on Computer and Communications Security – CCS '15* (pp. 1310–1321). ACM Press: New York, USA. https://doi.org/10.1145/2810103.2813687.

45. Bonawitz, K., Ivanov, V., Kreuter, B., Marcedone, A., McMahan, H. B., Patel, S., Ramage, D., et al., (2017). Practical secure aggregation for privacy-preserving machine learning. In: *Proceedings of the 2017 ACM SIGSAC Conference on Computer and Communications Security – CCS '17* (pp. 1175–1191). ACM Press: New York, USA. https://doi.org/10.1145/3133956.3133982.

46. Phong, L. T., Aono, Y., Hayashi, T., Wang, L., & Moriai, S. (2018). Privacy-Preserving Deep Learning via Additively Homomorphic Encryption. *IEEE Transactions on Information Forensics and Security, 13*(5), 1333–1345. https://doi.org/10.1109/TIFS.2017.2787987.

47. Smith, V., Chiang, C. K., Sanjabi, M., & Talwalkar, A. (2017). Federated multi-task learning. *Advances in Neural Information Processing Systems, 2017-Decem,* 4425–4435. Retrieved from http://arxiv.org/abs/1705.10467.

48. Lin, Y., Han, S., Mao, H., Wang, Y., & Dally, W. J. (2017). *Deep Gradient Compression: Reducing the Communication Bandwidth for Distributed Training.* Retrieved from http://arxiv.org/abs/1712.01887.

49. Pan, S. J., & Yang, Q., (2010). A survey on transfer learning. *IEEE Trans. Knowl. Data Eng., 22*(10), 1345–1359. https://doi.org/10.1109/TKDE.2009.191.

50. Liu, Y., Chen, T., & Yang, Q. (2018). *Secure Federated Transfer Learning.* Retrieved from http://arxiv.org/abs/1812.03337.

51. Sharma, S., Chaoping, X., Liu, Y., & Kang, Y. (2019). *Secure and Efficient Federated Transfer Learning.* Retrieved from http://arxiv.org/abs/1910.13271.

52. Chen, Y. R., Rezapour, A., & Tzeng, W. G., (2018). Privacy-preserving ridge regression on distributed data. *Inf. Sci. (NY)., 451, 452,* 34–49. https://doi.org/10.1016/j.ins.2018.03.061.

53. Nikolaenko, V., Weinsberg, U., Ioannidis, S., Joye, M., Boneh, D., & Taft, N., (2013). Privacy-preserving ridge regression on hundreds of millions of records. In: *Proceedings – IEEE Symposium on Security and Privacy* (pp. 334–348). https://doi.org/10.1109/SP.2013.30.

54. Hardy, S., Henecka, W., Ivey-Law, H., Nock, R., Patrini, G., Smith, G., & Thorne, B. (2017, November 28). Private federated learning on vertically partitioned data via entity resolution and additively homomorphic encryption. *ArXiv.* Retrieved from http://arxiv.org/abs/1711.10677.

55. Zhao, L., Ni, L., Hu, S., Chen, Y., Zhou, P., Xiao, F., & Wu, L. (2018). InPrivate Digging: Enabling Tree-based Distributed Data Mining with Differential Privacy. *Proceedings – IEEE INFOCOM, 2018-April,* 2087–2095. https://doi.org/10.1109/INFOCOM.2018.8486352.

56. Cheng, K., Fan, T., Jin, Y., Liu, Y., Chen, T., & Yang, Q. (2019, January 25). SecureBoost: A lossless federated learning framework. *ArXiv*. Retrieved from http://arxiv.org/abs/1901.08755.

57. Bonawitz, K., Eichner, H., Grieskamp, W., Huba, D., Ingerman, A., Ivanov, V., Kiddon, C., et al., (2019). *Towards Federated Learning at Scale: System Design*.

58. Yang, Q., Liu, Y., Chen, T., & Tong, Y., (2019). *Federated Machine Learning: Concept and Applications, 10*.

59. Lindell, Y. (2011). Secure Multiparty Computation for Privacy Preserving Data Mining. In *Encyclopedia of Data Warehousing and Mining*. https://doi.org/10.4018/9781591405573.ch189.

60. Ohrimenko, O., Schuster, F., Fournet, C., Nowozin, S., Vaswani, K., Costa, M., & Mehta, A. (2016). *Oblivious Multi-Party Machine Learning on Trusted Processors*. Retrieved from https://www.usenix.org/conference/usenixsecurity16/technical-sessions/presentation/ohrimenko.

61. Zhao, Y., Li, M., Lai, L., Suda, N., Civin, D., & Chandra, V. (2018). *Federated Learning with Non-IID Data*. Retrieved from http://arxiv.org/abs/1806.00582.

62. Du, W., Han, Y. S., & Chen, S., (2004). Privacy-preserving multivariate statistical analysis: Linear regression and classification. In: *SIAM Proceedings Series* (pp. 222–233). https://doi.org/10.1137/1.9781611972740.21.

63. Nakamoto, S. (2009). Bitcoin: A Peer-to-Peer Electronic Cash System. *Cryptography Mailing https://Metzdowd.Com*.

64. Dillenberger, D. N., Novotny, P., Zhang, Q., Jayachandran, P., Gupta, H., Hans, S., Verma, D., et al., (2019). Blockchain analytics and artificial intelligence. *IBM J. Res. Dev., 63*(2, 3), 5:1–5:14. https://doi.org/10.1147/JRD.2019.2900638.

65. Preuveneers, D., Rimmer, V., Tsingenopoulos, I., Spooren, J., Joosen, W., & Ilie-Zudor, E., (2018). Chained anomaly detection models for federated learning: An intrusion detection case study. *Appl. Sci., 8*(12). https://doi.org/10.3390/app8122663.

66. Kim, H., Park, J., Bennis, M., & Kim, S. L., (2019). Blockchain on-device federated learning. *IEEE Commun. Lett.* https://doi.org/10.1109/LCOMM.2019.2921755.

67. Decker, C., & Wattenhofer, R. (2015). A fast and scalable payment network with bitcoin duplex micropayment channels. *Lecture Notes in Computer Science (Including Subseries Lecture Notes in Artificial Intelligence and Lecture Notes in Bioinformatics), 9212*, 3–18. https://doi.org/10.1007/978-3-319-21741-3_1.

68. BigchainDB: The scalable blockchain database powering IPDB, 2017. – Google Search. (2017). Retrieved May 18, 2020, https://www.google.com.

69. *IPFS Powers the Distributed Web*. https://ipfs.io/ (accessed on 28th July 2021).

70. Yao, Y., Rosasco, L., & Caponnetto, A., (2007). On early stopping in gradient descent learning. *Constr. Approx., 26*(2), 289–315. https://doi.org/10.1007/s00365-006-0663-2.

71. Huang, J., Qian, F., Guo, Y., Zhou, Y., Xu, Q., Mao, Z. M., … Spatscheck, O. (2013). An in-depth study of LTE: Effect of network protocol and application behavior on performance. *Computer Communication Review, 43*(4), 363–374. https://doi.org/10.1145/2534169.2486006.

72. Smith, V., Chiang, C. K., Sanjabi, M., & Talwalkar, A. (2017). Federated multi-task learning. *Advances in Neural Information Processing Systems, 2017-Decem*, 4425–4435. Retrieved from http://arxiv.org/abs/1705.1046.

73. Li, J., Khodak, M., Caldas, S., & Talwalkar, A. (2019). *Differentially Private Meta-Learning*. Retrieved from http://arxiv.org/abs/1909.05830.

74. Nilsson, A., Smith, S., Ulm, G., Gustavsson, E., & Jirstrand, M., (2018). A performance evaluation of federated learning algorithms. *DIDL 2018 – Proc. 2nd Work. Distrib. Infrastructures Deep Learn. Part Middle, 2018*, 1–8. https://doi.org/10.1145/3286490.3286559.

75. Wang, Y., (2017). *Co-Op: Cooperative Machine Learning from Mobile Devices.* Univ. Alberta. https://doi.org/https://doi.org/10.7939/R32805C45.

76. Zheng, Z., Xie, S., Dai, H. N., Chen, X., & Wang, H., (2018). Blockchain challenges and opportunities: A survey. *Int. J. Web Grid Serv., 14*(4), 352–375. https://doi.org/10.1504/IJWGS.2018.095647.

77. Carlini, N., Liu, C., Erlingsson, Ú., Kos, J., & Song, D. (2019). The secret Sharer: Evaluating and testing unintended memorization in neural networks. *Proceedings of the 28th USENIX Security Symposium*, 267–284. Retrieved from http://arxiv.org/abs/1802.08232.

78. Dwork, C., & Roth, A., (2013). The algorithmic foundations of differential privacy. *Found. Trends Theor. Comput. Sci., 9*(3, 4), 211–487. https://doi.org/10.1561/0400000042.

79. Duchi, J. C., Jordan, M. I., & Wainwright, M. J. (2012). Privacy aware learning. *Advances in Neural Information Processing Systems, 2*, 1430–1438. Retrieved from http://arxiv.org/abs/1210.2085.

80. Geyer, R. C., Klein, T., & Nabi, M. (2017). *Differentially Private Federated Learning: A Client Level Perspective.* Retrieved from http://arxiv.org/abs/1712.07557.

81. Mohassel, P., & Zhang, Y., (2017). SecureML: A system for scalable privacy-preserving machine learning. In: *Proceedings – IEEE Symposium on Security and Privacy* (pp. 19–38). Institute of Electrical and Electronics Engineers Inc. https://doi.org/10.1109/SP.2017.12.

82. Kilbertus, N., Gascón, A., Kusner, M., Veale, M., Gummadi, K. P., & Weiler, A. (2018). Blind justice: Fairness with encrypted sensitive attributes. *35th International Conference on Machine Learning, ICML 2018, 6*, 4123–4137.

83. Araki, T., Furukawa, J., Lindell, Y., Nof, A., & Ohara, K., (2016). High-throughput semi-honest secure three-party computation with an honest majority. In: *Proceedings of the 2016 ACM SIGSAC Conference on Computer and Communications Security – CCS'16* (pp. 805–817). ACM Press: New York, USA. https://doi.org/10.1145/2976749.2978331.

84. Mohassel, P., Rosulek, M., & Zhang, Y., (2015). Fast and secure three-party computation. In: *Proceedings of the 22nd ACM SIGSAC Conference on Computer and Communications Security – CCS '15* (pp. 591–602). ACM Press: New York, USA. https://doi.org/10.1145/2810103.2813705.

85. Dwork, C., (2008). Differential privacy: A survey of results. In *Theory and Applications of Models of Computation* (pp. 1–19). https://doi.org/10.1007/978-3-540-79228-4_1.

86. Sweeney, L., (2002). K-anonymity: A model for protecting privacy. *Int. J. Uncertainty, Fuzziness Knowledge-Based Syst., 10*(05), 557–570. https://doi.org/10.1142/S0218488502001648.

87. Agrawal, R., & Srikant, R., (2000). Privacy-preserving data mining. In: *Proceedings of the 2000 ACM SIGMOD International Conference on Management of Data – SIGMOD '00* (pp. 439–450). ACM Press: New York, USA. https://doi.org/10.1145/342009.335438.

88. Rivest, R., Shamir, A., & Adleman, L. (1978). On Data Systems and Privacy Homomorphisms. *Foundations of Secure Computation, 4(11)*, 169–180. Retrieved from https://pdfs.semanticscholar.org/3c87/22737ef9f37b7a1da6ab81b54224a3c64f72.pdf%0Ahttp://files/834/22737ef9f37b7a1da6ab81b54224a3c64f72.pdf.

89. Giacomelli, I., (2018). *Privacy-Preserving Ridge Regression with Only Linearly-Homomorphic Encryption.*

90. Hall, R., Fienberg, S. E., & Nardi, Y., (2011). Secure multiple linear regression based on homomorphic encryption. *J. Off. Stat., 27*(4), 669–691.

91. Acar, A., Aksu, H., Uluagac, A. S., & Conti, M., (2018). A survey on homomorphic encryption schemes. *ACM Comput. Surv., 51*(4), 1–35. https://doi.org/10.1145/3214303.

92. Li, X., Jiang, P., Chen, T., Luo, X., & Wen, Q., (2018). A survey on the security of blockchain systems. *Futur. Gener. Comput. Syst., 107*, 841–853.

CHAPTER 3

Blockchain-Based Security Solutions for Big Data and IoT Applications

RAJDEEP CHAKRABORTY, ABHIK BANERJEE, and SOUNAK GHOSH

Department of Computer Science and Engineering,
Netaji Subhash Engineering College, Kolkata, West Bengal, India,
E-mails: rajdeep_chak@rediffmail.com (R. Chakraborty),
abhik.banerjee.1999@gmail.com (A. Banerjee),
sounakghosh.official@gmail.com (S. Ghosh)

ABSTRACT

For any distributed system, consensus plays a pivotal role in holding together the whole system. This is true not just for any blockchain (BC) peer-to-peer (P2P) Network but also for Cloud Computing where the Leader Election Problem is solved by the use of various consensus algorithms. No "one" consensus algorithm is universal or perfect. Depending upon the type of BC employed and the use case of the BC, there may be a consensus algorithm that would help the system reach optimum functioning without compromising the integrity of the system.

While proof of work (PoW) [2] became the first Consensus Mechanism to be used in BC, there have been many consensus algorithms to date. These are mechanisms which help the P2P network of a BC arrive at a decision. The decision taken most often is who is going to add the next block on the BC. This decision is crucial because if there exists a malicious entity in the network who gets in charge of the chain, then it would be detrimental to the overall functioning of the network.

The chapter discusses the following consensus algorithms-PoW [2], proof of stake, delegated proof of stake, Byzantine fault tolerance (BFT), Crash Fault Tolerance, Hashgraph Consensus Algorithm, Proof of Elapsed Time, and Proof of Authority (PoA). The pros of the consensus algorithms and main drawbacks against implementation such as low throughput and excessive consumption of bandwidth have been discussed as well. A comparative analysis of the aforementioned algorithms with a major focus on use cases has been done to help understand how such mechanisms may be used for designing solutions which surpass the initial application of cryptocurrency and delve into real-world solutions in the fields including but not limited to the Internet of Things (IoT) and big data.

3.1   INTRODUCTION

Blockchains (BCs) have evolved from being considered to be suitable for the only maintenance of cryptocurrency operations to being applied to almost every field from healthcare to identity management. It has been theorized that there are four generations of BCs (analogous to the generations of computers) with bitcoin (BTC) being the first and the earliest instance of it being successfully utilized. The fourth generation of BCs has a typical trait which is amalgamation with artificial intelligence (AI)-based solutions. This is slated to give rise to smarter and more scalable chains [28], which might even have the capacity to self-heal any disruptions caused by nodes. But even with this amalgamation, the role of consensus algorithms [37] would stay pivotal in the whole Distributed Ledger Ecosystem. Without a proper mechanism to reach a decision, the system would invariably collapse.

Big Data and IoT are two fields where the adoption of BC has been demonstrated with many Proofs-of-Concept. IoT and BC, in particular, have been shown to provide a secure and trusted way of supply chain manage-ment (SCM). But BC has its own set of limitations, which if not properly addressed, can lead to poor disaster recovery and cost management.

BC is a specific subclass of Distributed Ledger Technologies. But given that the two terms have become analogous to one another over the course of time, they have been used interchangeably. In this chapter, we provide a review of the most commonly used consensus algorithms and whether they are feasible for use in big data and IoT applications. The consensus algorithms discussed include proof-of-work [2], proof-of-stake, delegated proof-of-stake, proof-of-authority, proof-of-elapsed time (PoET), practical Byzantine fault tolerance (PBFT) [27]. An attempt has been made to intro-duce a concept of a new Consensus Algorithm with criticism as to why it might be a better alternative to the aforementioned consensus algorithms.

Before we discuss the consensus, algorithms and compare them, we need to understand two major concepts that govern the functioning of any distributed system- "Fallacies of a Distributed System" discussed in Section 3.2; and "CAP Theorem" [31] and how they relate to BC is discussed in Section 3.3. "Consensus Algorithms" are discussed in Section 3.4, and a short review is done in Section 3.5. Consensus algorithms applications in IoT and big data are discussed in Sections 3.6 and 3.7, respectively. "Current and Future Works" are discussed in Section 3.8, and finally, a conclusion is drawn in Section 3.9.

3.2   FALLACIES OF A DISTRIBUTED SYSTEM

The fallacies of a distributed system present a set of naive assumptions one might make with a distributed system. Since BC is a distributed system, these fallacies apply to it as well. The first seven of these fallacies were introduced by L Peter Deutsch, while the last was added by James Gosling. The following are the statements and their implications on a distributed ledger technology (DLT) which are discussed in Sections 3.2.1–3.2.8.

3.2.1   THE NETWORK IS RELIABLE

This fallacy is centered towards the infrastructure side of the network. This is where a BC ecosystem is assumed to be always available. This might not be the case always since all the nodes in the system may not be available due to fault in their own respective system or refusal to participate in the transaction or election process.

3.2.2   LATENCY IS ZERO

In the original version of the fallacy, this assumption is based on the Round Trip Time in the network. In other words, this fallacy assumes that the transfer of data through the network is instantaneous. This cannot be true in a BC Network for the following reason-the time is taken by data packets to travel from one node to another. These data packets may contain information about the next block in the BC maintained on that node or vote process for the next peer. Thus, latency in a BC network cannot be zero.

3.2.3   BANDWIDTH IS INFINITE

Zero-latency coupled with infinite bandwidth can lead to near-instantaneous transmission of data in the network. In the case of a BC network, the bandwidth does not measure with respect to the entire network but individual nodes. As such, the nodes themselves may not have the privilege of having a high bandwidth since a BC network can be heterogeneous (discussed in the eighth fallacy).

3.2.4   THE NETWORK IS SECURE

In any real-world scenario, it can never be assumed that the network is completely secure. In the case of the original fallacy, this assumption related to the absence of DDoS, MITM, and Injection attacks, etc., on the network. In the case of BC platform, this relates to the intention of the nodes in the network. Not all nodes may be healthy or good nodes. The presence of malicious nodes in the system is what propels the idea of BCs being able to operate in a "trust-less" environment. While there may be some level of trust present in a consortium BC, on a whole, a BC network cannot be taken to be completely secure.

3.2.5   TOPOLOGY DOES NOT CHANGE

The fifth fallacy in distributed systems relates to the overall network topology. This refers to the fact that during the course of being deployed on production, the overall network topology of the server can change over duration of time and may become more complex. In the case of a BC network, any node may come to join the network or leave it. There is little to no control over the number of peer nodes as we move from a private to public BC model.

3.2.6   THERE IS ONLY ONE ADMINISTRATOR

There is no one "administrator" in a BC network who can control the network. This may not be true depending on types of BC. For instance, in case of a public BC model which employs proof-of-work as a consensus algorithm, an entity which commands over 50% of the network's total compute resources may be assumed as the influencer but this has never occurred till date because of constraints placed over the peer nodes in the network and the structure of a BC network which promotes competitive interests.

3.2.7 TRANSPORT COST IS ZERO

The seventh fallacy may be interpreted in two ways-the cost to transfer data from the application to the transport level of the OSI model or the overall cost to initiate and maintain the network. Both of these interpretations would be wrong in case of a BC network since firstly, the cost required in this case can be measured in two ways analogous to the original fallacy:

- The cost to verify a transaction Block is never zero, compute resources are required for it. Also, for sending the decision arising from the verification process needs to be sent over the network which again invalidates the fallacy by invalidating the first original interpretation.
- The cost to initiate and maintain the network and individual node is not zero.

3.2.8 THE NETWORK IS HOMOGENEOUS

Satoshi Nakamoto's original design [1] for BC did not take into account the platform or operating system of the nodes. To this date, BC networks are platform agnostic in the sense that a node may run on Windows while another may run on Linux, however, as long as they can solve the cryptographic puzzle or take part in the leader election process, they are termed as healthy. Thus, the network may not be homogenous in this sense.

Furthermore, a network may not be homogenous in its intentions as not all nodes may think towards reaching a consensus or enhancing the overall network strength. There can be malicious peer nodes present in the network.

3.3 CAP THEOREM

One of the major theorems which govern any distributed storage system is the CAP theorem [38]. CAP stands for C-Consistency, A-Availability, and P-Partition Tolerance. The goal of any distributed web services or file storage system is to achieve proper balance among the three for optimum performance.

CAP theorem was conjectured by Brewer [11] when he stated that any Web Service must have these three properties. But it has been proved that it is not possible for any distributed storage system to have all three of the stated properties simultaneously.

To understand how BCs fall under this theorem, we need to understand the implication of each of the properties from the perspective of BC. If each property is considered to be a single vertex of a triangle, a distributed solution can only rest on any one of its edges. This concept has been explained in Figure 3.1. Section 3.3.1 discusses "consistency;" Section 3.3.2 discusses "availability;" and Section 3.3.3 discusses "partition tolerance."

3.3.1   CONSISTENCY

For a distributed system like BC to achieve the first property of consistency, it must be able to return the result of any transaction fast. Unfortunately, that is not the case with BC. BC is not completely consistent.

It might seem contradictory to the nature of BC, but BCs need to verify the transactions by peer nodes before they commit it to their ledger. This process is the same regardless of the kind of BC-private or public.

3.3.2   AVAILABILITY

BC does provide high availability though it also depends on the type of consensus algorithm. BC platforms which use PBFT [30] require that at most $1/3^{rd}$ of the network be faulty. This tolerance may increase or decrease depending on the type of consensus mechanism used. But for all intents and purposes, a BC is considered to satisfy the condition of availability.

3.3.3   PARTITION TOLERANCE

The ability of a distributed storage system to return results even after it has been divided is regarded as the partition tolerance. By design, BC is partition tolerant. If one were to look at the original design of BTC, it supports what is known as "forks" or "daughter chains." This condition primarily arises when two peer nodes in the network produce the right result at the same instant of time. As the size of the system grows, the chances of this happening reduce but do not become zero.

In case of such an event, both the chains are regarded as live chains. This gives rise to a "race." The chain where the next block is attached is regarded as the longer and correct chain. This is the chain that is maintained.

Thus, even in the event of a partition in the system, the BC can continue to function. This is what makes it partition tolerant.

Thus, in our analogy of CAP theorem and an equilateral triangle where each vertex would represent one of the properties of consistency, availability, and partition tolerance, the position of BC-based ledgers would be near the edge connecting the properties of availability and partition tolerance.

3.4 CONSENSUS ALGORITHMS

This section discusses the most used consensus algorithms in the industry. As mentioned earlier, the properties of a BC are heavily impacted by the consensus mechanism adopted by the network. There is no "one" perfect consensus mechanism suitable for every BC system. However, there are consensus algorithms which have been implemented in practice [26]. Section 3.4.1 discusses "Proof of Work;" Section 3.4.2 discusses "Proof of Stake;" Section 3.4.3 illustrates "Delegated Consensus Mechanisms;" Section 3.4.4 discusses "Proof of Elapsed Time;" Section 3.4.5 illustrates "Practical Byzantine Fault Tolerance" mechanism [32]; Section 3.4.6 discusses "Proof of Authority (PoA);" and finally Section 3.4.7 gives "Hedera Hashgraph Consensus Service."

3.4.1 PROOF OF WORK (POW)

PoW consensus protocol is the first consensus algorithm to be used in a BC [1]. As of writing, it accounts for more than 90% of the total market capitalization of existing digital currencies. The specialty of the function is that it is easy to check but extremely hard to compute. PoW is a consensus algorithm in which a production of a piece of data is highly expensive and time-consuming, but the verification of the data on whether it is correct or not is easy for others. BTC, the most popular cryptocurrency uses Hash cash PoW system.

For a network to accept a block, PoW has to be completed by miners for all the transactions to be verified in the block. The difficulty of this work keeps adjusting and not always the same, hence new blocks can be generated every 10 minutes. Users in a network send each other coins, and the decentralized ledger collects all the transactions into the block, but every block should be taken care of by someone and must be validated. In every BC, some nodes are always under the process of validation. In summary, PoW is a system which ensures security and consensus [14] throughout the BC network. The participant in the network who/which validates block evidently shows a huge amount of investment in computing power to do so.

The miners in the network try to guess a number at random, which should produce a right hash for the block of transactions. A nonce is a random number used only once, generally an integer between 0 and 4294967296 and hash is an algorithm which can also be defined as a very complex formula that converts any sequence of characters into a string of 64 numbers or characters. Every block in the BC has its hash(id) which is basically a string that a person got when the block was verified by him. So, when we need to verify the next block, we would need to take the hash and add to the current transaction blocks. We will be given a big block of text. The next step would be to take a nonce, i.e., a random number, and then we have to add it to the end of the block of that text. Now we have a big block of text which has a hash of the previous block, new transactions, and a completely random number. When we have all that, we can start calculations, and for the calculations, we use a hash function, and then we change the random number until we get a string that has a certain number of zeros in front of it [36].

The process seems to be easy, but we have to consider the fact that our computer has to perform around 10n21 computations to find the correct number, and almost 10 minutes is required to find the right number, which would provide us with the correct string of data. BTC has been using the SHA-256 hash algorithm as its hash function in its PoW.

A computer or a machine in the network must keep the increasing nonce value until the correct one is found. This means that brute force has to be applied by the computer to generate millions of hashes per second to generate that exact number with the same number of starting zeros which are defined. This whole process is expensive and time-consuming to execute a PoW block, but it is relatively easy for a user to verify if the block is correct or not. Figure 3.1 illustrates PoW flow diagram.

For example, a party in the network desires to check if Node A did the required work. They will simply use the block string which Node A received after validation and take its nonce digits. Over this data, they would have to apply a hash function and if the result has the correct number of leading zeroes. If it is the same, then the results are all right.

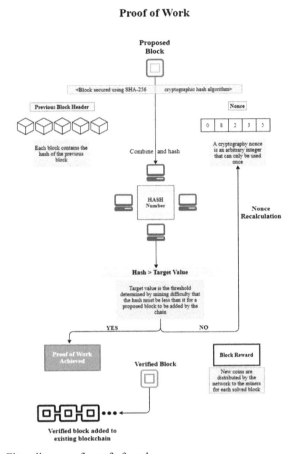

FIGURE 3.1 Flow diagram of proof of work.

PoW ensures that no new block can be added to the BC without the necessary work being executed. In this way, any malicious node can be easily avoided as they are very difficult to validate and hence any random block cannot be added to the BC and if tried, other participants in the network would just dismiss the block and it would be informed to everyone about the invalidity of the block.

Hash functions like SHA-256 give different result whenever any changes are made in a block of the text which was previously validated, which means that if one transaction is changed even for a small amount, the result will be different and everyone in the network will know about it and it will not be treated as a valid transaction. It gives the same result if the input is the same but never the same result for different input.

PoW is not perfect. To execute it, a huge amount of computation power has to be spent to brute force hashes and the power used is just a waste of resources. The other problem is centralization and mining pools [33]. A miner with a mining pool of 1,000 CPUs will have a better chance for a reward than one with one CPU. Currently, almost 50% of BTC hash power is met by few mining pools, which means that a handful of people have to agree on a 51% attack, and the BC can be changed which creates a centralized PoW and goes against the idea of decentralization. Also, more than 70% of mining power comes from China hence the pools are becoming highly centralized as the cost of electric power is different for different countries.

For a PoW system to work properly, at least 50% of the participants in the system need to have good intentions. If more than half of the validators have bad intentions, then someone in the network can put faulty transactions on the block, and someone else can have their fund stolen away from them. This would cause a loss in trust among the participants inside the system and hence would start another BC and if this happens, miners who were earning for validating blocks and invested money for the equipment would lose everything. The use of BC could decrease heavily, which would be unacceptable by miners who are actually validators, therefore if any faulty data has to be added to the BC, it can only be done with their help. This shows that the BC network needs to stay fair and honest for its economic reasons.

PoW is an inefficient system, and the cost of resources required to keep the network running is high. Criticism against it has existed for a long time. One of the first being described in Ref. [2].

3.4.2 PROOF OF STAKE

Proof of stake [3] is a consensus algorithm which was introduced in 2011 to solve the disadvantages of the PoW algorithm which was the most popular consensus algorithm back then, and it states that a user can mine block according to the number of coins being held by him/her which means that the more amount of coins (altcoin or BTC) owned by a miner, the more mining power is possessed.

Proof of Stake (PoS) has a lot of similarities with PoW, but they are fundamentally different from each other in many ways. In any BC consensus algorithm, achieving distributed consensus [13] is always the ultimate goal, i.e., a security system has to be created whereby validation of transactions of other people are to be incentivized by the users meanwhile maintaining integrity completely. Figure 3.2 shows the use case diagram of proof of stake.

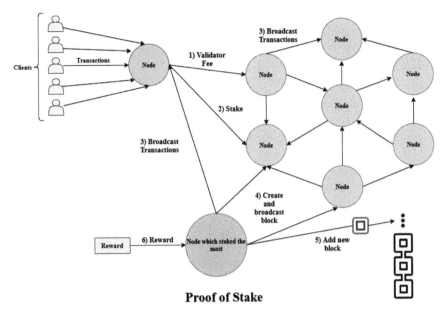

Proof of Stake

FIGURE 3.2   Proof of stake use case diagram.

In PoS algorithms, a consensus has to be achieved by requiring users to use the number of their tokens as a stake so that they can have a chance of getting selected to validate the transaction blocks and they get rewarded after doing it.

In PoS, the forger or the miner of the new block is semi-randomly chosen in a two-part process. User's stake is the first element that is to be considered in the process. A stake must be owned by every validator in the network. The process of staking involves the deposition of a certain amount of tokens and then locking that amount in a virtual safe and then, to vouch for the block it is used as collateral.

The more is the stake of the user, the more is their chances of getting selected to validate a block as they have invested a lot in the process and hence if they act maliciously, it would set them back by a lot greater amount than a user who has fewer stakes.

It can be thought that the PoS can be abused by the wealthy but the main point here is to have a degree of chance to get to the process of selection of the blocks so as a avoid a case where the users who are the richest are selected most of the time for the transaction validation, and take the rewards and grow even richer.

The second element is this algorithm adds to the randomness of its selection process and the method differs in every BC system. The two most common methods are Randomized Block Selection and Coin Age Selection.

In random block selection, selection of the users who are to be forgers is done by looking for people with a combination of highest stakes and lowest hash value. In the CoinAge Selection, forgers are chosen on the basis of the time the tokens of the user have been staked for. The longer the time, the larger the stake. There are also several other methods. Some currencies combine these methods, whereas others experiment with their own. Figure 3.3 compares reward potential versus investment between PoW and proof of stake.

PoS provide better security as the attackers have to put for their assets to attempt a 51% attack which is a big amount. It is also energy efficient when compared to PoW as it cuts out the extensive mining process of PoW. Large mining pools for PoW were a problem from the centralization of the BC in the case of PoW algorithm which causes an exponential increase in the reward per investment on PoW systems whereas the PoS has a linear increase.

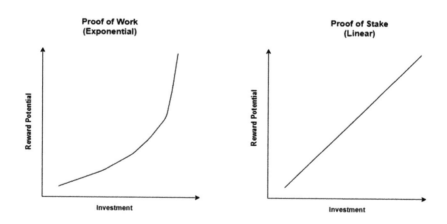

Proof of Work vs Proof of Stake

FIGURE 3.3   Comparison of proof of work and proof of stake.

3.4.3 DELEGATED CONSENSUS MECHANISMS

Delegated consensus mechanisms revolve around the simple mantra "if I cannot do it, you do it for me, and I shall give you a small portion of my rewards." The most famous delegated consensus mechanisms in BC include those of Proof-of-Work and Proof-of-Stake.

It is easily noticeable from the two that not most of the peer nodes in the BC network would have to compute or stake that match to the players in the system who have higher capacities discretion.

Therefore, delegated mechanisms offer the following advantages in a BC network:

- An equal footing to the lowest node in the system to take part in the consensus mechanism;
- Gain rewards from the system thus giving the nodes an option to stay incentivized;
- Prevent low-capacity peer nodes from becoming renegade nodes or malicious actors.

While it is possible for PoW to have a delegated version, a practical system with the said algorithm work has not been properly implemented due to the protocol's dependence on Compute Resources. What have been explored in practical implementation is Delegated Proof of Stake in BC platforms like Tezos and XCASH [6]. The total mechanism and implementation details are given from Sections 3.4.3.1–3.4.3.6.

3.4.3.1   DELEGATED PROOF OF STAKE

In case of delegated proof of stake, the next leader election process can take place through either voting or round-robin. While both procedures have their fair share of advantages, the former is preferred over the latter due to removal of predictability.

3.4.3.2   LEADER ELECTION PROCEDURE

DPoS-based BC platforms dictate that those who have a larger share in the network be given a higher preference over the next block upload on the BC. This is akin to owning the majority of computing power in a BC network that uses PoW. However, those who have a lower stake do get to participate as well. That is exactly where the "delegation" comes in.

The idea behind the delegation is that if you do not have enough stake in the network, delegate it to a peer node who has a higher stake. Should that peer node be selected as the next leader, then you get a portion of the reward as well. In this chapter, we would look at two implementations of the Delegated Proof of Stake algorithm in Tezos. Other notable BC platforms to utilize DPoS are EOS.IO, TRON, and LISK.

3.4.3.3   EOS.IO

EOS.IO [4] has been hailed as the "Ethereum (ETH)"-killer. In fact, a substantial percentage of decentralized apps which started out on ETH have explored and even shifted to EOS.IO. Just like ETH introduced the first Turing Complete scripting language in the form of solidity (which lead to the dawn of the second generation of BCs), EOS.IO too started a revolution with its implementation of Delegated Proof of Stake.

Unlike ETH which employed Proof-of-Work, EOS.IO focused on increasing the transaction throughput of the network. There were many reasons cited behind this motive. Firstly, a low transaction per second (throughput) of a BC network would lead to congestion of network traffic and higher block times. Secondly, an implication of the first would be that the network consensus would be delayed.

Due to the aforementioned reasons, EOS.IO implemented the Delegated Proof of Stake consensus algorithm described by Daniel Larimer, "Delegated Proof-of-Stake (DPOS)" 2014 in its network. DPoS is essentially a 2-phase process as described in the whitepaper. Figure 3.4 gives the work cycle of delegated proof of stake.

3.4.3.4   ELECTION OF BLOCK PRODUCERS

In the first phase of the DPoS algorithm, the network votes on a selection of 21 "producers." They are selected based on the amount of stake or ownership of tokens they have in the network. This is where the beauty of Delegated Proof of Stake comes in.

A network node who is unable to participate as a producer can vote another node. Essentially, they delegate their power to that node by voting for it. This helps every node in the network who has a stake in it participate in the network block formation process. Once, the election phase is over, the top 21 producers are chosen. This marks the start of the second phase.

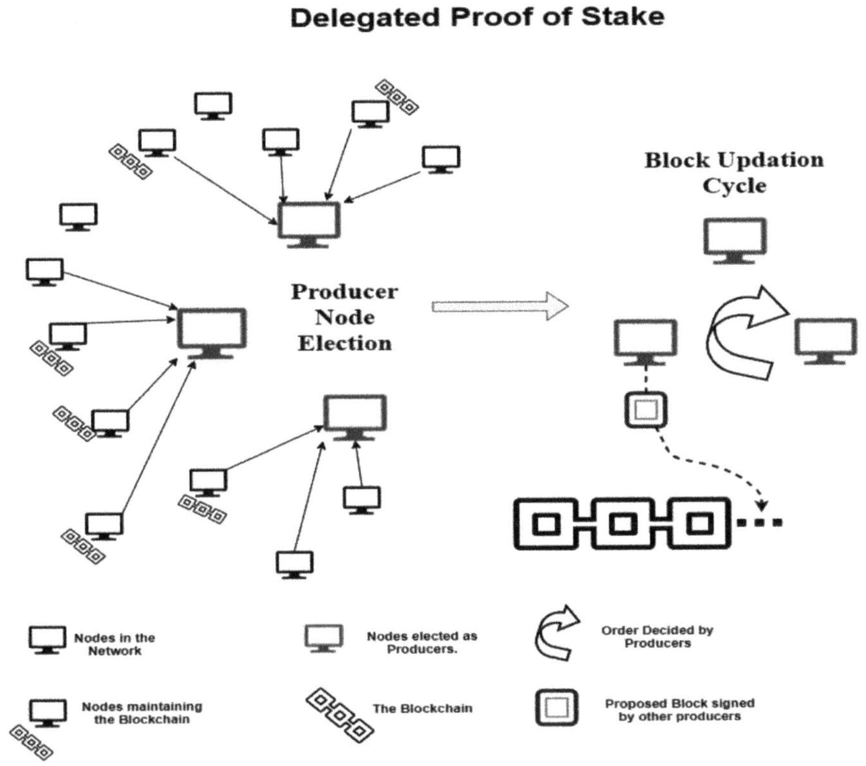

Delegated Proof of Stake

FIGURE 3.4 Delegated proof of stake work cycle.

3.4.3.5 PRODUCTION OF BLOCKS

In this phase of the algorithm, the producers come together to decide a particular order among them. This order is crucial as the role of producing a 'block" in the network is assigned to them in this order. At least 15 producers have to agree on the ordering before it is finalized.

After the order is finalized, the producers need to create a block in their own scheduled time. Failure to do so would result in their turn being skipped. If this happens repeatedly, it can also lead to that producer being voted out of the network. Once a block has been created by one producer, another 15 elected producers need to sign it. This makes the block "irreversible." In the case of ETH, this phenomenon is observed once at least 100 blocks have been added after the block in question. Use of DPoS is what allows the EOS. IO network to make the blocks irreversible faster than ETH.

Every producer needs to produce 6 blocks under normal conditions. Thus, if we take each of the "election-block creation" phases as one cycle then, we can say each cycle lasts 126 blocks at one go (21 producers, 6 blocks each).

3.4.3.6   TEZOS IMPLEMENTATION OF DPOS: A MISCONCEPTION

Like EOS.IO before it, TEZOS [5] too has been hailed as a revolutionary BC platform. But there is a misconception around it regarding its consensus algorithm. Tezos does not implement "delegated proof of stake." Tezos has a different version of the PoS model as its consensus mechanism-liquid proof of stake [5].

Under this mechanism, members of the network need to possess at least a specified amount of tokens before they are allowed to participate in the network's block creation process. Tezos functions on the Tez tokens. At the time of writing, a member of the network needs to have at least 10,000 Tezos (termed as 1 roll) before participating in the block creation process. In Tezos Ecosystem, the block creation process is termed as "baking" and the member nodes in the network who take part in the mechanism are known as "bakers."

If a member in the network does not have the required amounts of tokens, then that member may delegate the tokens to another member who has a higher chance of attaining the required amount of 1 roll. Since, in this case, the tokens of the member are kept in waiting (or delegate), this consensus mechanism is different from EOS.IO's and the official "Delegated Proof of Stake."

The delegation of tokens is what calls for the term "Liquid" to be added to the name of the mechanism. Liquid Proof of Stake, thus, prevents malicious behavior since the tokens of the network member are put at stake in the consensus mechanism. This is akin to the current system of banking where interested parties take a loan by keeping their belongings at mortgage. A similarity to the existing system coupled with the advantage of decentralization and achieving transaction verification faster and in a more secure manner has led to Tezos being favored by a number of Financial Institutions.

3.4.4   PROOF OF ELAPSED TIME

Proof of elapsed time [8] is a consensus mechanism which focuses on leader election for block upload in a secure environment and using z-test measures to test the validity of the leader election process. It has been used in Hyperledger Sawtooth-a Framework developed by Intel Kelly

[29] for designing Distributed Ledger Solution and later brought under the Hyperledger ecosystem [19]. Section 3.4.4.1 discuss various aspect to join a network; and Section 3.4.4.2 gives the "Ledger Election Process."

3.4.4.1 JOINING THE NETWORK

Hyperledger Sawtooth is mainly used for creating permissioned distributed ledger solutions [15]. This means that not everyone is free to join the network of nodes. But should a peer join the network as a "validator," then that peer has to download what is known as an "Enclave." Proof of Elapsed Time is essentially a 'lottery"-based consensus mechanism which requires execution in a secure environment. This environment is termed as "Intel Secure Guard Extension." Without this, it is not possible to be a validator node in the network or to run PoET.

After setting up the secure environment, the node requires ECDSA generated public and private key pairs to further generate the enclave's public and private key. The public keys are broadcasted to the network along with the joining request.

Once a node does this, the node has to wait for at least a specified amount of Blocks to be added to the chain before participating in the leader election process. This is what prevents a random node from gaining access into the network for a short duration just for the purpose of uploading malicious transactions onto the ledger and then leaving the network. In layman's terms, this is a proof of trust given by the peer.

3.4.4.2 LEADER ELECTION PROCESS

The leader election process in Proof of Elapsed Time consists of each validator node getting a random time interval from the Enclave function. The nodes need to wait for their allotted time. The node whose allotted time is up the first gets to be the leader and upload the next block on the BC. Figure 3.5 illustrates two phases of proof of elapsed time.

The node which gets elected as the leader needs to broadcast a set of values for verification. These include.

- WaitCertificate which is created by the enclave after consumption of the hash digest of the block of transactions;
- The block of transactions to be uploaded on the blockchain;

- The public key (public and private key pair is generated by the enclave function) and ECDSA public key signature of the node.

The verification procedure consists of checking the aforementioned data, the number of blocks that the node has waited and also conducting a z-test to find whether the winning the node is statistically permissible. Should the z-test or any of the above fail, then the node is revoked from the network.

What is notable here is that nodes do not have an infinite duration of time to consume their WaitTimer and get their certificate. Should they fail to do so, they are not allowed to take part in the leader election. This prevents erratic network traffic due to the presence or absence of transactions for blocks.

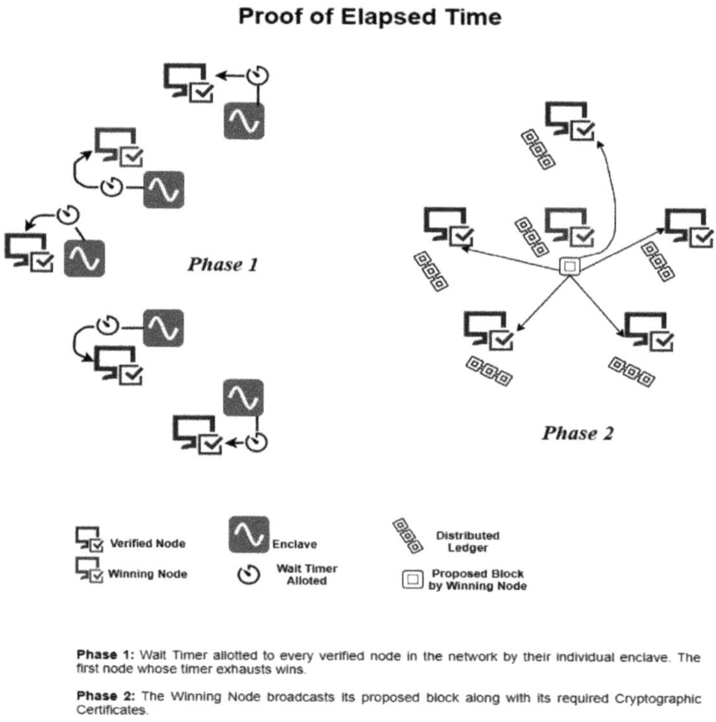

Proof of Elapsed Time

Phase 1

Phase 2

Verified Node   Enclave   Distributed Ledger

Winning Node   Wait Timer Alloted   Proposed Block by Winning Node

Phase 1: Wait Timer allotted to every verified node in the network by their individual enclave. The first node whose timer exhausts wins.

Phase 2: The Winning Node broadcasts its proposed block along with its required Cryptographic Certificates.

FIGURE 3.5   Proof of elapsed time phases.

3.4.5   *PRACTICAL BYZANTINE FAULT TOLERANCE (BFT)*

Practical Byzantine fault tolerance (PBFT) [9] is an applied solution of the Byzantine generals problem [10], which is used to reach consensus in a network of distributed systems in the presence of malicious actors. This requires at most 1/3rd of the network of nodes to consist of malicious actors. The algorithm fails to reach consensus if there are more bad actors in the system.

While the algorithm is comparatively robust, it is aforementioned limitation coupled with increasing complexity as the number of nodes in the system increases has been a pain point in practical implementation in BC.

It has been used in Hyperledger Fabric as a consensus algorithm, and many of the present consensus algorithms have been designed to be Byzantine Fault Tolerant and modeled after PBFT.

There are other variations of Byzantine Fault Tolerant consensus algorithms. But before we delve into it, a brief discussion on Byzantine Generals Problem – the problem which serves as a motivation for building a robust consensus mechanism – is required. The Byzantine general problem and mechanism are discussed in detail in Section 3.4.5.1.

3.4.5.1   BYZANTINE GENERALS PROBLEM

The Byzantine generals problem [10] describes a scenario whereby the presence of malicious actors in a group can have disastrous consequences in the overall action taken by the group. The solution seeks to make sure that even in the presence of such treacherous elements, the network can arrive at the right consensus.

The easiest method to understand the Byzantine generals problem is via visualization of four cases:

- When there is only 1 general leading the army;
- When there are 2 generals leading the army;
- When there are 3 generals leading the army;
- When there are 4 generals leading the army.

The army needs to attack or retreat, and only a unanimous decision can result in victory. So, if the army decides to retreat together, that scenario also counts as a win. In cases except case 1, there is one leader and the rest of the generals follow the leader.

> **Case I-1 Generals:** In the case of the only general, the outcome can be that either the general adheres to the orders or is malicious and takes the opposite decision. However, since there are no other

generals to lead the army, this action can be termed as being taken in "unison." Due to this, the scenario is a "win" in terms of consensus being achieved. This is a typical example of a centralized decision-making system.

➢ **Case II-2 Generals:** When there are two generals, there may be two scenarios. First, where both the generals are loyal and follow the command passed down and there is an overall 'win.' Second, when one general is treacherous (may be the leader or the follower). In such a scenario, one general may choose to follow the order given by the leader while the other does the opposite. This results in defeat. >50% malicious actor presence is not discussed since, in that case, the system would be able to cope.

The above 2 cases form the trivial cases of the problem. Figure 3.6 illustrates the attack and retreat mechanism.

➢ **Case III-3 Generals:** In the case of three generals, there may be 3 scenarios-all generals are loyal, 1 general is treacherous, 2 generals are treacherous. In the first and third scenarios, the system can work perfectly and become dysfunctional, respectively. The scenario may be discussed since if the number of traitors is a minority, then the system may function.

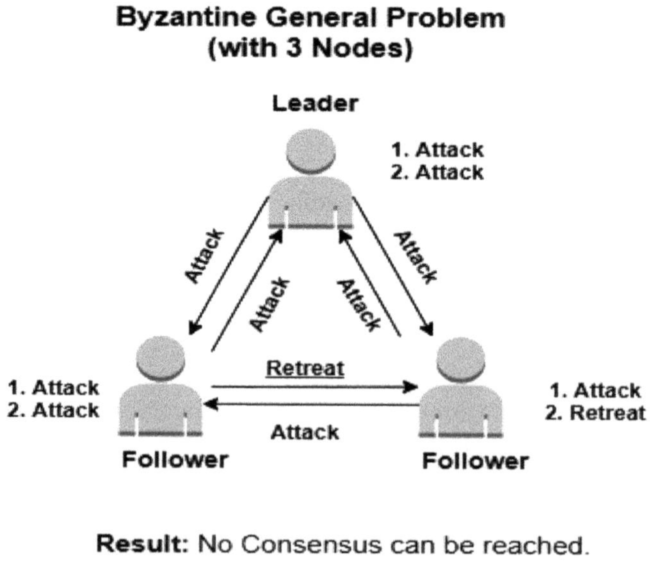

Byzantine General Problem (with 3 Nodes)

Result: No Consensus can be reached.

FIGURE 3.6   Attack and retreat mechanism.

In the absence of a medium of communication, the result may be a defeat since the army's decision would be in unison. If the leader relays the command to attack, the treacherous general may choose to retreat and vice-versa. Can this problem be solved with signed messages being passed between the leader and the follower generals? Supposing that the medium of travel of the messages is reliable, then let us assume two scenarios. First where a follower is treacherous and relays opposite instruction to another general. Second where the leader itself is corrupt. It is simple to observe that there cannot be any united decision in either case since at least one of the generals receives conflicting commands which cannot be solved via a simple majority (see Figure 3.7). Thus, the Byzantine Generals Problem is not solvable by just any "Odd" number of nodes as observed in CASE 1. Figure 3.7 illustrates Byzantine Attack and Retreat Mechanism.

➢ **Case IV-4 Generals:** In the last case, there are four generals-one leads while the other three follows. Here there may be three scenarios:

- **When One Following General is a Traitor:** Suppose the leader issues a command to all followers where he tells them to retreat. In this case, the treacherous general would get the command and relay it on to the two other remaining generals as an 'attack' command. All the generals in the army can then compare the commands received from the leader and relayed onto them by follower generals and via simple majority decide on a course of action. This course of action would be the true command issued by the leader, as seen in Figure 3.7.
- **When the Leader is a Treacherous General:** In this case, the leader would issue conflicting commands to the following generals. Let us suppose that the leader issues 'attack' command to two generals and 'retreat' to one. In this scenario, after receiving the commands, each general relays their commands to other generals, including the leader. As can be observed in Figure 3.7, even in this scenario, a simple majority of all the received commands yields a unanimous result, and the army can win.
- **When More than One General is Treacherous:** In this scenario, no matter the position of general (leader or follower), the army cannot overcome the treacherous nature of the generals, and thus this results in a failure.

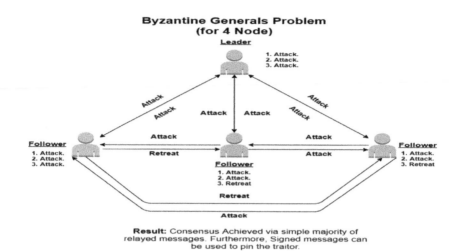

FIGURE 3.7   Byzantine attack and retreat mechanism.

From the above cases and scenarios, it can be seen that the Byzantine Generals Problem cannot be solved by the presence of nodes in any number. The general relation put forward in solving the Byzantine Generals Problem is that for 'm' number of traitors, there must be a total of 3m + 1 number of generals in total for a solution to exist.

Its limitation on the number of nodes notwithstanding, Byzantine Generals Problem is very important not just for BCs and distributed ledgers, but for every distributed system alike. It dictates how a system might be designed to tolerate malicious nodes called "Byzantine Nodes" as well as unanticipated failures in the distributed system. PBFT is an implementation of this which helps distributed systems attain a consensus. While the algorithm is robust and secure, it suffers from an inherent problem-PBFT is not scalable for a large number of nodes. Furthermore, there might be a message or token drops in the network, which can lead to delayed consensus. As a result, there has been a lot of research work into finding a better solution and implementation of the algorithm. Some of them are mentioned in the future works section of the chapter.

At the time of writing, PBFT [9] is not strictly used by any BC. Most BC platforms strive to make their consensus protocols "Byzantine Fault-tolerant." Variations have been brought up as well [18, 20]. The motivation behind discussing PBFT as a consensus mechanism in BC lies in the fact that there are BCs and DLT Frameworks which support pluggable "modules" for consensus mechanisms. Most of such platforms have support for PBFT. An example of such can be Hyperledger Sawtooth Framework which sports this feature.

Furthermore, PBFT was the consensus mechanism in Hyperledger Fabric (a DLT framework designed by IBM and hosted at Hyperledger) version 1. Hyperledger Fabric has seen a lot of changes in terms of the primary consensus mechanism used. Until Hyperledger Fabric 0.6, the consensus mechanism used was RAFT. This was downgraded in favor of PBFT in version 1.0 due to better robustness offered by PBFT. At the time of writing, the latest version of Hyperledger Fabric is 2.0 which uses Kafka (a messaging service)-based consensus algorithm.

3.4.6 PROOF OF AUTHORITY (POA)

The use of reputation as an alternative for computing power to block validation processes has been already tried out as consensus algorithms [23]. PoA [7] is a consensus algorithm which is based on reputation and can be used as a very efficient solution for, most importantly, private BC networks. The number of validators of blocks are very limited in the PoA algorithm, and hence its very scalable as the approval of blocks are done by participants who are already approved and have the roles of a moderator in the system. The validators stake their own personal reputation instead of coins in PoA.

Application of the algorithm is in various scenarios and is often chosen for logistical applications, like supply chains.

Maintaining privacy without taking away the advantages of a BC technology can be enabled by PoA. It is also implemented in Microsoft Azure's Managed BC solution. The platform does not use 'gas' currency; hence mining is not needed and is extensively used for private networks. The detail discussion and limitation of PoA is done from Sections 3.4.6.1–3.4.6.3.

3.4.6.1 PROOF OF AUTHORITY (POA) VS. PROOF OF STAKE

PoA [7] is considered to be a modified PoS [3] by some, as the identity of the validator is at stake than the amount of coins possessed by him. Since BC is decentralized in its nature, it is not always suitable to use PoS as a solution as large corporations and organizations can have a leverage on others on the financial terms. On the other hand, PoA can be a better solution for private BCs because its performance is considerably higher.

3.4.6.2   PROOF OF AUTHORITY (POA) CONSENSUS CONDITIONS

The PoA consensus algorithm is usually reliant upon:

- Identities which are trustworthy and valid: Identity confirmation is mandatory for the validators of the system.
- Validation difficulties: Money and reputation must be put to stake to be a validator. A tough process reduces the risks of selecting questionable validators and incentivizes a long-term commitment.
- A standard for validator approval: the method for selecting validators must be equal to all candidates.

The essence behind the reputation mechanism is the certainty behind a validator's identity. This cannot be an easy process or one that would be readily given up. It must be capable of weeding out bad players. Finally, ensuring that all validators go through the same procedure guarantees the system's integrity and reliability.

3.4.6.3   LIMITATIONS

The perception of the PoA mechanism is that it forgoes decentralization. So, one could say that this model of consensus algorithm is just an effort to make centralized systems more efficient. While this makes PoA an attractive solution for large corporations with logistical needs, it does bring some hesitation-especially within the cryptocurrency scope. PoA systems do have a high throughput, but aspects of immutability come into question when things like censorship and blacklisting can be easily achieved.

Another common criticism is that the identities of PoA validators are visible to anyone. The argument against this is that only established players capable of holding this position would seek to become a validator (as a publicly known participant). Still, knowing the validators' identities could potentially lead to third-party manipulation. For instance, if a competitor wants to disrupt a PoA-based network, he may try to influence public known validators to act dishonestly in order to compromise the system from within.

PoW [1], PoS [2], or PoA [7] all have their own unique advantages and disadvantages. It is well known that decentralization is highly valued within the cryptocurrency community and PoA, as a consensus mechanism, sacrifices decentralization in order to achieve high throughput and scalability. The inherent features of PoA systems are a stark contrast from how BCs have been functioning until now. Still, PoA presents an interesting approach and cannot be disregarded as an emerging BC solution, which may suit well for private BC applications.

3.4.7 HEDERA HASHGRAPH CONSENSUS SERVICE

The main network or the main net was launched by the Hedera platform [22] with cryptocurrency, file storage, and smart contracts (SCs) as its three main services. Licenses are not required when the software is written which uses any of the main services or uses the Hedera Hashgraph platform. All open-source projects or proprietary decentralized applications (DApps) that would be built on the platform would be owned by the developers. Only tokens are required to use the platform, also known as HBARs. A nominal fee is paid by the users to the platform when an API call is made to the services available.

Transactions, accounts, balances typically comprise the payment network. The most important thing that we can get from a platform like Hedera is a cheap native micro-transaction, on which Hashgraph functions on.

1. **Accounts:** Hedera Hashgraph cryptocurrency [22] was created to make microtransactions feasible and hence made fast and also performs with low network fees.

2. **Account ID:** A global account ID is assigned to every account on Hedera which in the end, decides the amount of money present on our account. HBAR transactions are made to and from this ID. A public/private key which would be assigned to our account can be used to sign and validate the transactions.

   All the ledger accounts have an HBAR balance and a key associated with it. A sign is required by the key holders for any transactions which cause coins to get transferred from the accounts.

3. **Proxy Staking:** For a fixed amount of time, our crypto-coins in our account would be connected to a node when the staking takes place, while we would get rewarded for running the network. It is important that the HBARs are stacked so that the network keeps running. Interests would be earned by stakes when nodes join the network, and this will, in turn, declare the associated accounts, which looks over the fact that the said nodes are paid to function properly.

   Staking is present in most PoS BC and this locks the tokens, thereby disabling their spending. But on the other side, Hedera provides users with proxy staking. Only the HBAR amount is proxied to a node, but circulation is not locked out. When coins in an account are connected to another node, proxy staking takes place, and it does not have to be run by the staking account to get the rewards. When our account is proxy staked, we can earn HBARs without running a node.

4. **Thresholds:** A value is set to which the Hedera platform compares the transaction and then generates records. The confirmation to the applied consensus to a transaction on the Hedera platform can be obtained by the clients as an array is provided by the platform and the records provided are a particular type of the confirmation of the transactions.

   Hedera generates records when a threshold for payment can be set up by us, and also the flag values are set to be true.

   Record threshold can be sent or received as per being set by the user which will generate records when a payment gets sent, which would be above the threshold. When a record threshold is received, it will, in turn, generate a record when an incoming payment is above that threshold.

5. **Receiver Signature:** Acceptance of the payment is made sure by the receive record threshold, and validation is hence allowed from the receiving end of the payment which is incoming. This feature allows certain foundations to take donations from particular individuals, which could not have been possible previously.

6. **Auto-Renew Period:** Retention of an account on the Hedera Platform costs money which allows renewal to be automatic when money is taken out. The fees are still not determined exactly.

7. **Transaction and Queries:** We can make transactions in the Hedera framework which lets the users access information or make changes regarding to the account.

   Accounts can be created and deleted if and only if the initial balance and public key are provided. When an account is deleted until the expiration is crossed, the ledger will contain the account but it will not be able to receive any transactions. All the HBARs from the deleted account would be removed to another alternate specified account.

8. **Adding and Deleting Claims:** There can be claims posted against the account, which in turn sets information upon the accounts by the users. Credentials or certificates or threshold key list can be among the information.

Each claim is given keys which is basically a protector for verification or removal of specified claims.

9. **Transferring:** Hedera Hasgraph is ideal for developers, ground-level users, and organizations as it allows micropayments on its platform, and the fees are at a fractional value of the amount to be transferred, which can be as low as $1/1000^{th}$ of a cent. Single or multiple accounts can also receive payment. The transaction can also be between the same accounts if the total adds up to be 0.

10. **Querying:** Information identification can be done by using an account querying method. It can also be used to retrieve information which is associated with a transaction like a receipt to simply check the account balance.

    Account properties are updated in Hedera, which is a benefit to the users like expiration date, time, signatures.

    All the different properties of its cryptocurrency are offered to the users and the developers by Hedera Hasgraph and the fact that it supports microtransactions at a minimal fee and fast speed are a huge advantage.

3.5 COMPARATIVE REVIEW OF CONSENSUS ALGORITHMS

This section gives a comparison of the various consensus mechanisms discussed in the chapter. Since the implementation of the algorithm can cause certain differences in BC platforms with the same consensus protocol, live (at the time of writing) BC platforms have been showcased.

This section gives a comparison of the various consensus mechanisms discussed in the chapter. Since the implementation of the algorithm can cause certain differences in BC platforms with the same consensus protocol, live (at the time of writing) BC platforms have been showcased. This comparative study is given in Table 3.1.

The comparison between the consensus mechanisms have been done based on the following criteria:

1. **Are there any live blockchain platforms currently employing the algorithm?** The consensus protocol's adoption in the industry is what makes it ideal. If the consensus mechanism has been devised but there has been no history of any BC platform-live or defunct using it, it can mean that the mechanism was not considered feasible in terms of security and infrastructure.

2. **Is the consensus protocol Byzantine fault-tolerant?** Byzantine Generals Problem was discussed in Section 3.4.5. The ability of a consensus mechanism to tolerate Byzantine faults is crucial. It is one of the factors which makes the BC robust. However, BFT-adherence can lead to scalability issues. It is, therefore, a credit to the architecture of a BC that can employ a BFT consensus mechanism while retaining scalability. It may be said here that it was due to the problems arising out of pBFT that led Hyperledger Fabric to consider another consensus mechanism. EOS.IO (discussed in Section 3.4.3.3) has been able to strike a balance between BFT and performance via its use of DPoS.

3. **Can the mechanism tolerate crashes in the network?** Crash Fault Tolerance is one of the measures of the BC's availability. If the network is prone to crashing, then it can put the integrity of the transactions at risk. A BC Platform which is not crash tolerant can even lead to double spending going unwatched. The basic tenets of 'immutability' and 'trustworthy' can be put to question if a network is not crash fault-tolerant by virtue of its consensus protocol.

TABLE 3.1 Comparative Study of Consensus Algorithms

Consensus Algorithm	Specific Use Case (If any)	Byzantine Fault Tolerant	Crash Fault Tolerant	Throughput (Transactions per Second)	Block Size	Block Time
Proof of work	Bitcoin	Yes	Yes	11	1 MB	10 minutes
	Ethereum	Yes	Yes	16	Dynamic	20 seconds
Proof of stake	Peercoin	Yes	Yes	8	1 MB	8 minutes
Delegated proof of stake	EOS	Yes	–	1,000	Dynamic	500 ms
Liquid proof of stake	Tezos	Yes	Yes	40	Dynamic	1 minute (cycle-4096 minutes)
Proof of elapsed time	Hyperledger sawtooth	Yes	Yes	Dynamic	Dynamic	Dynamic
Practical Byzantine fault tolerance	Hyperledger fabric 1.0	Yes	No	Dynamic	Dynamic	Dynamic
Proof of authority	–	Yes	Yes	Dynamic	Dynamic	Dynamic

TABLE 3.1 *(Continued)*

Consensus Algorithm	Specific Use Case (If any)	Byzantine Fault Tolerant	Crash Fault Tolerant	Throughput (Transactions per Second)	Block Size	Block Time
Hedera hashgraph consensus mechanism	Hedera hashgraph	Yes	Yes	10,000	Dynamic	3–5 seconds

4. **Throughput (number of transactions per second) of the use case/ live blockchain platform (if any):** Not every transaction recorded on the BC is deemed 'final.' The very concept of 'irreversibility' is practiced by most BCs by making sure that there are a specified 'c' number of blocks after the transaction. Moreover, the throughput of a BC platform affects the number of transactions that can be bundled together in a block. High throughput can mean reaching 'finality' for multiple transactions faster.

5. **Block size and block time of each block on the blockchain of the framework/blockchain platform:** Block Size and Block Times both can be varied based on the design of the BC platform and are not always an explicit function of the consensus protocol employed by that specific platform. For instance, even though both ETH and BTC both use PoW consensus mechanism at this time (ETH has been planning to move to Proof of Stake), the block size in the former is a limitation posed by the gas of that block while in the latter it is 1 MB. In the same manner, the block times in both are 15 seconds and 10 minutes, respectively.

The above criteria are used to determine the performance of a consensus algorithm implemented on a specific platform. The BC platforms and frameworks used for this purpose are as follows:

i. **Bitcoin (BTC):** First-generation BC which utilizes PoW and runs on BTC script-a scripting language implemented along with the platform and tokens.

ii. **Peercoin:** The first cryptocurrency which used Proof of Stake instead of PoW as its consensus algorithm.

iii. **Ethereum (ETH):** The first second-generation BC platform sporting a Turing complete scripting language-Solidity and running on SCs. This sparked the development of decentralized apps. At the time of writing, ETH uses PoW with the Ethash scheme for hashing the blocks. However, a long-overdue community plan has been to migrate the BC from PoW to Proof of Stake.

iv. **EOS.IO:** A BC platform designed after the release of ETH to enable high-throughput and help scale the BC better. Thus, it was made with the motive to replace ETH as the go-to BC platform for the design of decentralized apps.

v. **Tezos:** A distributed ledger which is focused on on-chain governance and enabling security via liquefaction of assets owned by the network members. Tezos has been particularly popular with financial institutions.

vi. **Hyperledger Sawtooth:** A framework for developing Distributed Ledger Solution conceived by Intel. It was termed as Sawtooth Lake while it was being developed by Intel. Later, it was contributed to the Hyperledger ecosystem where it graduated to a production-grade framework. Hyperledger Sawtooth employs pluggable consensus mechanisms. It is known for its use of Proof of Elapsed Time consensus mechanism and focuses on 'interoperability' with other BC platforms.

vii. **Hyperledger Fabric:** Made by Intel, Hyperledger Fabric is one of the most widely used frameworks for developing distributed ledger solutions for private and consortium class of BCs. This framework, too, has graduated from the Hyperledger Ecosystem.

Here special mention must be made to Managed Solution provided Microsoft Azure known as Azure Managed BC Service which implements a version of ETH using PoA consensus mechanism. But since no notable production class use cases or adoption from the industry were found, the said algorithm could not be reviewed with a framework/BC platform.

From Table 3.1, it can be observed that EOS.IO's implementation of Delegated Proof of Stake gives a very high rate of transactions and scales well even in the presence of a large number of nodes. The second feature is absent in Hyperledger Fabric 1.0 which uses pBFT to achieve network consensus. Tezos, on the other hand, provides better overall security as the assets of the network member are at stake in achieving consensus. Hedera Hashgraph provides a different architecture which utilizes DAGs or Directed Acyclic Graphs for information dissemination and consensus. But it has been shown to slow down in a network of many members. Thus, rendering it partially unscalable. However, all of the algorithms provide ways to reach a decision without straining too much compute resources which gives them an upper hand in consortium solutions.

3.6   CONSENSUS ALGORITHMS FOR BLOCKCHAIN (BC) SOLUTIONS IN IoT

The amalgamation of industrial Internet of Things (IIoT) and BC is one of the most hotly discussed topics in both domains. BC can build on IoT, identity management and decentralized identifiers [16, 17] to give privacy. Operations in an ecosystem consisting of the IoTs can be made reliable and transparent with the help of distributed ledger solutions. A flip side to this would be discarding initial blocks when the chain grows. This idea was advocated by Satoshi [1] in this BTC whitepaper as well. Figure 3.8 shows the solution of BC in IoT.

The discarding of initial blocks can be done under the following simple scheme:

- For every participant joining the network at or after a block of height 'h,' take the genesis block as the start of their copy BC.
- For the next 'h-1' blocks, create a Merkle tree hash of the blocks' individual hashes including the genesis block and use that hash in the header of the second block.
- For any other node in the network which has a memory storage problem, follow the same clipping procedure.

This procedure can be repeated after 'h' blocks in the BC (similar to BTC's 'halving' every 210,000 blocks). Any new node joining the system may or may not have a choice to replicate the whole or the clipped part of the BC depending on network architecture and the node's capacity.

Such a process would help make the system lighter but has its own disadvantages. One of the major disadvantages would be downtime in the system. While the clipping is in process, the system cannot verify any transactions. They would have to be put on hold to be included at the "h+1"th block.

On edge, IoT devices are tuned to have low compute power and memory storage. Thus, even in IIoT, storage is not really a strong suit for the 'things' in the network. They are optimized continuously for better compute capacity. Keeping this in mind, there can be nodes in the network whose sole purpose would be to store the original BC. All the other nodes can only maintain the clipped copy of the chain and focus on their tasks.

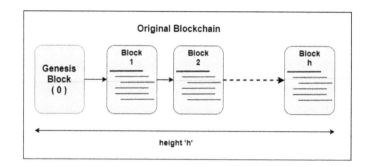

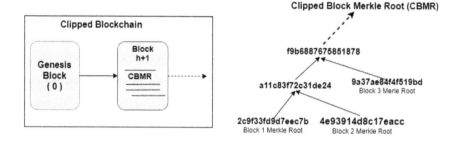

Blockchain in IoT - Clipping for better scalability

FIGURE 3.8   Blockchain in IoT.

Such a system would lead to the design of an autonomous and self-healing network of IIoT networks where every other node would keep track of a specific node's reliability. In case, a node (IoT device) stops functioning properly, its reputation points could be dropped, and it might be put out of commission. The system would also be able to arrive at a decision without human consensus and then keep a record of the decision as a 'transaction' on the chain.

The consensus mechanism used in such a network would need to ensure high throughput and scalability. It cannot be PoW [1] as a network composed of IoT devices would not be able to cope with the increase of difficulty of mining operations (refer to Section 3.4.1). Tolerating Byzantine Nodes would also be a priority for the network IIoT devices.

Keeping this in mind, a delegated consensus mechanism like DPoS [4] would be highly suitable for such a system. The asset to be put at stake in a BC network like this would be the roles of the network devices. Section 3.6.1 discusses the implementation of BC clipping.

3.6.1 IMPLEMENTATION OF BLOCKCHAIN (BC) CLIPPING

As mentioned earlier, the algorithm works based on the assumption that no transactions are recorded into the ledger while the clipping is being facilitated. This downtime can affect the system's robustness. However, a simple method to check the integrity of the nodes after the procedure has concluded is by announcing the latest hash of the block that results after the summary block to the network.

An implementation of the above procedure was made using PoW as the consensus mechanism. Please note that even though PoW was used in the Proof of Concept, in a real-life scenario, use of PoW is not ideal given its disadvantages discussed earlier. Sections 3.6.1.1–3.6.1.4 give the details about the implementation.

3.6.1.1 THE BLOCKCHAIN (BC) NETWORK

The network is meant to be formed between IoT devices which do not possess much compute resources. As such, it is of paramount importance that the whole network and its architecture remain as lightweight as possible. In the proposed implementation of the BC network, we have the following parts:

1. **Peers/Nodes:** These represent the IoT devices which are meant to form the network. The IoT devices use the network to record device telemetry, and this can be used by the devices to ascertain the behavior and, by implication, the trustworthiness of the IoT device in the network they are about to communicate with.
2. **Certifiers:** The Certifiers act as the entry point into the network. When a node joins a network, a X.509 Certificate is generated which is signed by the Certifier's Private Key. This acts as the new node's identity in the network.
3. **Ordering Service:** Ordering Service receives the transactions in the network and then after verification, makes a block out of them and distributes them among the nodes.

The minimum system requirements of a node to make it a committing node needs to be specified when bringing up the network. This prevents any node with low capacity to become a committing node. This is done so that the network is not frequently spammed with transactions which signal the start of the clipping process due to a committing node running low on resources.

The network which was used to demonstrate the concept of Clipping the BC was kept simple by use of the above entities only. In the implementation, we take 1 certifying node, 1 ordering node and 1 peer node. Since the network cannot function without the Certifier, the first node in the network, the one to bring up the network is the Certifier.

3.6.1.2 NETWORK ENTRY BY NEW NODE

Any new node wishing to become a part of the network and propose transactions has to follow the procedure given below:

- The entrant pings any node already part of the network to get the address of the certifier node in the network.
- The entrant then produces a ECDSA Key Pair. This ECDSA Key Pair is used in the Certificate that will be awarded to this entrant by the Certifier.
- The Certifier node is supplied with the public key of the ECDSA Key Pair of the entrant. The Certifier generates a X.509 Certificate and signs this certificate using its own Private Key from its ECDSA Key Pair. The unsigned digital certificate provided to the certifier also contains the system configuration of the entrant node. This is done to check whether the node conforms with.
- The signed X.509 Certificate is then passed to the entrant node. With this, the node is now enrolled in the network and can propose transactions for blocks.

With the generation of the signed digital certificate, the process of enrolling a node into the network is complete.

3.6.1.3 MAINTAINING THE MERKLE TREE VALIDITY/CRYPTOGRAPHIC INTEGRITY CHECK

Even though the BC network can now be used as a lightweight solution which is compatible with edge devices. However, the main problem that arises at this conjecture is how to maintain the cryptographic validity of the BC.

When checking the validity of the longest chain of the BC, calculating the Merkle tree is the easiest solution. However, with the aforementioned clipping procedure alone, this can prove to be detrimental as the Hash of the Genesis Block is available but the previous block is not available in the new BC. The following are the shortcomings which need to be accounted for:

- The new BC has to account for the previous genesis block, the previous block and the block which records the votes for hashes of the new clipped BC.
- The previous genesis block needs to be discarded in the new clipped chain; however, the validity of the blocks added up to the event of clipping needs to be checked as well.
- The new genesis block also needs to account for rare events of hash collisions which can be leveraged to change the facts recorded in a specific block on the BC without the need to change the whole BC even though the said block has been discarded in the clipped chain.
- If we take the new block which records the hash votes as the genesis block of the new chain, then there needs to be a provision such that nodes in the network which are storing the undiscarded blocks in the chain as well can also validate the blocks originating from here on out.
- The state of the whole network needs to be saved in the new genesis block as well in a summary since the previous blocks are discarded.

The last point is especially important in case the BC is being used to maintain tangible assets like cryptocurrency. Keeping in mind the aforementioned points, we propose the following design of the new BC:

- All the previous blocks in the BC are discarded in nodes which are incapable of storing the blocks.
- The block which records the hash votes forms the new genesis block of the clipped chain.
- The states of the network nodes are stored in the new genesis block as transactions.
- The hash of the new block is formed by using the Merkle tree of the hash of the votes and hashes the network nodes' state, along with the hash of the genesis block, any arbitrary block in the chain and the penultimate block to the block which recorded votes.

This would offer the following advantages in the network:

- The block can be validated by referring to the hash of the previous genesis block, the arbitrary, and the penultimate block in nodes in the network which are storing the BC.
- When the event of BC clipping occurs again the network, the original BC can be safely discarded by all nodes in the network while maintaining the state in the form of the Merkle tree of the new genesis block.
- Use of arbitrary blocks' hash in the new genesis block would reduce the chances of premeditated attack in the network.
- The whole procedure can be facilitated without deferring from the original method of calculation of Merkle trees and validation of the chain.

The proposed process has been discussed in Section 3.6.1.4.

3.6.1.4   NETWORK FUNCTIONING AND BLOCKCHAIN (BC) CLIPPING

The network nodes except the certifier and orderer continue to function normally as long as every committing node can hold the BC in its entirety. It has to be noted at this point, that not all nodes need to maintain a full BC. While the designed system allows every node in the network to propose transactions, it also allows the nodes to be either committing or non-committing nodes, meaning they may choose to store or not to store and maintain the BC.

Once a committing node reaches its limit of storage, it issues a 'LIMIT' Flag in the network using its Digital Certificate as evidence of identity. The other committing nodes and the Orderer acknowledges this and then begins the BC clipping.

Every committing node in the system calculates the clipped BC hash independently, after this they release their calculated hash along with their Digital Certificate onto the network for every other node to verify. In this process, the orderer needs to receive this hash from the committing nodes. It is the orderer in the network that then calculates the majority of the hashes and declares the clipped BC hash. This functions as a majority vote. The current states of the network nodes are also recorded in this block as transactions along with the hashes of:

- Previous genesis block;
- Penultimate block in the chain;
- Any arbitrary block denoted by the height of the block in the chain.

The Merkle tree along with the block is released to the network.

The majority vote is recorded in a block and released to all the committing nodes. The committing nodes can, thus, check the results of the majority after BC clipping. This block is added as the first block (new genesis) in the clipped BC if the committing node wants to record a shorter, clipped BC. In case of nodes which can store the full BC, this block becomes the next block in the BC. Either way, the cryptographic validity of the BC is preserved since the votes and states are treated as transactions in this case.

IoT devices on edge layer have low computation power and memory storage. So, even in IIoT this storage is not enough. They are optimized continuously for better computational capacity and memory, so that they can store the original BC. All the other nodes can only maintain the clipped copy of the chain and focus on their tasks.

The design of an autonomous and self-healing network of IIoT networks with tracking of node's reliabilityis leaded here. Reputation points could be dropped for misfunctioning node, and it might be put out of commission. This system makes a decision without human consensus and keep a record of the decision as a 'transaction' on the chain.

The consensus mechanism used in such a network would need to ensure high throughput and scalability. It cannot be PoW [1] as a network composed of IoT devices would not be able to cope with the increase of difficulty of mining operations (refer to Section 3.3.1). Tolerating Byzantine Nodes would also be a priority for the network IIoT devices.

DPoS [4], a delegated consensus, would be highly suitable for such a system. The role of network devices is to put the asset in a BC network.

3.7 CONSENSUS ALGORITHMS FOR BLOCKCHAIN (BC) SOLUTIONS IN BIG DATA

The use of a distributed database for big data solutions like healthcare, appliance, and user data has been a debated topic. A distributed database would provide better security as the data itself would not be stored in just one place. Given the current processing power and infrastructure offered by public cloud providers like Amazon and Microsoft, a big data solution for storage and processing of medical records of a whole country will not be impossible. The problem would then lie in maintaining the integrity of the data and putting a check on tampering of its copies by respective owners. Thus, it becomes an ideal use case of BC. Figure 3.9 shows the solution of BC in big data.

BC cannot be used to store and maintain voluminous data. Big data is famously characterized by its three properties, the three 'V's of Big Data-volume, velocity, and variety. Contrary to popular belief, a BC is not a distributed database that can be used for large amounts of data storage. If anything, using it in that manner would only make it unscalable faster [24].

Instead of using BC to maintain the data for guaranteeing 'integrity' and 'transparency,' the state of the data can be tracked on-chain. Big Data solutions like Hadoop store the data in three different copies. These would ensure that the data is not centralized. Maintaining a state of the data, for example, the hash of the contents on the BC and maintaining the copy of that chain on multiple peers would be helpful in keeping the network fast, scalable, and congestion-free.

A common ingestion pipeline can end in the hash being recorded and then stored on the BC. This would mark the data integrity irrespective of the partitions being stored. Thus, if the partitions of data are corrupted anyhow, then that would be easily detectable using BC. BC can be used for enabling fast data sharing as well. A smart contract-based role-based access control in the network could be used to limit and record the access in the system, particularly for update tasks.

As mentioned before, consensus algorithms would play a key role in determining the rate of adoption of BC in big data solutions. At first glance itself, a resource-intensive algorithm will adversely affect the whole system's performance. Thus, PoW [1] would be detrimental to the idea of providing 'real-time insights' into the data being ingested. Unlike the previous case (Section 3.6), proof of stake [2] mechanisms cannot be used since here the only element that can be delegated is one's 'vote' for leader election. This does not carry enough weight to provide an incentive to all the members of the network.

Even though it is designed to be used in private and permissioned BC solutions, proof of elapsed time [8] cannot be suitable here either since the velocity and variety of data arriving would make 'waiting' impossible. Under such circumstances, pBFT [9] or any variant of it is an ideal candidate. Though PoA is also Byzantine Fault Tolerant since it has not been deployed in production workloads, choosing a time-tested warrior in pBFT is a better option. Alternatively, a Tezos-styled liquid proof of stake [5] is also viable. In this case, the asset to be liquified and kept in holding would not just be limited to the role of the participant but also the data that the network member wants to introduce in the system.

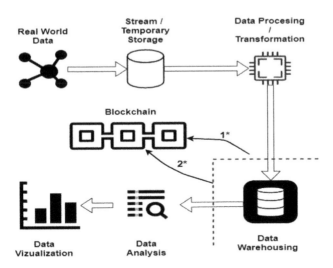

Big Data Ingestion and Visualization Pipeline with Blockchain

1.* Data Integrity Check (Uploading of the hash of the data to the blockchain).

2.* Blockchain Smart Contract-based Role-Based Access Control (RBAC).

FIGURE 3.9 Blockchain in big data.

3.8 CURRENT AND FUTURE WORKS

As noted before, the disadvantage with pBFT [9] lies with its limitation of at most $1/3^{rd}$ corrupt nodes. In a real-life scenario, this might not be the case. This begs for a consensus mechanism which can handle worse network corruption and is better scalable. An example of his has been discussed in Refs. [12, 18, 20]. There are also the case where faulty nodes need to be detected. Faster the detection, the more robust the network. There has been work done in this regard with respect to hardware distributed systems [25]. Such works can be modified and adapted into BC with ease.

Making faster consensus mechanisms and encryption standards would be crucial since cryptography is at the center of the BC revolution. Adoption in the industry can be propelled with use of lightweight cryptography algorithms [21], which can help make the overall procedure faster and less resource-intensive. Case in point, the hashing difficulty can be brought down using lightweight cryptography along with the time required to produce the digital signature of a new participant in the system. It can also help speed the adoption of IoT and on-edge solutions.

Since we have already achieved what is termed as 'quantum supremacy' [34], a special mention must also be made for making BC architecture more robust by introducing 'quantum-resistant [35] encryption and hashing schemes. A common scheme in this regard is to increase the keyspace of the algorithm being used. But such simple schemes can work only so long for noisy scale intermediate quantum computers [39] which exist today. There is a need for a better standard of encryption in BC to make sure that the oncoming Quantum Age does not break the BCs.

3.9 CONCLUSION

Industry 4.0 has sparked a widespread change in technology trends. One of the major driving principles of it is 'decentralization.' BC has the capability to make that happen. It has the potential to be a driving force behind many ongoing transformations. However, the belief that everything can be made better using BC is completely wrong. While there are certain areas like big data and IoT which would benefit from decentralization that BC provides, there are use cases where adoption of BC has led to higher costs. Furthermore, there is no 'one right consensus algorithm' to be used in a BC. At the end of the day, it comes down to the better judgment of the developer designing the system.

KEYWORDS

- blockchain
- Byzantine fault tolerance
- consensus algorithms
- cryptocurrency
- decentralization
- directed acyclic graph
- distributed ledgers
- peer-to-peer network
- proof of elapsed time
- proof of stake
- proof of work

REFERENCES

1. Satoshi, N., (2009). Bitcoin: A Peer-to-Peer Electronic Cash System (pp. 1–9). In https://bitcoin.org/bitcoin.pdf (accessed on 28th July 2021).
2. Ben, L., & Richard, C., (2004). Proof-of-work proves not to work. In: The Cambridge MIT Institute (CMI) Project: The Design and Implementation of Third-Generation Peer-to-Peer Systems (pp. 1–9).
3. Sunny, K., & Scott, N., (2012). PPCoin: Peer-to-Peer Crypto-Currency with Proof-of-Stake (pp. 1–6). In https://decred.org/research/king2012.pdf (accessed on 28th July 2021).
4. Ian, G., (2017). EOS – An Introduction (pp. 1–8). In https://iang.org/papers/EOS_An_Introduction-BLACK-EDITION.pdf (accessed on 28th July 2021).
5. Goodman, L. M., (2014). Tezos — A Self-Amending Crypto-Ledger (pp. 1–17). In https://tezos.com/whitepaper.pdf (accessed on 28th July 2021).
6. Guilhem, C., Paul, B., Zach, H., & Balthazar, G., (2019). DPoPS: Delegated Proof-of-Private-Stake, a DPoS Implementation Under X-Cash, a Monero Based Hybrid-Privacy Coin (pp. 34–37). In https://github.com/X-CASH-official/xcash-dpops (accessed on 28th July 2021).
7. Barinov, I., Baranov, V., & Khahulin, P., (2018). POA Network – White Paper (pp. 1–34). In https://github.com/poanetwork/wiki/wiki/POA-Network-Whitepaper (accessed on 28th July 2021).
8. Lin, C., Lei, X., Zhimin, G., & Weidong, S., (2017). On security analysis of proof-of-elapsed-time (PoET). In: International Symposium on Stabilization, Safety, and Security of Distributed Systems, SSS 2017: Stabilization, Safety, and Security of Distributed Systems (pp. 282–297).
9. Miguel, C., & Barbara, L., (1999). Practical Byzantine fault tolerance. In: The Proceedings of the Third Symposium on Operating Systems Design and Implementation (pp. 1–14). New Orleans, USA.
10. Leslie, L., Robert, S., & Marshall, P., (1982). The Byzantine generals problem. In: ACM Transactions on Programming Languages and Systems (Vol. 4, No. 3, pp. 382–401).
11. Seth, G., & Nancy, L., (2005). Brewer's conjecture and the feasibility of consistent, available, partition-tolerant web services. In: PODC-2005 (pp. 51–59).
12. Miguel, C., Giuliana, S. V., & Lau, C. L., (2010). Asynchronous Byzantine consensus with 2f+1 processes. In: SAC'10 (pp. 1–6). Sierre, Switzerland. Copyright 2010 ACM 978-1-60558-638-0.
13. Yang, X., Ning, Z., Wenjing, L., & Thomas, H. Y., (2020). A survey of distributed consensus protocols for blockchain networks. In: IEEE Communications Surveys & Tutorials (pp. 1–34). doi: 10.1109/COMST.2020.2969706.
14. Stefano De, A., (2017). Assessing security and performances of consensus algorithms for permission blockchains. In thesis of MSc. In: Engineering of Computer Science (pp. 4–42). Under Faculty of Information Engineering, Computer Science and Statistics Department of Computer, Control and Management Engineering.
15. Stefano De, A., (2017). Assessing security and performances of consensus algorithms for permissioned blockchains. In MSc. In: Engineering of Computer Science (pp. 14–44). Faculty of Information Engineering, Computer Science and Statistics Department of Computer, Control and Management Engineering, Sapienza, University Di Roma.

16. Yki, K., Dmitrij, L., Tommi, E., & Nikos, F., (2019). Improving the privacy of IoT with decentralized identifiers (DIDs). In: Hindawi Journal of Computer Networks and Communications (Vol. 2019, p. 1–10). Article ID 8706760. doi: https://doi.org/10.1155/2019/8706760, SCI-Indexed.

17. Fuchun, G., Yi, M., Willy, S., Senior, M., Homer, H., Duncan, W. S., & Vijay, V., (2015). Optimized identity-based encryption from bilinear pairing for lightweight devices. In: IEEE Transactions on Dependable and Secure Computing (pp. 1–11). doi: 10.1109/TDSC.2015.2445760.

18. Loi, L., Viswesh, N., Kunal, B., Chaodong, Z., Seth, G., & Prateek, S., (2015). SCP: A Computationally-Scalable Byzantine Consensus Protocol for Blockchains (pp. 1–16). In https://eprint.iacr.org/2015/1168 (accessed on 28th July 2021).

19. Qassim, N., Ilham, A. Q., Manar, A. T., & Ali, B. N., (2018). Performance analysis of Hyperledger fabric platforms. In Hindawi Security and Communication Networks (Vol. 2018, 1–14). Article ID 3976093. doi: https://doi.org/10.1155/2018/3976093.

20. Libo, F., Hui, Z., Yong, C., & Liqi, L., (2018). Scalable dynamic multi-agent practical Byzantine fault-tolerant consensus in permissioned blockchain. In: Appl. Sci. (Vol. 8, pp. 1–21, 1919). doi: 10.3390/app8101919, www.mdpi.com/journal/applsci (accessed on 28th July 2021).

21. Job, N., Jo, V. B., et. al., (2017). Sancus 2.0: A low-cost security architecture for IoT devices. In: ACM Transactions on Privacy and Security (Vol. 20, No. 3). Article 7.

22. Leemon, B., Mance, H., & Paul, M., (2019). Hedera: A Public Hash Graph Network and Governing Council-The Trust Layer of the Internet (pp. 1–97). In whitepaper of https://hedera.com/hh_whitepaper_v2.1-20200815.pdf.

23. Kai, L., Qichao, Z., Limei, X., & Zhuyun, Q., (2018). Reputation-based Byzantine fault-tolerance for consortium blockchain. In: 2018 IEEE 24th International Conference on Parallel and Distributed Systems (ICPADS) (pp. 604–611). 978-1-5386-7308-9/18/$31.00 ©2018 IEEE doi: 10.1109/ICPADS.2018.00084.

24. Fredik, S., & Oliver, P., (2019). A comparison of the password- authenticated key exchange protocols, SRP-6a and PAKE2+. In: Degree Project in Technology (pp. 6–26). First cycle, 15 credits Stockholm, Sweden 2019, Kth Royal Institute of Technology, School of Electrical Engineering and Computer Science.

25. Ben, H. H. J., & Timo, D. H., (2004). Accelerating the Secure Remote Password Protocol Using Reconfigurable Hardware (pp. 2–10). In: CF'04. Ischia, Italy. Copyright 2004 ACM 1581137419/04/0004. doi: 10.1145/977091.977157.

26. Andy, R., Foteini, B., Gesine, H., & Christof, P., (2015). Cryptographic theory meets practice: Efficient and privacy-preserving payments for public transport. In: ACM Trans. Info. Syst. Sec., 17(3), 1–31. Article 10. doi: http://dx.doi.org/10.1145/2699904.

27. Mozaffari-Kermani, M., Kai, T., Reza, A., & Bayat-Sarmadi, S., (2014). Fault-resilient lightweight cryptographic block ciphers for secure embedded systems. In: IEEE Embedded Systems Letters (pp. 1–4). doi: 10.1109/LES.2014.2365099.

28. Avi, A., Gad, C., Ido, G., & Maya, L., (2018). Helix: A Scalable and Fair Consensus Algorithm Resistant to Ordering Manipulation (pp. 1–21). https://eprint.iacr.org/2018/863.

29. Kelly, O., Mic, B., James, M., Shawn, A., Dan, M., & Cian, M., (2018). Sawtooth: An introduction. In: Hyperledger – Blockchain Technologies for Business (pp. 1–7).

30. Fatemeh, B., & Andr'e, S., (2010). A leader-free Byzantine consensus algorithm. In: ICDCN 2010 (pp. 1–12, 67–78). LNCS 5935. Springer-Verlag Berlin Heidelberg 2010.

31. Seth, G., & Nancy, A. L., (2012). Perspectives on the CAP Theorem (pp. 1–9). In https://groups.csail.mit.edu/tds/papers/Gilbert/Brewer2.pdf. Whitepaper (accessed on 28th July 2021).

32. Xinwei, G., Jintai, D., Jiqiang, L., & Lin, L., (2018). A Leader-Free Byzantine Consensus Algorithm (pp. 1–18). In https://eprint.iacr.org/2017/1196.pdf. E-print archive (accessed on 28th July 2021).

33. Blockstack, P. B. C., (2020). PoX: Proof of Transfer Mining with Bitcoin (pp. 2–8). In Proof of transfer whitepaper draft v.0.2.

34. Shreya, B., Arghya, M., & Prasanta, K. P., (2020). Quantum blockchain using weighted hypergraph states. In: Physical Review Research (Vol. 2, pp. 1–7, 013322). doi: 10.1103/PhysRevResearch.2.013322.

35. Chuntang, L., Yinsong, X., Jiahao, T., & Wenjie, L., (2019). Quantum blockchain: A decentralized, encrypted and distributed database based on quantum mechanics. In: Journal of Quantum Computing, JQC (Vol. 1, No. 2, pp. 1–15, 49–63). doi: 10.32604/jqc.2019.06715.

36. Ignat, K., & Eugene, P., (2016). Secure Comparator: A ZKP-Based Authentication System (pp. 1–8). In 2015–2016; Cossack Labs Limited, www.cossacklabs.com. Whitepaper. (accessed on 28th July 2021).

37. Shehar, B., et. al., (2019). SoK: Consensus in the age of blockchains. In: 1st ACM Conference on Advances in Financial Technologies (AFT '19) (pp. 16, 183–195). Zurich, Switzerland. ACM, New York, NY, USA. doi: https://doi.org/10.1145/3318041.3355458.

38. Lars, F., et. al., (2014). The CAP theorem versus databases with relaxed ACID properties. In: IMCOM (ICUIMC)'14 (pp. 1–7). Siem Reap, Cambodia. Copyright 2014 ACM 978-1-4503-2644-5.

39. Fernández-Caramés, T. M., & Fraga-Lamas, P., (2020). Towards post-quantum blockchain: A review on blockchain cryptography resistant to quantum computing attacks. In: Special Section on Emerging Approaches to Cyber Security, IEEE Access (pp. 21091–21111).

CHAPTER 4

Intelligence on Situation Awareness and Cyberthreats Based on Blockchain and Neural Network

S. PORKODI[1] and D. KESAVARAJA[2]

[1]*Scholar, Department of Computer Science and Engineering, Dr. Sivanthi Aditanar College of Engineering, Tiruchendur, Tamil Nadu, India, Tel.: 7339464560, E-mail: Ishwaryaporkodi6296@gmail.com*

[2]*Associate Professor, Department of Computer Science and Engineering, Dr. Sivanthi Aditanar College of Engineering, Tiruchendur, Tamil Nadu, India, Tel.: 9865213214, E-mail: dkesavraj@gmail.com*

ABSTRACT

In today's world, the critical infrastructure (CI) is much harder to protect against cyber threats, increasing day-by-day. To identify cyber-attacks and efficiently perform as a cyber analyst, a completely new reaction method and distributive detection approach based on information security technology is required. This method automatically analyzes the report of incidents and shares the result securely and secretly with other CI stakeholders. The major goal is to use cyber triage for the classification of cyber incidents reports instead of human inputs, to remove irrelevant information, and filter the related reports much faster in a scalable way and automate the process of managing report life cycle process. Artificial intelligence (AI) is for to generate fast and efficient incident management techniques which can help the cyber analyst to broadcast cyber situation awareness to quickly adopt the necessary countermeasures if an attack arises in the system. In this chapter, Blockchain technology in support of autoencoder deep neural networks is evaluated for managing and classifying the incidents and validate its performance and accuracy. The cyberthreat classification and management system using neural network and blockchain (CCMSNNB) minimizes the use of the storage area and reduces the number of manual operations. A trusted automated system is built with smart contact blockchain technology to manage the workflow of incidents that gives access to automatic acquisition of data, classification, and removal of irrelevant information data enrichment. Centers of the security operation can perform this technique in supporting incident handling. There are 28,000 parameters used totally by the neural network during the training of the autoencoder, where the neural network consists of a single input layer and five hidden layers. The parameters that are related to the smart contract of the blockchain are blockchain id, consumed time, number of significant terms, and level of threat (high or medium or low). The random forest classifier and neural network are for classifying the cyber incident threats at an accuracy of 91.3%.

4.1   INTRODUCTION

4.1.1   TODAY AND TOMORROW

In the current world, the higher officials and leaders need a meaningful awareness in the cyber situation to safeguard the sensitive data, sustain in fundamental operation, and to protect the national infrastructure. In an organization, the critical intellectual property or infrastructure is protected by security experts in information technology by using cyber threat intelligence (CTI). The cyber threats and vulnerability are increasing day-by-day, intelligence in situation awareness is necessary for data protection, or to take proper action against cyber threats. CTI and Situational Awareness helps the organization analyze future risk status to protect data from threats.

4.1.2   VISION

The vision is to see the innovative development in CTI and situational awareness by using the blockchain and neural network, which can be used for the analysis and prediction of threats and to safeguard the private data of the user. The smart contracts (SCs) of blockchain are used for cyber incident management, whereas the autoencoder of the neural network can perform cyber incident classification [9]. A detailed description of cyber situational awareness, autoencoder, smart contract, and system management follows in the upcoming sections.

4.1.3   CYBER SITUATIONAL AWARENESS

The perception of threat and security coupled with impact assessment present and future can be termed as cyber situational awareness [1, 16]. As the world of technology develops, there is much development in complex tools by the researchers in the situational awareness field across many application domains. The real-time cyber situational awareness of cyber operation gets difficult for evaluating due to the overload of data, insufficient meaning [2], and events speed. The data are often not precise and vague. This imperfect information has to be trusted to find the real attacks and prevent the attack before it is done by suitable risk management. A vast amount of threat reports is received daily by analysts of the security operation center (SOC). The challenge faced by SOC analysts is to find the relevant information in a complicated and huge data set. When exploring the data, insights, and patterns are discovered. Also, the business process of the organization is followed, such as acquisition, archive, usage, and disposing of the threat reports. To identify and defeat the cyber-attacks, the cyber analyst highly needs new tools that can merge the gap between situation and cyber data comprehension. In this chapter, a system which can give automatic support to a cyber analyst in analysis and classification of the incoming incidents, to search for the high similarity of cyber incidents which can affect the cyber situational awareness and to manage the life cycle of the cyber incident has been developed [4, 35].

4.1.4   AUTO ENCODER (AE)

When a stakeholder in a critical infrastructure network generates a cyber-incident report, the incident analysis gets triggered. With the help of incident analysis tools and a strong and huge knowledge base (KB), incident analysis is performed on a large set of data. Deep autoencoder (AE) analyzes a single colossal data set or existing KB. The basic need to design a deep autoencoder for situational awareness is to maximize the sharing speed of the severe high data, to get a trustworthy and fast classification of the cyber incidents, and to avoid any need for human involvement. In this chapter, a study on existing tools and techniques available for intelligence on the cyber threat, intelligence on automatic cyber analysis is described with the help of a neural network based on the deep autoencoder, and evaluation results are shown.

This proposed chapter's main contribution is that the inputs of humans can replace a real-time solution for a large set of tasks on cyber incident analysis. To improve the organization of information and to improve the access of the cybersecurity system where the cybersecurity documents are classified automatically to its threat level. The SCs application, based on the block-chain of ETH [26, 27], could be used to solve cyber situational awareness problems. The SCs are mainly designed for cyber situational awareness for building a trusted and rapid cyber incident management and cyber incident classification without any centralized authority. The smart contract is purely a decentralized system that can minimize the cost of manual analysis and also minimize the incident life-cycle management effort. Novel techniques can make automatic decisions in SCs, where the cyber-attacks are identified and defeated [8, 38].

4.1.5   SMART CONTRACT

A smart contract is a protocol that verifies and fixes negotiated behavior, which cannot be manipulated as it has been distributive and also executed on many different nodes of a single blockchain [13, 19]. A benefit in using a smart contract is that it is only necessary to be deployed once, and then it functions automatically even with no human interaction involved. In the intelligent threat analysis system, the procedure for incident handling and instructions for using the programming language of the smart contract (Solidity) and method to upload smart contract into private blockchain network (ETH) is described [29].

The overview of cyber situational awareness establishment using smart contract and neural network for life cycle management and information classification is illustrated in Figure 4.1. In which SCs such as acquisition, archive, usage, and disposing, these smart contract's source code defines the rules and instruction need to be processed. The smart contract's state is stored in blockchain, which is transparent and can be accessed by all the registered members of the community. Smart contract's source code can undergo parallel execution by miners in the network under the consensus regarding execution outcome. When there is a result of smart contract execution, it leads ways to update the state of contract (Block n+2) in blockchain. Then it is synchronized among every user (Critical Infrastructure 1 to Critical Infra-structure n) participating in the network via poof of work-based consensus mechanism and a peer-to-peer (P2P) standard tool. When any security expert (user) who protects critical infrastructures produce an incident report, it is directed to go through SCs. Then it is automatically handled based on the instruction program.

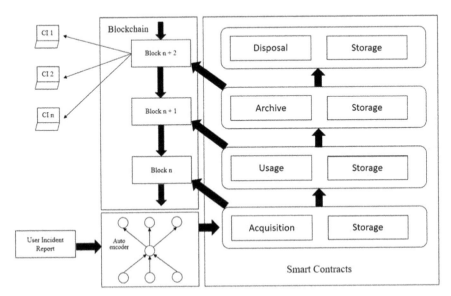

FIGURE 4.1   Overview of cyberthreat classification and management system using neural network and blockchain.

4.1.6   MANAGEMENT SYSTEM

Management system focuses on, some tools for analysis of threats: IntelMQ, MISP (malware information sharing platform), or CAESAIR [12, 32], providing the automatic threat reports management and gives efficient decision support to the operator of SOC. The automatic classification done based on the threat level is much faster, can provide significant support, and accelerates the SOC analyst's reaction time compared to the time taken by the manual classification. For example, a tool which is known as collaborative analysis engine for situational awareness and incident response (CAESAIR) that gives support to various secure techniques of information correlation and a multitude of security-relevant sources provides importing capabilities which can be customized. The sources consist of open-source intelligence (OSINT), custom repository, IT-based security bulletins, and a vulnerable standardized library, which consists of common vulnerability and exposure (CVE). This CVE is essential, especially in SCs, regarding the likelihood assessments of the game theory [10, 34], which is implemented in risk scoring [3]. CAESAIR is employed with the CVE scoring [25] and is extended by tagging automatically, and it can thus provide many useful inputs for life cycle management and information classification. Such type of system is designed for a single organization by implementing SCs. An institution can have much variety of profile definitions according to the network, cyber analyst role, and critical infrastructure.

4.2   RELATED WORKS

The number of tools used in the cyber incident analysis provides intelligence about the threat in cybersecurity. Some examples of the related works include analytical support provided by the CAESAIR tool (http://caesair.ait. ac.at) to security experts, who carry out incident handling tasks at national and international levels. Also, it gives the facility to identify the implicit relation among the available information pieces. MISP tool (https://github. com/MISP/MISP) is an open-sourced tool with a platform of threat intelligence that can do data correlations automatically when a relationship is found between the indicators and attributes from analysis or malware or attack campaigns. The indicator database stores all the information regarding both technical and also non-technical, including incidents, malware samples, intelligence, and attackers, then to do data exchange there is sharing functionality with various distribution models. IntelMQ tool (https://github. com/certtools/intelmq) is also an open-source tool that is developed by the Austrian CERT in collaboration with other parties, aiming to correlate and parse cyber incidents.

The majority of various analytical tasks uses autoencoder. The recursive autoencoder-based framework of machine learning (ML) [39–43] is used to perform sentence-level prediction for the distribution of sentiment labels [20, 28]. Deep autoencoders have been employed for the image retrieval process based on its content, whereas auto coder can learn automatically [6, 37] from the existing examples. The autoencoder uses a neural network that has been used by analysis of latent semantics for better performance, categorizing text [22], and minimizing dimensionality. In an application [36], an artificial neural network [39] is used for improving the scalability of text classifiers.

The implementation of the classification method in the tools of threat intelligence suffers by the massive size of the vector, and the effectiveness decreases as the number of incidents increases. The drawback found in the present existing methods of text classification includes word embedding neural network, Gensim tools or support vector machine (SVM) [21]. SVM is, these methods need a large database to perform training to get meaningful results; however, the database of SOCs are not enough to perform tasks based on semantics. Also, there is no transparency in the results since a vector with real numbers is used. The explanation for the results calculated by the tools is difficult, and also specifically, the SVM has a limitation in choosing the kernel. The words that are not in previous training vocabulary or merely the unknown words cannot be handled. In this chapter, the threat incident classification in a use case of SOCs, autoencoder with better scalability, small vector size, and along with high-level accuracy is suggested.

Automatic technologies are developed by many researchers to make information classification system automated. Principles of the ontology are used to classify the relationship between concepts and documents [15]. An automated hierarchical classification method for constructing project documents based on the project components is developed for easily accessing the construction management and for improving more organized information [7, 17]. A survey-based on cyber-attacks and its classification are taken to create a new ontology design for the cybersecurity incidents, which are classified by the motivation, purpose, and the characteristics. The classification of the cyber-attacks can also be based on its type of network, scope, or involvement severity. But in this chapter, the classification is only done by the threat level, which varies in each organization. The goal of this chapter is mainly to focus the human experts on incidents that are most important to a specific organization. The information life cycle model explained in Refs. [11, 33] can also be applied in the cybersecurity domain. The reports of cyber incidents are acquired, then analyzed later turns outdated. Efficient retention, disposal policy, and automated classification can reduce risks on data, thus effectively managing information. Data classification makes SOC or a company focus its resources to handle urgent or most essential incidents first and then moves to the incidents with less importance automatically, to save cost and time. Since the critical infrastructure network members have no central authority, they have no trust in each other to share the incident's state of the life cycle. Thus, to maintain a life cycle management blockchain is suggested.

Blockchain technology is proposed along with its potential in smart contract design, the transaction of money, digital assets, and automatic banking ledger [18, 24, 30]. A comparison of five blockchain platforms is made and studied about how the blockchain works and used outside the bitcoin (BTC), to develop a customized application on top [5]. So currently, ETH is a well-established and most suitable platform. Thus, ETH blockchain technology (Pyethereum implementation) is used for the analysis of cyber incidents. It supports a testing environment for SCs without mining. In the proposed chapter, a smart contract is applied for the life cycle of cyber incident management and autoencoder of cyber incidents classification for the novel given domain.

4.3   CYBERTHREAT CLASSIFICATION AND MANAGEMENT SYSTEM USING NEURAL NETWORK AND BLOCKCHAIN (CCMSNNB)

4.3.1   NEURAL NETWORK

Neural networks are an algorithm set that is mainly developed for recognizing patterns. The data from the sensors are interpreted through machine perception, either clustering or labeling the raw input. Then the patterns are identified in numerical form in vector form where every data in the real world, such as text, sound, time series, or image is translated [23]. Neural networks do the clustering and classification process on top of stored and managed data. The data is grouped according to the similarities in them. For example, the inputs are acquired, then the data are clustered based on the similarities and label it to create a training data set.

4.3.2   NEURAL NETWORK ARCHITECTURE

All the elements present in the neural network are nodes, which are called units [31]. These units are linked and connected via links. Every link consists of numeric weights. The architecture of the neural network is shown in Figure 4.2.

4.3.3   NEURAL NETWORK USE IN BLOCKCHAIN

The autoencoder of the deep neural network supports the blockchain technology to manage and classify incidents and then validate the system's performance and accuracy. The use of storage areas is minimized in the system, the amount of manual work reduces, and the security increases in the system. An automated trusted system is built with

SCs of the blockchain technology. It is used for managing the workflow of incidents, which gives for the automatic data acquisition, SCs are triggered for analysis and classification of incident report automatically and data enrichment by removal of all the irrelevant information in the classified data. The center of security operations performs this technique for supporting incident handling. The timestamps can be defined by the smart contract to automate the disposal and archival of the incident data. There is no need to worry about the life cycle of the incident by a cyber analyst.

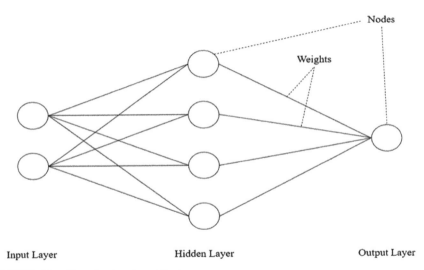

FIGURE 4.2 Neural network architecture.

4.3.4 AUTO-ENCODER FOR CYBER INCIDENT CLASSIFICATION IN THE CCMSNNB SYSTEM

Cyber experts are in charge of critical infrastructure as well as for deducting suspicious behavior of a system. A lot of information is needed for an expert to choose the right mitigation strategy, so all available data relevant to the previous and ongoing attacks are collected and analyzed for a specific use case model and then transformed into actionable intelligence. Information on security is basically in a semi-structured document of text such as vulnerability alerts, bulletins, incident reports, advisories, etc. Manual review or judgment is needed to acquire the CTI present in the documents to find the required information and identify the implicit correlation present in the document. The outline of any practical mitigation strategies and their impacts can be found.

These manual efforts can be avoided by generating incident reports by critical infrastructure experts. The incident reports can be given as input to deep autoencoder, and if it has any sufficient severity, then the output will be a threat report. This intuitive approach reduces personnel expenses while comparing to handle manual cyber incidents, and this can help the analyst to be updated with the situational awareness status for scalable and fast information enrichment and exchange. The main idea to use autoencoder is to map the N-dimensional data with the M-orthogonal directions, so that the data will have greater variance and thus form a less dimensional subspace. The drawback in this conversion, which can also be acceptable, is that there will be information loss regarding the location of original data points in remaining orthogonal directions.

A deep autoencoder is trained as illustrated in Figure 4.3 workflow of the CCMSNNB. The first step of the workflow is to read the incident report as input and then report content is parsed. In the second step, convert the input data with that particular organization's settings of the expert profile into a binary vector with the help of a technique called the bag of words. The third step is normalization; the output of this step is passed to the fourth step in an encoded form to an autoencoder. The most used words of the document are compiled. The remaining vector is left with the word count, which is irrespective of its order. In simple words, 0 can be marked in the first step if there is no word in the original document, and 1 can be indicated if there are words that can be counted greater than 0. Stop words (prepositions, articles, etc., that are not necessary for analysis) can also be ignored. The vector is reduced to minimum size, still holding enough document's data content for achieving scalability and excellent reasonable performance. Then the neural network is trained to produce input vectors, where information is compressed to its maximum as possible, producing only 10 numbers of a central bottleneck as deep autoencoder training result. Cosine similarity is used in step five for scalable and fast document comparison. Then deducted similar incidents are merged with settings of specific institutions, and priority levels are decided (refer Eqn. (1)) [14], which are applied for a given incident. Compressed vectors can be later stored in the neural network hidden layer.

$$P = f(I_r, W_r, W_o, T_s, V_s) \tag{1}$$

where; P is the priority level of incident; I_r is the total count of related incidents; W_r is the total count of related words; W_o is the total count of original words; T_s is the deducting significant terms; V_s is the score of vulnerability.

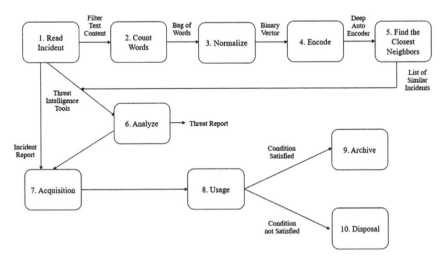

FIGURE 4.3 Workflow of the cyberthreat classification and management system using neural network and blockchain.

The P in Eqn. (1) shows the priority level of the incident, which will return the value as 0 or 1 corresponding to low priority or high priority, respectively. This priority level can be defined as a function that consists of metrics of incident evaluation depending on the basis of indicators, including I_r, W_r, W_o, T_s, V_s.

4.3.5 SMART CONTRACTS (SCS) FOR CYBER INCIDENT MANAGEMENT IN THE CCMSNNB SYSTEM

The SCs are used for classification and management of incident reports, which is labeled as a threat with high priority by autoencoder. The SCs are used to find high priority cyber threat incidents, tag incidents with the acquisition, searching incident using the tag, removing the incidents after a setup predefined amount of time. Then to check the data integrity in a particular interval of time to avoid hardware or manual corruption, to find data origin and to allocate access rights (such as public, private, confidential, sensitive). The goal of the CCMSNNB system is to improve the performance, for saving storage space and for keeping updated information in a trustable way by using the blockchain's distributive nature profitably. Trigger a smart contract to automatically share the analysis report to all the participants in the block-chain network. The major advantage is that the SCs of the Blockchain cannot be compromised or changed. As any such changes can be deducted easily via hashed transactions and messages, they are verified that they originated only from trusted sources with the help of public-key encryption. As soon as the incidents are acquired, incident report classification based on threat level is carried out by smart contract, and the obtained result is stored on blockchain. Then, the process of life cycle management is initiated for given incidents. The report can be archived or used or disposed of in the next step.

Cyber incident processing includes four different SCs, as illustrated in Figure 4.2. In the workflow, autoencoder execution of the steps of cyber incident management comes after the implementation of the steps of cyber incident classification by SCs. As the last step of classification has proceeded as the incident report analysis is read and then parsed. The content of the report is enriched with results classification in step six. In step seven, the input data with that particular organization's expert profile settings are shared with the first coming smart contract for acquisition that allows tools of threat intelligence. The incident text classification is done by splitting works and phrases based on threat relevance such as low, middle, or high. Risk points are compared by counting the number of terms presented in the cyber incident report for each given threat level. The threshold is applied for every level, or else the weighted method, as seen in formula 2 is used to calculate the threat level. In formula 2 [14], calculated points based on threat level is multiplied additionally with constant, representing the related level of threat weight. The range of the threat level is from 1 to 3, which represents a low threat to 1 and a high threat to 3. RP is the Risk point, which is the total sum of H_{rp} as high-risk points are multiplied by HT_w as high threat weight, M_{rp} as a middle-risk point is multiplied by MT_w as middle threat weight, and L_{rp} as a low-risk point is multiplied by LT_w as low threat weight.

$$RP = H_{rp} * HT_w + M_{rp} * MT_w + L_{rp} * LT_w \qquad (2)$$

where; RP is the risk point; H_{rp} is the high-risk points; HT_w is the high threat weight; M_{rp} is the middle-risk point; MT_w is the middle threat weight; L_{rp} is the low-risk point; LT_w is the low threat weight.

$$T_l = \begin{cases} 3(high) \, if \, RP > HT_t, \\ 2(middle) \, if \, RP \geq MT_t, \\ 3(low) \, else \, RP < MT_t, \end{cases} \qquad (3)$$

where; RP is the risk point; HT_t is the high threat; MT_t is the middle threat. where; $LT_w = 1$, $MT_w = 2$, $HT_w = 3$ and $MT_t = 3$, $HT_t = 10$. T_l is the threat level which is inferred by using the RP, weighted points of risk and HT_t high threat and MT_t middle threat from the above formula 2 and formula 3. In step seven, the acquisition step is divided into many different tasks. Then according to the threat level, classification is automatically done, which will be among the 3 levels. The high-level threat needs a faster reaction, triage process, and mediation steps. The medium-level threat dedicates metrics that point out the possible vulnerabilities or indicators of corruption (IoC), which also require a software update. The low-level threat tells about the cybersecurity information regularly. This step needs attention, but it is not always a threat. To find, remove, or shift a report easily, specific tags are assigned for reports. Personal data can be protected by the removal of personal data present in the incident report, and this step may be needed when a normalized incident version is saved. In step eight, automatic similarity search, provenance retrieval, status, data integrity (checking metadata periodically), and data enrichment are supported. The incidents are archived in step nine or removed (by tag or date) on step 10 based on threat level after a considering amount of time.

The incident response is fed into the deep autoencoder when the new incident is read into the deep autoencoder. Threat report is generated based on the knowledge acquired from the past attacks and ongoing attacks. The threat report is analyzed by the smart contract and stored in blockchain, which is also given as an input to deep autoencoder that can learn from the attack and develop intelligence. When a new incident occurs, threat reports can be generated more quickly and efficiently, which helps the cyber analyst to broadcast cyber situation awareness to adopt the necessary countermeasures swiftly if an attack arises in the system; this is the novelty of the CCMSNNB system, as shown in Figure 4.4.

An automatic approach based on smart contract and autoencoder for cyber incident classification and management can be easily used by any analyst for critical infrastructure defense, to quickly adopt the necessary countermeasures if an attack arises in the system. The CCMSNNB system, which is suggested in this chapter, would analyze situational awareness with high throughput and low cost-intensive performance. But in some areas, the human approach and performance will have high accuracy.

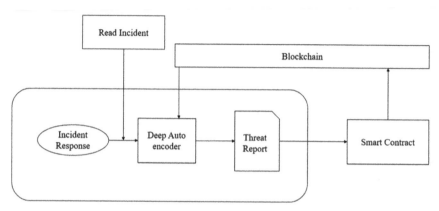

FIGURE 4.4   Flowchart of the cyberthreat classification and management system using neural network and blockchain.

The data set is imported in the r-tool, and the required packages that are necessary for incident classification are installed, data manipulation, and visualization of data. Then an analysis of data exploration is performed. The training data set is trained, and then n the test data, when a new cyber threat incident is fed into the system, the classification is done based on various elements. Random forest is to classify the cyber incident threats, and the neural network is to improve the accuracy of classifying the cyber incident threat classification.

4.4   LIST OF BENEFITS ON INTEGRATING NEURAL NETWORK AND BLOCKCHAIN IN THE CCMSNNB SYSTEM

The benefits of integrating the Neural Network and Blockchain are listed in Figure 4.5. The auditability is increased if there are proper authentication changes as required. The security can be improved by using the cryptographic techniques in the blockchain network. The fraud is reduced as all the data are stored in the blockchain ledger across the network. The failures in the system are reduced by avoiding tampering at any point. The effectiveness is increased by the secure sharing of data throughout the system.

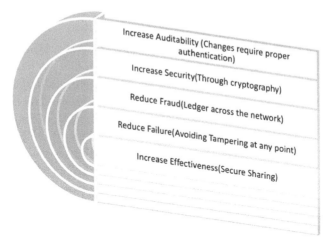

FIGURE 4.5   Benefits of integrating neural network and blockchain.

4.5   RESULTS

4.5.1   EVALUATION OF DATA SET

The goal of a cyber-analyst is to perform a deduction of cyber incidents for mitigating it or for performing cyber incident response. The data set consists of 2,000 training data with 14 columns each and two testing data sets with 2,000 test data, each with 14 columns. The neural network consists of one data input layer and five hidden layers. Various report categories are used to evaluate the cyber incident report, such as Bugtraq, full disclosure, Nmap-dev, and Pen-test. The Bugtraq is a mailing list specially designed to address security. Full disclosure is used to analyze the vulnerability of software by discussing exploitation techniques and vulnerability to the ordinary people at the public forums. Nmap-dev consists of a forum for discussing suggestions based on the proposed change in the Nmap, new ideas, etc. Pen-test is to find weak spots in the system where an attack can happen, and it gives strategies and techniques useful to maintain security. All these threat categories are trained in the CCMSNNB system that consists of a neural network and blockchain. The test set is given to classify the threat based on various threats and with many other elements. The autoencoder and the smart contract is deployed for performing cyber incident classification among a massive set of incident reports further incident management is done after analyzing the incidents.

4.5.2   EXPERIMENTAL RESULTS

The r-tool is used to perform the classification of threats. All the data sets are imported, and the required packages are installed that are necessary for incident classification, manipulation of data, and visualization of data. Then an analysis of data exploration is performed. There are one data input layer and five hidden layers in the neural network, which consist of neurons from 10 to 2,000. The activation function in the neural network is ReLu (rectified linear unit), and the last layer uses sigmoid. The training data set is trained by 100 iterations for autoencoder until it reaches a stable value. The autoencoder's accuracy in training improves for every iteration. The smart contract parameters in blockchain includes id of the blockchain, number of significant terms, time consumed and threat level (low, medium or high). The other parameters are similarity with the past incident that is a similarity score between the new incident and the relevant incident that is found. The categories of threats are 'Bugtraq' as shown in Figure 4.6, 'full disclosure' as shown in Figure 4.7, 'Nmap-dev' as shown in Figure 4.8 and 'Pen-test' as shown in Figure 4.9 and incident threat generated by a person or an organization.

On receiving the new incident report, the smart contract has been triggered automatically to perform analysis and then incident report classification. Then the weight of the threat is estimated. If the weight is high, then it should be handled soon. So that incident is tagged automatically and enriched. Similar incidents are linked. Then analysis steps are performed manually by a cyber-analyst. Based on the classified level evaluated, as shown in Figure 4.10, a timestamp is defined by the smart contract for the archive or disposing of the incident data. A cyber-analyst can then focus on the incident that has been triggered. The significant terms that are deducted include terms like 'phishing,' 'hack,' and 'attack' are listed as threat level high of an incident. The 'encode,' 'authentication,' and 'access' are listed as threat level medium of an incident. The 'investigation,' 'capability,' and 'key' are listed as threat level low of an incident are shown in Figure 4.11. The smart contract ID is used to retrieve data regarding a specific incident from the blockchain with the help of SCs, and they are owner, hash, time, provenance, tag, etc. The threat generated by an individual person or an organization, as shown in Figures 4.12 and 4.13 show that similarity with past incidents that is similarity score between the new incident and the relevant incident is shown in Figure 4.14. Random Forest classifier is used to find the importance of the variable, as shown in Figure 4.15.

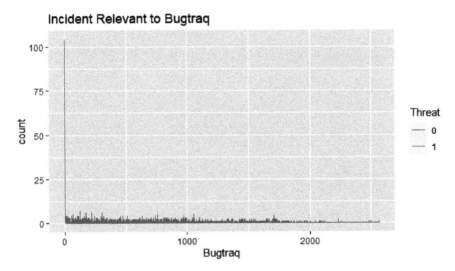

FIGURE 4.6   Incident relevant to Bugtraq.

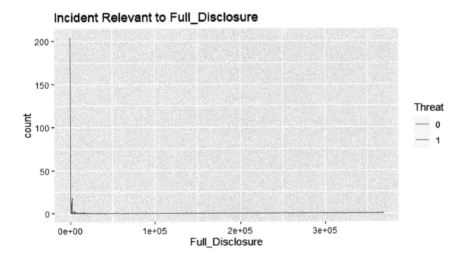

FIGURE 4.7   Incident relevant to full disclosure.

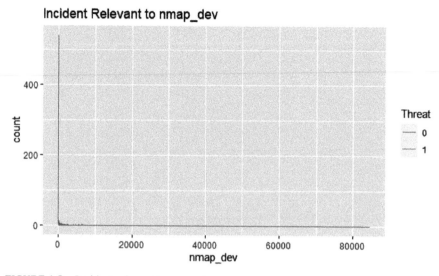

FIGURE 4.8   Incident relevant to nmap_dev.

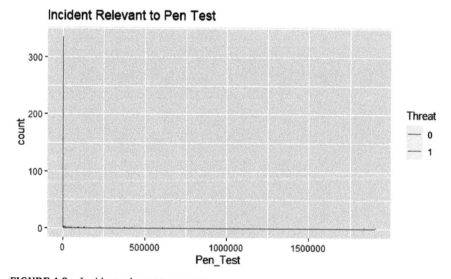

FIGURE 4.9   Incident relevant to pen test.

The overview of the category of the threats in the experimental results that combines test 1 and test 2 datasets are shown in Table 4.1 describes the distribution of the incident into 'high,' 'medium' and 'low' levels of threats. From Table 4.1, it can be observed that the least incident threat report comes from the Pen test threat category (62), and the most incident threat report comes from the Nmap-dev threat category (1916). It can also be observed that most number of threats comes under low-level threat (2383). But cases that come under high-level threats also seem to be high (715), which needs to be handled soon by incident management or a cyber-analyst. The graphical representation of Table 4.1 is shown in Figure 4.10, where the threats belonging to each category of threats can be easily seen.

TABLE 4.1   Overview of the Category of Threats in the Experimental Result

Threat Category	High-Level Threat	Medium-Level Threat	Low-Level Threat	Total
Full disclosure	226	185	447	858
Bugtraq	298	176	490	964
Nmap-dev	157	318	1,441	1,916
Pen-test	34	23	5	62
Sum	715	702	2,383	3,800

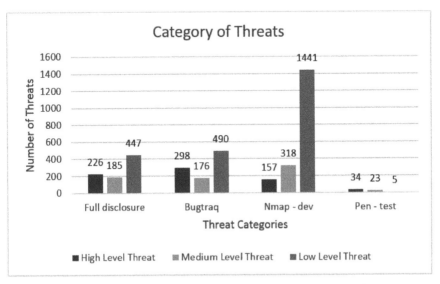

FIGURE 4.10   Overview of the category of threats in the experimental result.

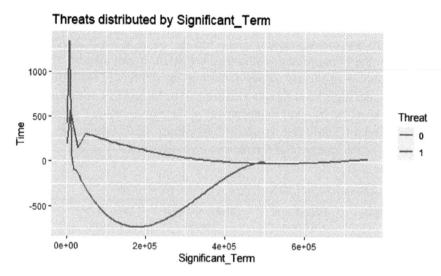

FIGURE 4.11   Threats distributed by significant terms.

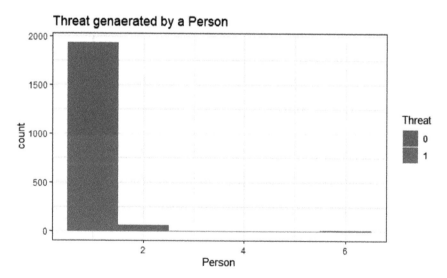

FIGURE 4.12   Threat generated by a person.

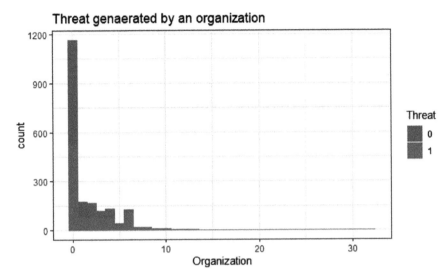

FIGURE 4.13   Threat generated by an organization.

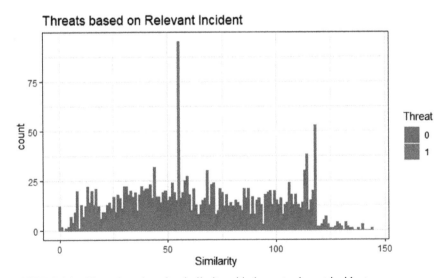

FIGURE 4.14   Threat based on the similarity with the past relevant incident.

The classification tree of cyber incident threat based on the observed elements is shown in Figure 4.16. The threats are also classified based on levels as high, medium, and low. The Recursive Partitioning and Regression Trees of the graphs shows a probability between 0 and 1, where 0 represents false, and 1 represents true, showing the possibility belongs to a positive class, thus confirming it as a threat. The accuracy of the Recursive Partitioning and Regression Trees algorithm is found to be 76.7%.

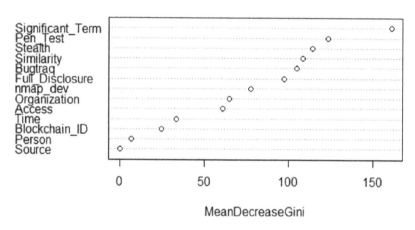

FIGURE 4.15   Importance of the variable thread classification.

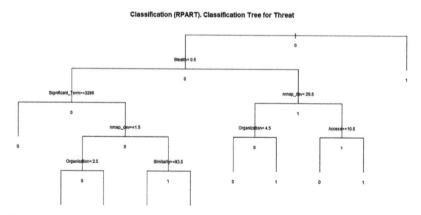

FIGURE 4.16   Cyber incident threat classification tree based on RPART.

The random forest iteratively performs to reduce the impurities in classification. The confusion matrix of the random forest classifier shows that 86.95% accuracy is achieved in the classification of the cyber incident threat, as shown in Figure 4.17.

Two deep neural networks are created to perform classification, such as a convolution network and a fully connected network, as shown in Figure 4.18. Both the networks are constructed by using the nnet library, and the classification of the cyber incident threat takes place.

```
> caret::confusionMatrix(pred_forest, test2$Threat)
Confusion Matrix and Statistics

          Reference
Prediction    0    1
         0 1664   25
         1  236   75

                Accuracy : 0.8695
                  95% CI : (0.8539, 0.884)
     No Information Rate : 0.95
     P-Value [Acc > NIR] : 1

                   Kappa : 0.313

 Mcnemar's Test P-Value : <2e-16

             Sensitivity : 0.8758
             Specificity : 0.7500
          Pos Pred Value : 0.9852
          Neg Pred Value : 0.2412
              Prevalence : 0.9500
          Detection Rate : 0.8320
    Detection Prevalence : 0.8445
       Balanced Accuracy : 0.8129

        'Positive' Class : 0
```

FIGURE 4.17 Confusion matrix for random forest.

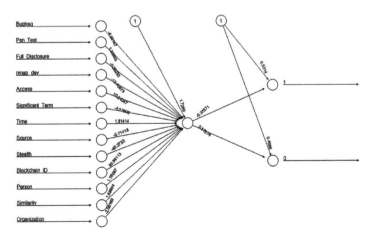

FIGURE 4.18 Cyber incident threat classification based on neural network.

The confusion matrix of the neural network classifier shows that 91.3% of the accuracy is achieved in the classification of the cyber incident threat, as shown in Figure 4.19.

```
> caret::confusionMatrix(pred_nnet, test2$Threat)
Confusion Matrix and Statistics

          Reference
Prediction    0    1
         0 1795   69
         1  105   31

               Accuracy : 0.913
                 95% CI : (0.8998, 0.925)
    No Information Rate : 0.95
    P-Value [Acc > NIR] : 1.00000

                  Kappa : 0.2176

 Mcnemar's Test P-Value : 0.00797

            Sensitivity : 0.9447
            Specificity : 0.3100
         Pos Pred Value : 0.9630
         Neg Pred Value : 0.2279
             Prevalence : 0.9500
         Detection Rate : 0.8975
   Detection Prevalence : 0.9320
      Balanced Accuracy : 0.6274

       'Positive' Class : 0
```

FIGURE 4.19   Confusion matrix for neural network.

4.5.3   PERFORMANCE ANALYSIS

The comparison of various algorithms such as Rpart (Recursive Partitioning and Regression Trees), Native Bayes, Neural Network, and Random Forest is shown in Table 4.2. The elements that are compared are 'sensitivity,' 'specificity,' 'positive predictive value,' 'negative predictive value (NPV),' 'prevalence,' 'detection prevalence,' 'balanced accuracy' and 'accuracy' of the algorithm. The graphical representation of Table 4.2 is given in Figure 4.20. Sensitivity is the true positive rate, as shown in Eqn. (4); it measures the actual positive proportion, which is identified correctly. The graph shows that the sensitivity of the neural network is high compared to other algorithms. Whereas the specificity is the true negative rate, as shown in Eqn. (5), it measures the proportion of actual negative, which is identified correctly. The graph shows that the specificity of the neural network is high compared to other algorithms except for random forest.

TABLE 4.2 Comparison of the Various Algorithm Used in CCMSNNB System

	Sensitivity	Specificity	Positive Predictive Value	Negative Predictive Value	Prevalence	Detection Rate	Detection Prevalence	Balanced Accuracy	Accuracy
Rpart	0.7705	0.2	0.8799	0.1383	0.95	0.732	0.747	0.5253	0.767
Native Bayes	0.9026	0.28	0.8988	0.1463	0.95	0.8575	0.8585	0.5413	0.8065
Neural Network	0.9447	0.31	0.963	0.2279	0.95	0.8975	0.932	0.6274	0.913
Random Forest	0.8758	0.75	0.9852	0.2412	0.95	0.832	0.8445	0.8129	0.895

$$Sensitivity = \frac{number\ of\ true\ positives}{Total\ number\ of\ Threats\ in\ the\ incident\ reports} \qquad (4)$$

$$Specificity = \frac{number\ of\ true\ negative}{Total\ number\ of\ Non-Threats\ in\ the\ incident\ reports} \qquad (5)$$

The positive predictive value (PPV) is a probability where a subject with positive is confirmed to be a threat. PPV can be calculated with the formula shown in Eqn. (6). The NPV is a probability where a subject with negative is confirmed to be a non-threat. NPV can be calculated with the formula shown in Eqn. (7).

$$PPV = \frac{Sensitivity \times Prevalance}{Sensitivity \times Prevalance + (1-Specificity) \times (1-Prevalence)} \qquad (6)$$

$$NPV = \frac{Specificity \times (1-Prevalence)}{(1-Sencitivity) \times Prevalance + Specificity \times (1-Prevalence)} \qquad (7)$$

Prevalence is the proportion of a certain amount of incidents, which are found to be a threat. It is a number of threats present in the total incident reports. 0.95 denotes that 95% of the incidents are threats in the total incident report data set that are taken to perform threat classification and management in the CCMSNNB system. The detection rate represents the no. of positive threats that are predicted correctly, which is made as a proportion among all predictions made. The deduction rate can be calculated by the formula given in Eqn. (8). The graph shows that the detection rate of the neural network is high. The deduction prevalence represents the no. of positive threats that are predicted, which is made as a proportion among all predictions made. The deduction prevalence can be calculated by the formula given in Eqn. (9). The graph shows that the detection prevalence of the neural network is high.

$$Detection\ Rate = \frac{True\ Positive}{True\ Positive + False\ Positive + False\ Negative + True\ Negative} \qquad (8)$$

$$Detection\ Prevalence = \frac{True\ Positive + False\ Positive}{True\ Positive + False\ Positive + False\ Negative + True\ Negative} \qquad (9)$$

The balanced accuracy gives the average of true positive and true negative rates. The balanced accuracy can be calculated by adding the sensitivity and specificity of the algorithm and divided by 2 as shown in Eqn. (10). The actual accuracy of the algorithm is accuracy which defines how accurate the algorithm works. It can be calculated by using the Eqn. (11). From the graph; the accuracy of the neural network is found to be high.

$$Balanced\ Accuracy = \frac{Sensitivity + Specificity}{2} \qquad (10)$$

$$Accuracy = \frac{True\ Positive + True\ Negative}{True\ Positive + False\ Positive + False\ Negative + True\ Negative} \qquad (11)$$

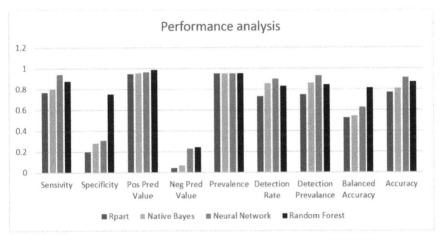

FIGURE 4.20   Performance analysis of various algorithms used in CCMSNNB system.

4.6   CONCLUSION AND FUTURE DIRECTIONS

In this chapter, an intuitive approach with intelligence to handle the classification and management of incident reports for situation awareness and cyber threat analysis is established by building the CCMSNNB system. A deep autoencoder neural network is used for cyber incident classification, and SCs in the blockchain technology are used for cyber incident management. This proposed system could assist the cyber analyst is also giving protection to the CI against the growing cyber threats. The significant contributions of the proposed chapter are to find a real-time solution that can replace the inputs of humans for the task of a massive amount of incident analysis to make some process easier. Such as eliminating irrelevant data, incident classification, and only focusing on the most critical data to perform the mitigation steps quickly. SCs are used to design an automatic incident life cycle management trusted system, with which data can be acquired, used, classified, disposed, or archived according to the need. The CCMSNNB system is built with the neural network and blockchain, and its accuracy is found to be 91.3% on predicting and classifying the threats correctly. The main advantage of this chapter is the human cost for analysis is reduced completely. In the future, this chapter could lead to the development of automatic tools for security assessment and increased effectiveness in handling cyber incidents.

KEYWORDS

- **blockchain**
- **cyber threats intelligence**
- **neural network**
- **proof-of-stake**
- **security**
- **situation awareness**

REFERENCES

1. Aggarwal, P., Moisan, F., Gonzalez, C., & Dutt, V., (2018). *Understanding Cyber Situational Awareness in a Cyber Security Game Involving Recommendation, 4*, 11–38. doi: 10.22619/IJCSA.2018.100118.
2. Auria, L., (2008). *Support Vector Machines (SVM) as a Technique for Solvency Analysis* (p. 811). DIW Berlin.
3. Barford, P., Dacier, M., Thomas, G., et al., (2010). Cyber SA: Situational awareness for cyber defense, in cyber situational awareness. *Advances in Information Security, 46,* 3–13. Springer, Boston, MA.
4. Berman, D. S., Buczak, A. L., Chavis, J. S., & Corbett, C. L., (2019). A survey of deep learning methods for cybersecurity. *Information, 10*(4), 122. doi: 10.3390/info10040122. 2019.
5. Bo, Y., Zong-ben, X., & Cheng-Hua, L., (2008). Latent semantic analysis for text categorization using neural network. *Knowledge-Based Systems, 21*(8), 900–904.
6. Buterin, V., (2013). *Ethereum: A Next-Generation Smart Contract and Decentralized Application Platform.* http://blockchainlab.com/pdf/Ethereum_white_paper-a_next_generation_smart_contract_and_decentralized_application_platform-vitalik-buterin.pdf (accessed on 28th July 2021).
7. Caldas, C. H., & Soibelman, L., (2003). Automating hierarchical document classification for construction management information systems. *Automation in Construction, 12*(4), 395–406.

8. Carrio, A., Sampedro, C., Rodriguez-Ramos, A., & Campoy, P., (2017). A review of deep learning methods and applications for unmanned aerial vehicles, Hindawi, *Journal of Sensors, 1–13.* Article ID: 3296874.

9. Choo, K. R., Dehghantanha, A., & Parizi, R., (2020). *Blockchain Cybersecurity, Trust and* Privacy. doi: 10.1007/978-3-030-38181-3.

10. Cleveland, F., (2010*). White Paper Cyber Security Issues, the Smart Grid Xanthus Consulting International.* https://www.scribd.com/document/182064845/Cyber-Security-Issues-for-the-Smart-Grid (accessed on 28th July 2021).

11. Collobert, R., & Weston, J., (2008). A unified architecture for natural language processing: Deep neural networks with multitask learning. In: *Proceedings of ICML* (pp. 160–167). https://ronan.collobert.com/pub/matos/2008_nlp_icml.pdf (accessed on 28th July 2021).

12. D'Amico, A., Whitley, K., Tesone, D., O'Brien, B., & Roth, E., (2005). Achieving cyber defense situational awareness: A cognitive task analysis of information assurance analysts. In: *Proceedings of the Human Factors and Ergonomics Society Annual Meeting* (Vol. 49, pp. 229–233).

13. Gopie, N., (2018). *What Are Smart Contracts on Blockchain?* IBM. https://www.ibm.com/blogs/blockchain/2018/07/what-are-smart-contracts-on-blockchain/ (accessed on 28th July 2021).

14. Graf, R., & King, R., (2018). Neural network and blockchain based technique for cyber threat intelligence and situational awareness. In: *10th National Conference on Cyber Conflict.* https://ccdcoe.org/uploads/2018/10/Art-21-Neural-Network-and-Blockchain-Based-Technique-for-Cyber-Threat-Intelligence-and-Situational-Awareness.pdf (accessed on 28th July 2021).

15. Harris, S., & Maymi, F., (2016). *CISSP All-in-One Exam Guide Book.* New York: McGraw-Hill Education.

16. Horneman, A., (2019). *Situational Awareness for Cybersecurity: An Introduction.* Software Engineering Institute. https://insights.sei.cmu.edu/sei_blog/2019/09/situational-awareness-for-cybersecurity-an-introduction.html (accessed on 28th July 2021).

17. Joachims, T., (1998). Text categorization with support vector machines: Learning with many relevant features. *Proceedings of ECML-98* (pp. 137–142). Chemnitz, Germany, Springer, Berlin.

18. Kanoun, W., Cuppens-Boulahia, T., Cuppens, F., Dubus, S., & Martin, A., (2009). Success likelihood of ongoing attacks for intrusion detection and response systems. In: *Proceedings IEEE CSE 2009, 12th IEEE International Conference on Computational Science and Engineering.* Vancouver, Canada. IEEE Computer Society.

19. Kishigami, J., Fujimura, S., Watanabe, H., Nakadaira, A., & Akutsu, A., (2015). The blockchain-based digital content distribution system. In: *2015 IEEE Fifth International Conference on Big Data and Cloud Computing* (pp. 187–190). doi: 10.1109/BDCloud.2015.60.

20. Kosba, A., Miller, A., Shi, E., Wen, Z., & Papamanthou, C., (2015). *Hawk: The Blockchain Model of Cryptography and Privacy-Preserving Smart Contracts.* Tech. rept. Cryptology ePrint Archive, Report 2015/675. https://eprint.iacr.org/2015/675.pdf (accessed on 28th July 2021).

21. Kott, A., & Wang, C., (2014). *Cyber Defense and Situational Awareness* (Vol. 62). Switzerland: Springer International Publication. ISBN 978-3-319-11391-3.

22. Krizhevsky, A., & Hinton, G. E., (2011). Using very deep autoencoders for content-based image retrieval. *Proceedings ESANN*. Bruges, Belgium. http://www.cs.toronto.edu/~fritz/absps/esann-deep-final.pdf (accessed on 28th July 2021).

23. Sharma, D., Gaur, L., & Okunbor, D., (2007). Image compression and feature extraction using Kohonen's self-organizing map neural network. *Journal of Strategic E-Commerce, 5*, 25. Jordan Whitney Enterprises, Inc.

24. Liu, Y., Comaniciu, C., & Man, H., (2006). A Bayesian game approach for intrusion detection in wireless ad hoc networks. In: *Proceedings from the 2006 Workshop on Game Theory for Communications and Networks*. GameNets, NY, USA. ACM.

25. Macdonald, M., Liu-Thorrold, L., & Julien, R., (2017). The blockchain: A comparison of platforms and their uses beyond bitcoin. *COMS4507 – Advanced Computer and Network Security*. University of Queensland.

26. Maghrabi, L., Pfluegel, E., Al-Fagih, L., Graf, R., Settanni, G., & Skopik, F., (2017). Improved software vulnerability patching techniques using CVSS and game theory. *International Conference on Cyber Security and Protection of Digital Services (Cyber Security)* (pp. 494–505). London.

27. Maghrabi, L., Pfluegel, E., & Noorji, S. F., (2016). Designing utility functions for game-theoretic cloud security assessment: A case for using the common vulnerability scoring system. In: *International Conference on Cyber Security and Protection of Digital Services (Cyber Security)*. London: IEEE. https://ieeexplore.ieee.org/document/7502351 (accessed on 28th July 2021).

28. Peters, G. W., (2016). *Understanding Modern Banking Ledgers through Blockchain Technologies: Future of Transaction Processing and Smart Contracts on the Internet of Money* (pp. 239–278). Springer International Publishing.

29. Praitheeshan, P., Pan, L., Yu, J., Liu, J., & Doss, R., (2020). *Security Analysis Methods on Ethereum Smart Contract Vulnerabilities — A Survey* (pp. 1–21). arXiv: 1908.08605v2.

30. Samarji, L., (2015). Coordination and concurrency aware likelihood assessment of simultaneous attacks. *Third International Conference on Security and Privacy in Communication Networks Secure Communication, 152*, 524–529.

31. Sharma, D. K., Gaur, L., & Okunbor, D., (2007). Image compression and feature extraction with neural network, allied academies international conference. *Academy of Management Information and Decision Sciences* (p. 33). Proceedings.

32. Settanni, G., Skopik, F., Graf, R., Wurzenberger, M., & Fiedler, R., (2016). Correlating cyber incident information to establish situational awareness in critical infrastructures. In: *14th Annual Conference on Privacy, Security and Trust (PST)* (pp. 78–81). Auckland, New Zealand. doi: 10.1109/PST.2016.7906940.

33. Socher, R., Pennington, J., Huang, E. H., Ng, A. Y., & Manning, C. D., (2011). Semi-supervised recursive autoencoders for predicting sentiment distributions. *Proceedings of the Conference on Empirical Methods in Natural Language Processing* (pp. 151–161). Stroudsburg, Edinburgh, Scotland, UK. https://www.aclweb.org/anthology/D11-1014 (accessed on 28th July 2021).

34. Uma, M., & Padmavath, G., (2013). A survey on various cyber-attacks and their classification. *International Journal of Network Security, 15*, 390–396. Coimbatore.

35. Vanderburg, E., (2018). *What is Cybersecurity Situational Awareness and Why Should it Be a Critical Part of Your Security Strategy?* DELL Technologies. https://www.delltechnologies.com/en-us/perspectives/

what-is-cybersecurity-situational-awareness-and-why-should-it-be-a-critical-part-of-your-security-strategy/ (accessed on 28th July 2021).

36. Weng, S. S., (2006). Ontology construction for information classification. *Systems with Applications, 31*(1), 1–12.

37. Wood, G., (2014). *Ethereum: A Secure Decentralized Generalized Transaction Ledger, EIP-150 Revision.* https://gavwood.com/paper.pdf (accessed on 28th July 2021).

38. Yousefi-Azar, M., Varadharajan, V., Hamey, L., & Tupalula, U., (2017). Autoencoder-based feature learning for cybersecurity applications. In: *2017 International Joint Conference on Neural Networks (IJCNN)*. doi: 10.1109/IJCNN.2017.7966342.

39. Pandey, S., & Solanki, A., (2019). Music instrument recognition using deep convolutional neural networks. *Int. J. Inf. Technol., 13*(3), 129–149.

40. Rajput, R., & Solanki, A., (2016). Real-time analysis of tweets using machine learning and semantic analysis. In: *International Conference on Communication and Computing Systems (ICCCS2016)* (Vol. 138 No. 25, pp. 687–692). Taylor and Francis, at Dronacharya College of Engineering, Gurgaon.

41. Ahuja, R., & Solanki, A., (2019). Movie recommender system using K-means clustering and K-nearest neighbor. In: *Accepted for Publication in Confluence-2019: 9th International Conference on Cloud Computing, Data Science & Engineering* (Vol. 1231, No. 21, pp. 25–38). Amity University, Noida.

42. Tayal, A., Kose, U., Solanki, A., Nayyar, A., & Saucedo, J. A. M., (2019). Efficiency analysis for stochastic dynamic facility layout problem using meta-heuristic, data envelopment analysis, and machine learning. *Computational Intelligence.*

43. Tayal, A., Solanki, A., & Singh S. P. (2020). Integrated framework for identifying sustainable manufacturing layouts based on big data, machine learning, meta-heuristic and data envelopment analysis. *Sustainable Cities and Society.* https://doi.org/10.1016/j.scs.2020.102383.

CHAPTER 5

WVOSN Algorithm for Blockchain Networks

NADA M. ALHAKKAK

Computer Science Department, Baghdad College for Economic Science University, Baghdad, Iraq, E-mails: nadahakkak@hotmail.com; dr.nada@baghdadcollege.edu.iq

ABSTRACT

Blockchain networks have two main scenarios; the first one is about nodes join the network, and the other one is about the mining process inside that network which is also known as consensus algorithm; which is used in blockchain strategy before adding the new block to the network's nodes. Those algorithms help in agree of network for one particular block, to be added to the network. Those algorithms are the solution to decentralized systems, where there is no trust on the network; we call this Byzantine Generals Problem (PG). There are multiple types of consensus algorithms, choosing the best one depends on the transmission type and data transferred. This proposed a hybrid algorithm that manages the decentralized network starting from joining the network as a new node till adding a new authorized block to the blockchain of network nodes. The hybrid algorithm built on the basis of tournament consensus algorithm (TCA) for node join to the network, on proof of work (PoW) for mining process, on Proof of Luck to assign priority for specific types of blocks, and on Chain Perfect Hashing (CPH), this algorithm should take advantage from good features of that basis and try to reduce the bad features.

5.1 INTRODUCTION

The consensus algorithm is an agreement between all nodes about the block that should be added inside the node's chain [16]. Each block in the blockchain contains multiple transactions, and blockchain solves the problem of low-trust centralized by one entity to high-trust decentralized by multiple entities. There are two types of blockchains; public and private. Public blockchain means anyone can join the network. All types of blockchains need consensus algorithms [1]. A consensus algorithm decides the agreement type that should be made for appending new blocks to the network's nodes. Before any addition of block to chain; verify sender's identity, by using public and private keys, also called digital signature, also check the validity of the transaction for the block which called (check public ledger) [15, 18]. This chapter is organized as follows; background for introducing basic ideas, literature review for previous attempts, methodology for presenting the main idea of the proposed hybrid algorithm, discussion, and future direction for last overview about the proposed idea and its relation to its basics, and finally the references section.

5.2   BACKGROUND

Blockchain is a sequence of blocks or list of blocks as in Figure 5.1 where the connection point between the blocks is "parent block hash," each block has two parts, as in Figure 5.2:

- Header of the block includes:

  o   Block version: for validation rules to be followed.
  o   Merkle tree root hash: hash value.
  o   Timestamp: current time in seconds.
  o   nBit: threshold of hash value.
  o   Nonce: its increasing length depends on calculation of hash value, and it is related to consensus algorithms in blockchain, i.e., proof of work (PoW) and proof of stack (PoS).
  o   Parent block hash: points to the previous block.

- Body of block includes:

  o   Transaction counter;
  o   Transactions [7, 33, 35].

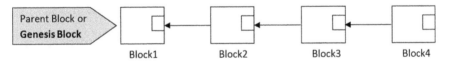

FIGURE 5.1   Blockchain with four blocks.

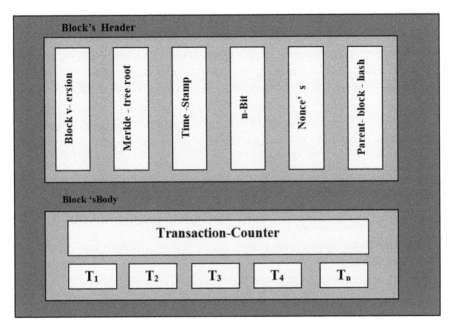

FIGURE 5.2   Block structure of blockchain.

Blockchain's characteristics:

1. **Decentralization:** No third party is needed with the help of consensus algorithm.
2. **Persistency:** Checking transaction's validation.
3. **Anonymity:** Users access blockchain using fake addresses, not allowed, this is an interstice constraint in blockchain.
4. **Audit Ability:** Easy verification and track of transactions [3].

There are multiple types of blockchains, regarding accessibility mode, those are:

1. **Public Blockchain:** All blocks are available to the public for the consensus process.

2. **Private Blockchain:** Only pre-selected blocks available for consensus process for only one organization.
3. **Consortium Blockchain:** Only a small group of organizations can access a pre-selected block, for the consensus process [23, 32].

In Figure 5.3, any node connects to the network can be any electronic device like computer, phone, printer, bridge; in other words, any device that can connect to the internet. All nodes have equal importance on the blockchain. Depending on the role of the node, there are three types of nodes: light node, full node, and mining or forging node. Any node could add to the network, will get a copy of the blockchain available on each node on the network [18, 19, 36, 37]. We need consensus algorithm to ensure that the system is fully decentralized, and because of having many types consensus algorithms we need to compare between them using some features or criteria, as below:

- Consensus standard;
- Hash function computation quality;
- Power consumption;
- Maintenance cost;
- Network security;
- Expected problem [36].

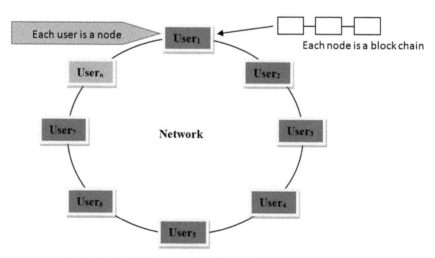

FIGURE 5.3   Network of users (nodes) as blockchains.

Another study showed different kinds of comparative criteria:

- Energy-efficiency;
- Modern-hardware;
- Forking-process;
- Double-sending's attack;
- Block-creating speed;
- Pool's mining;
- Example [12, 22].

5.3   LITERATURE REVIEW

There has been numerous attempts and studies that written on blockchain from different directions; i.e., review; security enhancement; integrating with e-applications; etc. Most important type of studies those related to consensus algorithms; which explain managing the decentralization feature of blockchain network. The prior studies on multiple kinds of consensus algorithms are described in subsections.

5.3.1   *PROOF-BASED CONSENSUS ALGORITHM*

Works on public blockchain networks, the joining is freedom to the chain; the results have to be collected with all details from all nodes before the final decision, which is difficult. There should be too many nodes, and the exchange of agreement between them would be difficult. For appending a block, all nodes in the chain should be qualified, and there is a composition between them, because of receiving rewards, and finally suitable for networks with lots of nodes [12, 22].

5.3.2   *CONSORTIUM BLOCKCHAINS (PRIVATE BLOCKCHAIN)*

Works on private blockchain networks, only permitted nodes could join the verified system (in some cases, also work with proof-based consensus algorithm) [12, 22].

5.3.3   VOTING-BASED CONSENSUS ALGORITHMS

For any node that wants to append a block to this chain, then T of nodes would append the same block (T is the threshold limit, where greater than 50% of nodes should append the same block).
Fault-tolerant techniques:

- Crush fault tolerance;
- Byzantine fault tolerance (BFT) [12, 22].

5.3.4   PROOF-OF-BURN (POB)

The miners send their coins to be burned and could not be used by others. The miner that burns the largest coins would have the right to add (or mine) the new block [12, 22].

5.3.5   PROOF-OF-SPACE (POSP)

Investing the money on hard disk, generating datasets (plots), the node that have more plots will add block (mine) to the chain [12, 22].

5.3.6   PROOF-OF-ELAPSED TIME (POET)

Used with sawtooth lake-blockchain platform executed with trusted execution environment (TEE). The node (miner) having the shortest wait-time will be the one to add (mine) a new block.
Process steps:

- All nodes (at the same time) request a wait-time, from enclave;
- All nodes receive the replay from enclave;
- Nodes will wait until their wait-time finish;
- When nodes finish waiting, it will check with others, if it was the first one, then it will have the right to add the new block (mine).

Two functions are used: create time () and check times (). It is difficult with cheating, because of the above two functions [12, 22].

5.3.7   PROOF-OF-LUCK (POL)

Also work with TEE and XGS devices, synchronize the ledgers from all miners. Each node when have new block will randomly assign a number to that block the value is between (0,1), also called 'lucky value.' All nodes should agree about the chain that have the largest sum, for a specific node, will be the main chain. It is fair for all nodes, not like POW, and finally hard to be attacked by double-spending attack [12, 22].

5.3.8   MULTICHAIN

Another type of consensus algorithms, similar to POW, it uses a round-robin schedule to use rotation for the creation of node blocks. Depend on partner (0 < P < 1) called mining diversity. In each phase or round all nodes should wait for a specific time, then start check its right to append the new block to the chain. For node A, as an example, for blocks total number equal P×N (P→ 0 < P < 1, N→ nodes number) let us say X= P×N, if x-blocks does not to node A, then this node could append its block to its current chain, and then broadcast it to other nodes in network. But if any fork happens (like POW), then the longest chain would be chosen. There is an obvious difference between voting-based consensus algorithm and proof-based consensus algorithm; where in the first type nodes should be known and updatable in order to make message exchangeable easy. While the second type deal with network type that give ability to nodes join and leave the network in flexible way [12, 22].

5.3.9   PROOF-OF-AUTHORITY (POA)

Modified from PoS, uses set of authorities instead of puzzles; PoW uses mining mechanism while PoA uses identity for verification. Some nodes are authorized to create new blocks and add to the system, the rest nodes sometimes are authorized depending on the network's judgment and sections, also it need more processing time [36].

5.3.10   ROCK-SCISSOR-PAPER ALGORITHM (RSP)

Reduce processing-power and minimize maintenance cost, it depends on balancing the values of (R, S, P), where each user will have a unique number, by giving values to R, S, P like R = 0, S = 1, P = 2 then add all those values (SUM). The sum is divides by total user number (N), if the remainder matches to user-unique value, then user is selected as delegator, but it will change when new block is created, and uses random function to construct RSP value, and then broadcast to the network. The delegator is excluded from any consensus process because its job is only monitoring. After mining the new block, the whole above process is done again to select another delegator. RSP depends on the computing devices' performance, does not use hash function, improved network security, and defend against malicious participant's attack. For performance evaluation, it needs to work outside the lab environment (expanding development-environment scale) [36].

5.3.11   DELEGATED PROOF-OF-STAKE (DPOS)

Like PoS, the riches node chooses its representatives and puts them in circle way. Each representative will have the right to sign block/blocks to the network after council's agreement. Inside the circle, if the representative misses his turn, then he should leave the circle. The owner node should not lose its control on circle's members because it should then leave and lose its representatives. DPoS need less processing than PoS [27, 30].

5.3.12   LEASED PROOF-OF-STACK (LPOS)

Leased proof-of-stack (LPoS) is another type of PoS, it is supported by waves platform, any rich node could rent his balance to mining nodes, and the mining nodes could share the profit with users. It allows receiving income from mining activities without leading to those mining nodes [27, 30].

5.3.13   PROOF-OF-CAPACITY (POC)

Also called proof-of-space (PoSpace) supported by Rurstcoin platform, each miner calculates (Parcel) which is huge data stored in the cloud storage system of the node and related to old nodes. In order to create a new block, the miner reads a small part of data, let us say 0.024% from total data, and measure the time taken, called deadline, in seconds. From reading data till the last block is created, then it will be able to create the new block. The miner that had minimum deadline in seconds will get a reward for the transaction and signs the unit. The computation time is small also its mining is efficient. Better than PoW, because miners compete with each other's in data-size, and equipment speed [27, 30].

5.3.14 PROOF-OF-IMPORTANCE (POI)

Supported by NEM platform, it takes into consideration: amount of the currency (like PoS), and user's activity in the blockchain network [27, 30].

5.3.15 PROOF-OF-ACTIVITY (POA)

PoA is a combination of PoW and PoS, to increase protection level. Each miner creates empty block that contains blank headlines, with: hash (previous block), public address for miner, index of current-block in blockchain, and nonce. The node sends the generated block to the blockchain network, the network's nodes will use (follow-the-Satoshi) algorithm using hash-value of sent block-headline and hash of any previous block +N presets; in order to select the stakeholder. Each online stakeholder checks the blank headline block, for correctness. Also, if the stakeholder finds itself in the list of (N-1) of this block, then it assigns the blank headline block with its secret key then that key is sent to the network of blockchain. For stakeholder number = N, if he found that he should do two things:

- Sign the block; and
- Sign headline of blank block.

Then should add a block that contains: some amount of transaction chose by him, any signature with value N-1 from other stakeholders then gives a sign to the block. The stakeholder (with number N) sends the above work (as a new unit), the nodes receive this block and verify its validity, and then add it to their blockchain. The list of N-stakeholders is called "lucky ones" [27, 30].

5.3.16 PROOF-OF-BURN (POB)

It is the nearest alternative to PoW, it needs lower energy than PoW needs, it does not need high computing currency, it does not depend on high mining tools because it works with virtual mining tools. When the user burn coins in the network (virtually) this would increase its powerful that results in the increase of choosing him as a validation of a block [27, 30].

5.3.17   BYZANTINE FAULT TOLERANCE (BFT)

Related to Byzantine error, and complicated, used in ripple network and stellar, solve the scalability problem with low communications [27, 30].

5.3.18   PROOF-OF-WORK (POW)

Network nodes make some calculation to confirm the transaction, the first node that complete its calculation successfully will get a reward from blockchain network, need huge power for fast-calculations, all nodes make calculations but only one get rewarded. PoW, works in high speed and with a large number of users. User's weight depends on the coin's numbers in his account. This is called weight proof and it: reduce risk of fork, reduce risk of attack, and keeps the network safe. There are no rewards; hence it is used for creative projects rather than commercial ones [5, 27, 30].

PoW, Solves the trust issue between the nodes of decentralized system, the user collects the new transactions, then modify nonce-value in the block until the calculated value of the hash for the block is less than a specific called difficulty hash value, hence the block can be broadcast. Then block is verified by others, so it can be added to the chain, finally that user can get the reward.

The block contains B = <h′; txs; nonce>

- h′ = hash (previous-block);
- txs= transaction – record in the block;
- nonce=32-bit integer.

D = define number of teaching zeros are needed for current block-hash's value. Its difficulty value is calculated by using the node's construction block. The node should use all its functions or resources to calculate all possible values of nonce in order to get H (B) < D.

In other words:

- The new transaction is broadcast to all nodes or miners in the network;
- Each miner constructs its Merkle tree, from collecting transaction records;
- The miner calculates its nonce and compare with "difficulty-value;"
- When miner finds the solution, it will broadcast the block to the network;
- Verification from other miners should be done.

Check transaction record in this block: if valid then: Hash (block) ≡ value(difficulty) + this block is longer than other froks + other honest (nodes) construct their blocks after this block [9, 11].

5.3.19   PROOF-OF-STACK (POS)

Choose the node with the creator balance (rich) or oldest one, the choice done by the one who is responsible of creating the next block in blockchain's network [30]. Also, it Uses currency-age to affect mining of hashing (difficulty), the nodes should submit its transaction for ownership enhancement among blockchain assets.

When blockchain has more assets:

- Increase holding time;
- Easy of mining.

Users must make transfers to reach some level of blockchain assets. Because: the assets affect mining difficulty for blockchain's miners. More assets = more official nonce, and hashing work: proof hash< coins.age [9, 11].

5.3.20   DELEGATED PROOF-OF-STACK (DPOS)

Proposed by Bitshares community; it allows three types of people to vote: witnesses, delegates, and workers. Witnesses deal with transactions and blockchain maintenance, and then paid. Delegated does not contain any payment, but ask for Bitshares' update. Workers should be paid when they start working on desired projects.

DPoS does not require mining and full node verification and it is more centralized than (PoS, WoS). But it suffers from collision attacks.

This algorithm contains two main processes: witness's election and witness's block. The witness responsibilities are: witness the transaction, verify the signature, and time stamp of the transaction, and not participating in the transaction [9, 11].

5.3.21 PRACTICAL BYZANTINE FAULT TOLERANCE (PBFT)

The algorithm requires nodes in total to be $n > = 3F + 1$ (F=number of evil nodes). Failed nodes no less than Y3 of network nodes; Main steps of the algorithm: client request service from primary nodes (or master), master or primary node multicast the request to the secondary node, secondary node execute the request then replay to the primary node, the clients receive replays equal to $(f+1)$ with the same answer, and it gets the data it requests (f = max failed replicas) [9, 11].

5.3.22 TOURNAMENT CONSENSUS ALGORITHM (TCA)

It is based on proof-of-luck (PoL), any node wants to join private block-chain, it should send a request to all nodes, if authorized then it will receive blockchain with right to create transactions and add next block. After specific time, 15 seconds as example, request is broadcast to network nodes to choose (0 till 1), the node choose its number and broadcast it. And wait to receive numbers from other nodes. Nodes receive multiple random numbers, and choose the biggest one and send it back to its original node, called (winner vote). Any node receives maximum number of (winner votes) then the (winner-node), and he is chosen to add the next block. Wined node should mine the block using PoW-policy, but with very low difficulty to find (nonce-value) in less than a second. This algorithm avoids loss of time and energy [18, 25].

5.3.23 A LIGHT WEIGHT BLOCKCHAIN CONSENSUS PROTOCOL

The algorithm does not depend on a large number of calculation neither on depends on currency large amounts, it depends on voting schemes and uses announcements system for registering the miners, also uses scoring systems. Announcement system: (1) miner must announce its intention to mine (by announcement notification or transaction); (2) the miner wait for a specific value of time = N blocks then multiple blocks will added; (3) the miner in (1) called "gensis " does not included in the count of N blocks because its work focused on chain-grow. Minner message components:

- Miner's public key;
- Unique address identifier (UAI): IPv_4, IPv_6, MAG, domain name;
- Hash (1,2) using SHA256 (like one-way, unpredictable output, collision-proof); the hash produces miner identifier number (MIN).

Scoring system: mainly any consensus algorithm has three main actions:

- Function (sequence of blocks) with output as chain-score;
- Use the above function to returned scores; and
- Select the chain with the highest score.

Any new block should have a timestamp (creation time) in order to manage the time blocks are created. Selected blocks will have MIN close to:

- Hash value of preceding block;
- Hash (concatenation of L previous blocks);
- Root hash (Merkle tree).

Complete score for a chain blocks=total sum of score for each block in the chain. The score of each chain is computed, check all scores for all chains and choose the max-score for one chain and call it blockchain [13].

5.3.24 PROOF-OF-SPACE (POSP)

It was an alternative to PoW, used in public blockchain, it asks for a prove about a request disk space, and whether it will be used for storing specific information. It depends on two phases: initialization and challenge-response phases. It needs a method that keeps data fresh in each round. The prover must interact with any upcoming challenge, which is complicated with blockchain environment [26, 38].

5.3.25 PROOF-OF-ELAPSED TIME (POET)

Based on Intel platform and its secure execution, better than PoW in saving energy. Uses 'wait-time' concept to choose the node that is able to create the next block. All members wait for a random (wait time) and the winner will be the one with the least wait time; will be able to create the block. This random time, comes after a random distribution with specific mean value. It uses attestation in its verification, it means that the member provides sufficient information that make other member be sure it is created in the same community and the member has waited the correct time specified for it. PoET solves conflict in the same way PoW does. Its mean waiting time is calculated depending on membership size, target wait time). Target waiting time is calculated depending on network features. It uses z-test for cheater's detection for winner's who win multiple times more than others [26, 38].

5.3.26   PROOF-OF-LUCK (POL)

It built on TEE-based PoL primitive. It consists of two functions: POLROUND and POLMINE at begin of each round, the node makes the TEE to mine a particular chain bypassing the block into the called POLROUND. After passing the ROUND_TIME, the node calls POLMINE to mine a new block. The node passes new-block-header with the extended block, also known as "previous block," round block and previous block should have the same parent. The POLMIN function generates random value (1,0) using uniform distribution. Every time or round, nodes use a specific algorithm to collect pending transactions, execute them then put them into the new block. Also put PoL that is generated inside TEE for current chain extended. PoL prefers the chain with the highest luck.

When new block is appended, that mean preferring new chain, when network splits into heals, then luck goes to larger chain, any chain that attached more than others will not be the major one. Every node starts with: empty blockchain (ϵ), set of pending transactions (ϵ), and initial value to round block (NULL). Every time nodes listen to the network for network transactions, if found then add them to their current transactions (without duplicate). If the chain was new then assign it as (VALID) and consider it as (high-LUCK), then all nodes switch to the lucky chain and broadcast it, using NETWORK-BROADCAST. If there was another chain with different parent, then nodes can mine before broadcasting. Every node mine new block every round time, when node receive from its TEE a chain worse then what nodes have then no broadcasting happen [20, 34].

5.3.27 PROOF-OF-PROBABILITY

Solve some issues of PoW, which have been solved partially by PoS. those issues are high electricity because of overheated mining also it suffers from high cost of mining tools. Each in the network has its own hashing algorithm that makes sorting also. When translation happened in node, that node sends encrypted hash to the network; also, it sends some fake hashes. The receiving nodes, arrange received hashes depending on their priority, to mine with its own hash, using a sorting algorithm inside each node. Inside each node; it will take the input value and put it inside (sorted hash) in order to calculate the nonce, If the node's hash the fake value and that did not match the real hash, then it should wait for some time and take another received hash and make the calculation again with matching; Any node have nonce that matches real hash get rewarded [14, 21].

5.3.28 PROOF-OF-VOTE (POV)

There are four rules (also called four nodes types): commissioner, butler, candidate, and ordinary:

1. **Commissioners:** This can recommend, vote, and evaluate any activity. Those nodes also can verify and forward for any block or transaction. For new block generation, in the network of blockchain; it will be sent (the new block) to any kind of commissioners that is verified as votes, when more than 51% votes for it, this new block will be marked as valid and added to the network.

2. **Butlers:** Limited in numbers and are responsible for producing new block. Butler nodes are like miners in bitcoin (BTC), but they do not need high power for computations. Butler gather information from the blockchain network and put into one block, then sign the block. Any node can become a butler in two steps: become butler candidate, and win election for butler. A node can be both (commissioner and a butler) at the same time.

3. **Butler Candidate:** If butler candidate los in election (to be butler) then it can wait to another election (stay online for next election).

4. **Ordinary User:** Those nodes can join (exit the network with authorization) if they do not have permission to be part of the block process, then they can either distribute blocks or forward the message

(or both). If they are in the system, then they can see all steps of the consensus process.

The consensus process has some important steps; each consensus around a valid-block generate and sign. At the end of the consensus around, node butler uses a special function to generate a Random number between $(0, N_b)$, N_b is a number of butler nodes. If the random number is equal to R, then this butler is able or can generate the next new block. The generated block must assign as valid-block this happen when it gets signature from commission's nodes that are $(N_c/2+1)$, N_c is number of commissioner nodes. If no new block is generated after time T_b, then the butler node that have the value/number, R+1 block will re-generate. There are two types of voting: voting for block production, and voting for the butler candidate. Commissions do vote process by returning or sending their signatures. Commissioner's voting composite of two tickets: score ticket, and designated tickets. The random number (R) that is generated by each block to determine the next butler identity, this ensure random order generation, and there is a specific algorithm that uses signature and time stamp [28, 39].

5.3.29 PROOF-OF-EXERCISE (POX)

Used with BTCs for mining rationalize rationalization in cryptocurrencies. PoX improve in PoW's sustainability, and it replaced the hash function by matrix-based scientific computation problems: (1) If a node (employee or student) called E have a complex problem or exercise called X, when E-stores X in a highly available database called XDB. (2) Get from XDB; corresponding credentials, and hash digest H(X). (3) E creates: eXercise Transaction (XT) with its components: PoX version, H(X), meta-data about X. Then deposit credit-BTC for specific time where E gives up. (4) E components H(XT), called hash digest, and submit the value to the shuffling service that mix up this value several times in order to make it possible to have any relation with H(XT) value and E. (5) the shuffling service publish SH(XT), to be able selected by miners and prevent any fork in XBoard, the miner (M) collects multiple transactions in one block, for adding into blockchain, M need to solve eXercise chose from XBoard to give PoX. (6) M assigned to X randomly, this done by matching $H(B_h)$ to X in XBroad. (7) M solve X in XT, called eXecrcise Transaction by creating a Deal Transaction DT, where DT contains: PoX version, SH(XT), and $H(B_h)$. And get period of time by mapping a credit in BTC. (8) M miner work in a similar way to E employee by claiming credit in case X eXercise was incorrect or become unaccessable. (9) E uncovers meta-data of the eXercise in XT. (10) give corticated to X in XDB to start working on it. (11–12) if miner M finds a solution Y' to X, then if follows the same steps to get rest of solutions and make it available for verifies, this is called (Auditors). (13) VXT verify XT is created which is similar to XT, but without the need to credit, then use shuffling service that produce SH(VXT') then verify the results [24, 29].

5.4 METHODOLOGY

The main focus of this work is to design a technique that manage decentralized blockchain network, from joining the network to adding new block to chains of nodes.

When a node creates a new block; a copy of the new block will be distributed to all nodes on the network; each node will check the new block and then added to its chain block; all nodes in the network will vote on the new block (good/bad) this is called consensus; bad blocks are rejected by nodes of the network [18].

Consensus algorithms helps in agree of network for one particular block, to be added to that network. It is a solution to decentralized systems, where there is no trust on the network; it works on distributed computer systems; this is called "Byzantine generals problem, or (BG) problem."

There are multiple types of consensus algorithms with different scenarios, as discussed previously, this study proposes a new scenario for adding nodes and blocks in private networks of blockchains. This scenario is called WVOSN algorithm, where each capital letter refers to a specific meaning, as follows:

- **W:** Winner node;
- **V:** Vote node;
- **O:** Original node;
- **S:** Slotted node;
- **N:** Node.

However, the following sections describe the main idea about the proposed WVOSN algorithm and how it works when a new node would like to join the network, and on what level it will join depending on the votes it gets from other nodes on that network. Also, the proposed algorithm has another side related to the mining process when adding new transactions as blocks to the nodes of the blockchain network. As follows:

➤ **Step 1: Join the Network:**
  • Any node who wants to join the network will send a message (random number to network nodes) between (0,1) on a specific period of time.
  • The network's nodes will receive a request message in the format (Node-id, Random number), and compare between all received requests then choose the max-one, a replay is sent to the original node.
  • If the number of replays to original node (called agreement-replays) was:
    ○ Equal to a total number of networks nodes, then this new node will join the network as winner-node; that is mean it can vote, and create translations to be added as new blocks to the blockchain.
    ○ Equal or greater than 50% of network's nodes, then this new node will join as vote-node with only a voting ability and cannot add blocks to the network's blockchain.
    ○ Less than 50% of the network's nodes, then the requests are rejected to join the network.

➤ **Step 2: Node Classification:**
  • When a node joined the network as (vote-node) then it should work hard to be promoted as (winner-node).
  • This will happen when the vote-node resend requests to the original nodes above the threshold limit.
  • The original nodes then promote the vote-nodes to be winner-nodes.
  • Any node in the network has a table about other nodes (vote or winner node) to use it for accepting the transaction from winner nodes only.

➤ **Step 3: Mining Process:**
  • Any node wants to add a block, it will send (block, number of slots used in its private chain, random number as priority feature).
  • The node with the highest used-slots would add its block.
  • If two or more nodes have the same used-slots then compare the number of chains blocks in each slot.
  • If two or more nodes are having the same slots and chains, then compare using the random number (priority feature) [2].

This work tested and implemented using a written code much like SimBlock simulator which is an event-driven simulator, that simulator has the following parameters:

1. **Block Parameters:** It contains two different types of parameters. The first one is the size of the block that node generates and the second one is the interval of generating that block.
2. **Node Parameters:** Those are four different types of parameters, contain: total number of nodes in the blockchain's network; number of neighbors for each node; node location in blockchain network; capacity of block generation in blockchain network as it is presented by "computing power" in PoW consensus algorithm.
3. **Blockchain Network Parameters:** The bandwidth of blockchain network calculated from upstream bandwidth for sender region and downstream bandwidth for receiver region. Also, there is the blockchain network propagation delay that determines as propagation as average value between different regions.

The following sections describe how to calculate some parameters:

- Message arrival time = (propagation delay between nodes, bandwidth).
- Transmission time = (message size, bandwidth between regions), this study suggest that message size is equal to zero byte.
- Mining time = sum (capacity of all nodes for block generation, difficulty for block generation).

The simulation continues working until 10,000 blocks are generated, starting from 10 blocks, distributed in different regions with randomly selected neighborhoods.

In order to evaluate the proposed WVOSN's work, we should check two important things, affecting the network delay on the number of transmitted blocks, and the mining process over block transmission time [4, 6, 8, 17, 31].

Figure 5.4 represents the effect of network delay on block transactions in blockchain networks, and Figure 5.5 represents the effect of an increasing number of blocks on the mining process.

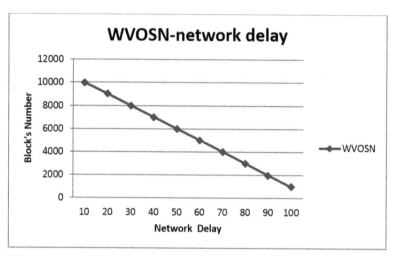

FIGURE 5.4   WVOSN-network delay.

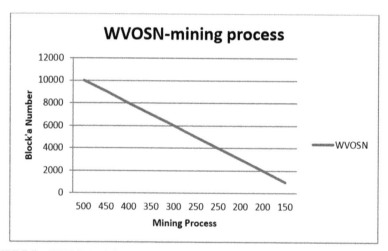

FIGURE 5.5   WVOSN-mining process.

5.5   DISCUSSION AND FUTURE DIRECTIONS

This section summarizes the main idea behind this work, where blockchain networks manage its work by itself without any centralized party. However, there are multiple types of those networks; i.e., public, private, and hybrid. This work presented a new mechanism for blockchain network used for private blockchain networks, the mechanism builds on tournament consensus algorithm (TCA) for node join to the network, on PoW for mining process, on Proof of Luck to assign priority for specific types of blocks, and on Chain Perfect Hashing CPH for constructing private chain for each block. The proposed mechanism is called WVOSN algorithm for adding new blocks to the network of blockchain and doing mining process as a consensus mechanism. The work implemented using simple written software in a specific programming language. The results showed network delay does not affect block increase or mining. So, this is a hybrid algorithm, but it works on private networks, hence, it may suffer from security issues when working with public networks.

KEYWORDS

- **blockchain**
- **Byzantine Generals Problem**
- **chain perfect hashing**
- **consensus algorithms**
- **decentralized systems**
- **proof of work**
- **tournament consensus algorithm**

REFERENCES

1. Ahmed, M., Elahi, I., Abrar, M., Aslam, U., Khalid, I., & Habib, M. A., (2019). Understanding blockchain: Platforms, applications and implementation challenges. In: *Proceedings of the 3rd International Conference on Future Networks and Distributed Systems* (pp. 1–8).

2. Al-Hakkak, N. M., (2013). Performance evaluation of encrypted data inquiries using chained perfect hashing (CPH). *Journal of Baghdad College of Economic Sciences University, 2013*(4), 511–520.

3. Alharby, M., & Van, M. A., (2019). Blocksim: A simulation framework for blockchain systems. *ACM SIGMETRICS Performance Evaluation Review, 46*(3), 135–138.

4. Alsahan, L., Lasla, N., & Abdallah, M., (2020). Local bitcoin network simulator for performance evaluation using lightweight virtualization. In *2020 IEEE International Conference on Informatics, IoT, and Enabling Technologies (ICIoT)* (pp. 355–360). IEEE.

5. Andrey, A., & Petr, C., (2019). Review of existing consensus algorithms blockchain. In 2019 *International Conference "Quality Management, Transport and Information Security, Information Technologies" (IT&QM&IS)* (pp. 124–127). IEEE.

6. Aoki, Y., Otsuki, K., Kaneko, T., Banno, R., & Shudo, K., (2019). SimBlock: A blockchain network simulator. In: *IEEE INFOCOM 2019-IEEE Conference on Computer Communications Workshops (INFOCOM WKSHPS)* (pp. 325–329). IEEE.

7. Eklund, P. W., & Beck, R., (2019). Factors that impact blockchain scalability. In: *Proceedings of the 11th International Conference on Management of Digital EcoSystems* (pp. 126–133).

8. El-Hindi, M., Heyden, M., Binnig, C., Ramamurthy, R., Arasu, A., & Kossmann, D., (2019). BlockchainDB-towards a shared database on blockchains. In: *Proceedings of the 2019 International Conference on Management of Data* (pp. 1905–1908).

9. Feng, L., Zhang, H., Chen, Y., & Lou, L., (2018). Scalable dynamic multi-agent practical Byzantine fault-tolerant consensus in permissioned blockchain. *Applied Sciences, 8*(10), 1919.

10. Finlow-Bates, K., (2017). *A Lightweight Blockchain Consensus Protocol.* Computer Security Resource Center.

11. Gupta, S. S., (2017). *Blockchain.* John Wiley & Sons, Inc.

12. Jeong, J. W., Kim, B. Y., & Jang, J. W., (2018). Security and device control method for fog computer using blockchain. In *Proceedings of the 2018 International Conference on Information Science and System* (pp. 234–238).

13. Kim, J., Kang, S., Ahn, H., Keum, C., & Lee, C. G., (2018). Architecture reconstruction and evaluation of blockchain open-source platform. In: *Proceedings of the 40th International Conference on Software Engineering: Companion Proceedings* (pp. 185, 186).

14. Kim, S., & Kim, J., (2018). POSTER: Mining with proof-of-probability in blockchain. In: *Proceedings of the 2018 on Asia Conference on Computer and Communications Security* (pp. 841–843).

15. Liu, Y., Zhang, Y., Zhu, S., & Chi, C., (2019). A comparative study of blockchain-based DNS design. In: *Proceedings of the 2019 2nd International Conference on Blockchain Technology and Applications* (pp. 86–92).

16. Lunardi, R. C., Michelin, R. A., Neu, C. V., Zorzo, A. F., & Kanhere, S. S., (2019). *Impact of Consensus on Appendable-Block Blockchain for IoT.* arXiv preprint arXiv:1912.11043.

17. Malik, H., Manzoor, A., Ylianttila, M., & Liyanage, M., (2019). Performance Analysis of blockchain based smart grids with Ethereum and Hyperledger implementations. In: *IEEE International Conference on Advanced Networks and Telecommunications Systems* (pp. 1–5).

18. Maroufi, M., Abdolee, R., & Tazekand, B. M., (2019). *On the Convergence of Blockchain and Internet of Things (IoT) Technologies.* arXiv preprint arXiv:1904.01936.

19. Mendki, P., (2020). Blockchain-enabled IoT edge computing: Addressing privacy, security and other challenges. In *Proceedings of the 2020 The 2nd International Conference on Blockchain Technology* (pp. 63–67).

20. Milutinovic, M., He, W., Wu, H., & Kanwal, M., (2016). Proof of luck: An efficient blockchain consensus protocol. In: *Proceedings of the 1st Workshop on System Software for Trusted Execution* (pp. 1–6).

21. Miraz, M. H., & Donald, D. C., (2019). LApps: Technological, legal and market potentials of blockchain lightning network applications. In: *Proceedings of the 2019 3rd International Conference on Information System and Data Mining* (pp. 185–189).

22. Nguyen, G. T., & Kim, K., (2018). A survey about consensus algorithms used in blockchain. *Journal of Information processing systems, 14*(1).
23. Zheng, Z., Xie, S., Dai, H., Chen, X., & Wang, H., (2017). An overview of blockchain technology: Architecture, consensus, and future trends. In: *2017 IEEE International Congress on Big Data (BigData Congress)* (pp. 557–564). IEEE.
24. Nofer, M., Gomber, P., Hinz, O., & Schiereck, D., (2017). Blockchain. *Business & Information Systems Engineering, 59*(3), 183–187.
25. Páez, R., Pérez, M., Ramírez, G., Montes, J., & Bouvarel, L., (2020). An architecture for biometric electronic identification document system based on blockchain. *Future Internet, 12*(1), 10.
26. Qasse, I. A., Abu, T. M., & Nasir, Q., (2019). Inter blockchain communication: A survey. In: *Proceedings of the ArabWIC 6ᵗʰ Annual International Conference Research Track* (pp. 1–6).
27. Salimitari, M., & Chatterjee, M., (2018). *A Survey on Consensus Protocols in Blockchain for IoT Networks.* arXiv preprint arXiv:1809.05613.
28. Samaniego, M., & Deters, R., (2016). Blockchain as a service for IoT. In: *2016 IEEE International Conference on Internet of Things (iThings) and IEEE Green Computing and Communications (GreenCom) and IEEE Cyber, Physical and Social Computing (CPSCom) and IEEE Smart Data (SmartData)* (pp. 433–436). IEEE.
29. Shoker, A., (2018). Brief announcement: Sustainable blockchains through proof of exercise. In: *Proceedings of the 2018 ACM Symposium on Principles of Distributed Computing* (pp. 269–271).
30. Sultan, A., Mushtaq, M. A., & Abubakar, M., (2019). IoT security issues via blockchain: A review paper. In: *Proceedings of the 2019 International Conference on Blockchain Technology* (pp. 60–65).
31. Toapanta, S. M. T., Quintana, T. F. P., Arellano, M. R. M., & Gallegos, L. E. M., (2020). Hyperledger technology in public organizations in Ecuador. In: *2020 3ʳᵈ International Conference on Information and Computer Technologies (ICICT)* (pp. 294–301). IEEE.
32. Wang, H., Cen, Y., & Li, X., (2017). Blockchain router: A cross-chain communication protocol. In: *Proceedings of the 6th International Conference on Informatics, Environment, Energy and Applications* (pp. 94–97).
33. Werner, R., Lawrenz, S., & Rausch, A., (2020). Blockchain analysis tool of a cryptocurrency. In: *Proceedings of the 2020 The 2ⁿᵈ International Conference on Blockchain Technology* (pp. 80–84).
34. Yang, D., Long, C., Xu, H., & Peng, S., (2020). A review on scalability of blockchain. In: *Proceedings of the 2020 The 2ⁿᵈ International Conference on Blockchain Technology* (pp. 1–6).
35. 35. Yang, X., Liu, J., & Li, X., (2019). Research and analysis of blockchain data. In: *Journal of Physics: Conference Series* (Vol. 1237, No. 2, p. 022084). IOP Publishing.
36. Zhang, R., Xue, R., & Liu, L., (2019). Security and privacy on blockchain. *ACM Computing Surveys (CSUR), 52*(3), 1–34.
37. Zhange, S., Yao, L., Sun, A., & Tay, Y., (2019). Deep learning-based recommender system: A survey and new perspectives. *ACM Computing Surveys (CSUR), 52*(1), 1–38.
38. Zhao, W., Yang, S., & Luo, X., (2019). On consensus in public blockchains. In: *Proceedings of the 2019 International Conference on Blockchain Technology* (pp. 1–5).
39. Zheng, X., Zhu, Y., & Si, X., (2019). A survey on challenges and progresses in blockchain technologies: A performance and security perspective. *Applied Sciences, 9*(22), 4731.

CHAPTER 6

Blockchain-Based Authentication and Trust Computation Security Solution for Internet of Vehicles (IoV)

SUNILKUMAR S. MANVI[1] and SHRIKANT TANGADE[2]

[1]School of Computing and Information Technology, REVA University, Bengaluru, Karnataka, India, E-mail: ssmanvi@reva.edu.in

[2]School of Electronics and Communication Engineering, REVA University, Bengaluru, Karnataka, India

ABSTRACT

A vehicular ad hoc network (VANET) is one of the applications of mobile ad hoc network (MANET). VANET is used to reduce road traffic and accidents to save millions of lives by providing safety applications. Whereas, the current evolution of the Internet is the Internet of Things (IoT), where billions of smart computing devices, including smart vehicles are connected to the Internet to transfer data without human intervention. The existing two technologies VANETs and IoT define new technology called internet of vehicles (IoV). The IoT enables various value-added services; however, IoV is susceptible to various security threats from malicious entities. The collective efforts of researchers enhance security against internal and external attacks. To secure the IoV against the attacks from malicious nodes, most of the existing technologies provided centralized and computation overhead-based solutions. Many researchers have proposed cryptography and trust management-based security solutions. These schemes have their own limitations. The cryptography-based schemes provide security over external attacks, whereas trust management schemes provide security over internal attacks. The new Blockchain technology enables decentralized and distributed operations. The chapter has proposed blockchain-based security solution for IoV to authenticate vehicles, calculate reward points and compute new trust value. The proposed Blockchain networks employ both private and public Blockchain networks. The first uniqueness of proposed scheme, it employs the combination of both public and private Blockchains, it is called consortium blockchain. And second uniqueness, the transmission and computation overhead in IoV are overcome by making use of deep neural networks at RSUs and ATA. The simulation results and comparative analysis of V2V, V2I, private, and public Blockchain networks are presented and it shows performance of the proposed scheme is comparably enhanced by minimizing transmission and computation overheads.

6.1 INTRODUCTION

In the field of intelligent transportation systems (ITSs), the vehicular ad hoc network (VANET) has become the most prominent technology [1, 2]. VANETs promote improvements in road safety, road traffic, driver assistance, infotainment, and entertainment [3, 4]. VANET encompasses vehicles as mobile nodes, stationary units at roadside known as roadside units (RSUs), wireless communication channel, and government trusted entity called trusted authority (TA). An on-board-unit (OBU) is embedded within each vehicle, which enables Vehicle-to-Vehicle communications and RSUs to provide Vehicle-to-Infrastructure (V2I) communications. The IEEE standard IEEE 802.11p which is known as dedicated short-range communication (DSRC) [5] and new IEEE standards 1,609 called wireless access in vehicular environment (WAVE) are dedicated for V2V and V2I communications in VANETs [6]. The main aim of VANET is to provide road safety to vehicle passengers through safety message which comprises of life critical information, these safety messages are broadcast between neighboring vehicles and RSUs during V2V and V2I, respectively [7, 8].

IoT is one of the applications of the Internet of Things (IoT). The IoV is considered as extension of V2V and V2I communications in VANETs. IoV main objective is to magnifying vehicle driving aids with the help of other emerging new technologies like artificial intelligence (AI), machine learning (ML) and deep learning (DL) [32–35]. In IoV, vehicles of VANETs are connected to the public Internet in addition to other neighboring vehicles and RSUs. This salient feature enables vehicles to instantaneously communicate with other vehicles in IoV. In today's world, increase in number of vehicles, the IoV has become most popular in both industry and academia [9, 31].

In recent days IoV got attracted significant attention due to some limitations in VANETs. The unique characteristics of VANETs leads to certain flaws like restrained processing capacity to analysis and take decision on received global safety and non-safety messages which halt the vehicles to become smart and intelligent nodes. These limitations in VANETs enabled new technology known as IoV [10]. It is a new technology revolution and provides larger extent benefits in the ITS as listed below [11]:

- **Reducing Accidents and Saving Lives:** IoV provides instantaneous traffic and road conditions information to nearby and surrounding vehicles to avoid accident prone situations, which saves millions of lives.
- **Cost Effective:** The road traffic facts and safety messages will lead to reducing maintenance rate, public health rate, life, and vehicle insurance rates.
- **Time Efficient:** The well inspected road traffic will save valuable time of many riders, drivers, and the consumers.
- **Expansion of Smart Vehicles and Cities:** IoV broad applications organize urban by providing smart parking, accident alert messages, the shortest path and lesser traffic route navigation.
- The passengers smart pick and drop services.
- IoV provides smart driving non-safety services like entertainment on the road, meals on wheels and vehicle service.

The limitations of VANETs are overcome by IoV by employing the vehicular cloud which provides an open platform for many smart applications. In IoV each smart vehicle broadcasts safety messages to its surrounding other smart vehicles about road traffic conditions and road safety alerts [7, 12]. These safety messages are broadcasted via open wireless communication medium which leads to security attacks like Sybil, modification, impersonation, identity-disclosure attacks and so many. These security attacks are very dangerous and may damage the entire ITS. Hence, providing security over these attacks becomes the primary objective before we implement IoV applications into real scenario. In order to deliver security, the primary security requirements must be achieved, they are: vehicle authentication, message integrity, non-repudiation, and privacy [13, 14].

6.1.1 SECURITY REQUIREMENTS IN IOV

As per Ref. [9], there are two available security standards for the V2V and V2I communications, they are: IEEE 1609.2 WAVE standard and another one European Telecommunications Standards Institute (ETSI) Intelligent Transport Systems (ITS) G5. These two standards differ in their message format, authentication schemes, and trust models [9, 15].

Many researchers have presented various security schemes to provide security against attacks in IoV. The literature of these schemes shows that these are broadly classified as cryptography and trust management-based schemes. In cryptography-based schemes, each smart vehicle registers with central authority and obtains its private key and public key. It digitally signs every broadcast safety message with its private key. These schemes have limitations, ensures only node authentication and message integrity. Whereas these schemes do not provide security against external attacks like broadcasting bogus messages, instead handles only external attacks [7, 12]. In order to provide security against internal attacks, trust management-based schemes are presented by many researchers [13, 16–19].

The objective of trust management is to detect dishonest peers and malicious data. It ensures the receiver accepts only trustworthy data from honest peer which has maintained good trust within the network. In these schemes' incentives are given dishonest peers to discourage self-interested behavior and behave honestly [20–22]. In IoV, the vehicles high mobility and random distribution constructs new challenge to establish trust between vehicles during V2V communications. Trust computation of vehicles depends on numerous factors, they are: neighboring vehicles recommendations, past interactions history, reward points, etc. [23, 24].

Most of industry and academia researchers are attracted towards Blockchain technology because of its extensive benefits in wider fields, such as, most popular banking, supply chains in agriculture and medicine, etc. In brief, a Blockchain is a shared, immutable ledger of records of data (block) that are secured and connected to each in a sequential order using cryptographic principles (i.e., chain) to form a blockchain [25]. This technology builds trust between vehicles and solves critical information dissemination issues in IoV. Blockchain has removed the central trusted third party and made distributed decentralized technology. This exceptional flexibility feature of Blockchain can also be employed for secure communications in IoV. Hence, IoV with new Blockchain technology provides decentralized and distributed network infrastructure, which also extends network scalability and security [26, 27].

The existing Blockchain-based security schemes are not sufficient, since there are additional end-to-end delay, computation, and communication overheads because of cryptographic operations. These influence the scalability and timeliness delivery of safety messages in IoV.

6.1.2 PROBLEM STATEMENT

This chapter proposes a Blockchain-based authentication and trust competition scheme to provide a larger extent secured communication in IoV. The proposed scheme aims to establish trust between vehicles and infrastructure in addition to providing basic security requirements. It employs DL-based private Blockchain networks for communications between Agent Trusted Authority (ATA) and RSUs, whereas public Blockchain network for V2V and V2I communications. RSU authenticates the sending vehicles, computes reward points and verifies vehicle using deep neural network models-1 and deep neural network models-2, respectively. Finally, ATA computes the trust values of smart vehicles using its reward points.

6.1.3 OUR CONTRIBUTIONS

In the proposed architecture, IoV consists of Agent Trusted Authority (ATA) which is connected to various RSUs in VANETs using private blockchain network and RSUs of network of these RSUs and smart vehicles are connected through public blockchain network. The chapter proposed Blockchain-based vehicle authentication and trust competition scheme for secured communications in IoV. Where, driver behavior-based reward-points are calculated using deep neural network model-1 for smart vehicle's trust-value computation and deep neural network model-2 verifies safety-messages are fraudulent or non-fraudulent. In the proposed work, RSU first authenticates the sending vehicles using a public blockchain. Finally, ATA computes the trust values of smart vehicles using private blockchain and deep neural network. The proposed scheme is unique, since it employs consortium blockchain (combination of private and public blockchains) and deep neural networks at RSUs and ATA. This leads to reduction in the transmission and computation overhead in IoV. And second uniqueness, the computation overhead in IoV is overcomes by making use of deep neural networks at RSUs and ATA. The performance analysis of transmission and computation overheads is deliberated through simulating a substantial set of experiments. Hence, the proposed approach leads to secured communications between smart vehicles with reduced computation overheads and decentralized network architecture of IoV.

6.1.4   ORGANIZATION OF THE CHAPTER

The rest of the chapter is organized as follows. In Section 6.2, IoV architecture and its background are described. Section 6.3 presents "Proposed Blockchain-Based Security Solution for IoV." The Section 6.4 describes the results of proposed schemes and performs comparative analysis. Lastly, Section 6.5 concludes the chapter and gives future work of this chapter.

6.2   NETWORK ARCHITECTURE AND BACKGROUND

The IoV network architecture is presented in this section and discussed basic security requirements and privacy preservation.

6.2.1   NETWORK ARCHITECTURE

Figure 6.1 presents a IoV network architecture model, this comprises of five fundamental components, they are: (i) cloud server (CS); (ii) regional transport office (RTO) as TA; (iii) agent of trusted authority (ATA); (iv) RSU; and (v) OBU. The working of these components is described as follows:

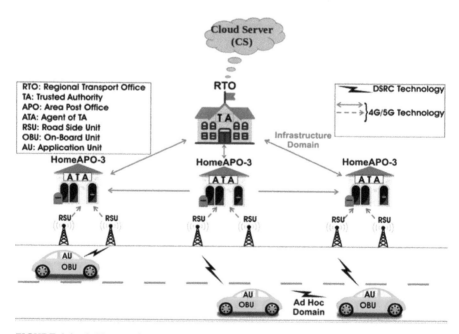

FIGURE 6.1   IoV network architecture model.

1. **CS:** The CS connects vehicles of the current cluster to other cluster vehicles through Internet.
2. **RTO:** It is TA which generates and issues cryptography private and public keys to all the entities those register off-line within IoV networks, like: ATAs, RSUs, and vehicles.
3. **ATA:** Each cluster of IoV is called area post office (APO) which is embedded with ATA. The new trust value of each vehicle is calculated by ATA through RSUs during V2V and V2I communications. ATA is connected to TA through secured wireless media.
4. **RSUs:** These entities are fixed on roadside to enable V2I communications between vehicles and RSU itself using public Blockchain network. It verifies received safety message and trustworthiness of sending vehicle and forwards verified safety message to the ATA via private Blockchain network for computation of new trust value of sending vehicle.
5. **OBUs:** It is hardware entity embedded within each smart vehicle of IoV for V2V and V2I communications.

6.2.2 SECURITY AND PRIVACY REQUIREMENTS

Many researchers concluded their literature survey on security in IoV and presented basic security and conditional privacy preservation, which must be implemented before IoV put into real practical scenario. These basic security requirements are follows [28]:

1. **Integrity:** The verification of data sent by legitimate nodes and data received at the receiver should be the same. This is sometimes also called message authentication.
2. **Authentication:** In order to avoid imitation of vehicles, authentication of vehicles ensures the legitimate of sending vehicles.
3. **Non-Repudiation:** It ensures that the sending vehicle does not deny of sent data.
4. **Confidentiality and Privacy:** The data and the privacy of vehicle or passenger are the most important part of non-safety applications. Hence, sensitive data and privacy of vehicles like real identity should not be known by illegitimate vehicles.

6.3 PROPOSED BLOCKCHAIN-BASED SECURITY SOLUTION FOR IOV

The proposed scheme objective is to verify the trustworthiness of smart vehicles and safety messages by employing public and private Blockchains. The public Blockchain network is used during V2V and V2I communications, whereas messages between ATA and RSUs are secured through private Blockchain network. Table 6.1 lists all notations used in this chapter and gives a description.

The proposed scheme has two phases, phase-I: system initialization and registration of entities and phase-II: authentication and trust computation in public and private blockchains. This section first present general Blockchain network in IoV and then we discuss the phase-I and phase-II steps of proposed scheme.

6.3.1 PROPOSED BLOCKCHAIN NETWORK IN IOV

A proposed Blockchain consists of a public and private Blockchain networks, together called consortium Blockchain. Figure 6.2 presents a public Blockchain network which stores and manages all nodes trustworthiness data within a given cluster. The data blocks are hashed and are chained in sequential order to construct a Blockchain, where each vehicle verifies and updates each block of the Blockchain.

TABLE 6.1 Notations and Description

Notation	Description
RTO_{ID}	RTO real ID
RTO_{Pr}	RTO private key
RTO_{Pu}	RTO public key
RTO_{Crt}	RTO certificate
RTO_{Srk}	RTO secret key
$PRBC_{Hk}$	Private blockchain network hash key
$PUBC_{Hk}$	Public blockchain network hash key
ATA_{ID}	ATA real ID
ATA_{Pr}	ATA private key

TABLE 6.1   *(Continued)*

Notation	Description
ATA_{Pu}	ATA public key
ATA_{Crt}	ATA certificate
ATA_{Srk}	ATA secret key
RSU_{ID}	RSU real ID
RSU_{Pr}	RSU private key
RSU_{Pu}	RSU public key
RSU_{Crt}	RSU certificate
SG_{RPr}	Signature of RSU
V_{ID}	Vehicle real ID
V_{Pr}	Vehicle private key
V_{Pu}	Vehicle public key
V_{PsID}	Vehicle pseudo-ID
V_{Rp}	Vehicle reward points
SG_{VPr}	Signature of vehicle
SMG_{V}	Safety message of vehicle
MSG_{Alt}	Alert message
DR_{Br}	Driver behavior
EX_{Fr}	External factors
H_{Code}	Hash code

6.3.2   PHASE I: BLOCKCHAIN-BASED IOV SYSTEM INITIALIZATION AND REGISTRATION OF ENTITIES

The IoV system initialization starts with the highest trusted entity *RTO*, which is called *TA*. It has to register itself using its real identity RTO_{ID} and generates its private key, public key, secret key, certificate, and private Blockchain network hash key (RTO_{Pr}, RTO_{Pu}, RTO_{Srk}, RTO_{Crt}, and RTO_{Hk}). Later, the primary components of IoV such as vehicles *V*, *RSUs*, and *ATA* has to register with local *TA* before they take part in IoV as legitimate nodes. In Phase I all these components register with *TA* as given the following steps:

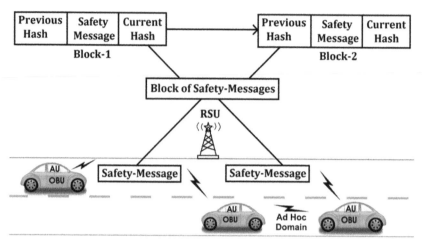

FIGURE 6.2   Public blockchain network.

1. **ATA Registration:** *ATA* and *RTO* are present in the same cluster and are connected with private Blockchain secured network. *ATA* sends registration request with its real ATA_{ID} to *RTO*. In response to the request by *ATA, RTO* generates private key, public keys, and certificate $(ATA_{Pr}, ATA_{Pu}, and ATA_{Crt})$ of *ATA* and then upload these credentials with $PRBC_{Hk}$ in to *ATA* as shown in Eqn. (1).

$$ATA = (ATA_{ID}, ATA_{Pr,} ATA_{Pu,} ATA_{Crt,} RTO_{Pu,} RTO_{Srk,} and PRBC_{Hk}) \qquad (1)$$

2. **RSU Registration:** Each *RSU* within the coverage area of *ATA* has to register with *ATA*. In order to register, *RSU* sends registration request with its real RSU_{ID} to *ATA*. Once *ATA* receives the *RSU's* registration request, it computes the private key, public key, and certificate $(RSU_{Pr}, RSU_{Pu}, and RSU_{Crt})$ of *RSU* and uploads these credentials with $(PRBC_{Hk} and PUBC_{Hk})$ into *RSU* as shown in Eqn. (2).

$$RSU = (RSU_{ID}, RSU_{Pr,} RSU_{Pu,} RSU_{Crt,} ATA_{Pu,} ATA_{srk,} PRBC_{Hk} and PUBC_{Hk}) \quad (2)$$

3. **Vehicle 'V' Registration:** Finally, each *'V'* has to register first with *RTO* through the nearest *RSU* and *ATA*. The vehicle *'V'* request for registration to *RTO* by sending its real identity V_{ID} through registration request. *RTO* then computes private key, public key, and certificate $(V_{Pr}, V_{Pu}, and V_{Crt})$ and uploads these credentials with initial trust value $TR_{TV} = 0$ and $PUBC_{Hk}$ in to 'V' as exhibited in Eqn. (3).

$$V = (V_{ID}, V_{PsID}, V_{Pr,} V_{Pu,} TR_{TV} = 0, HMAC\ (RTO_{Srk,}\ TR_{TV}),\ ATA_{Srk,}\ RTO_{Pu,}\ and$$
$$PUBC_{Hk}) \tag{3}$$

After successful registration of 'V,' RTO distributes V's credentials to all ATA.

6.3.3 PHASE II: AUTHENTICATION AND TRUST COMPUTATION IN PUBLIC AND PRIVATE BLOCKCHAINS

In IoV, each vehicle broadcasts safety message SMG_V during V2V communications within the private Blockchain network. The proposed work considers the safety message format as shown in Eqn. (4).

$$SMG_V = (MSG_{Alt} || DR_{Br} || EX_{Fr} || TRT_{Cur} || TRT_{Code} || HMAC\ (RTO_{Srk,}$$
$$MSG_{Alt} || TRT_{Cur}) || V_{PsID} || SG_{Vpr}\ TRT_{Cur}\ (MSG_{Alt} || TRT_{Cu\ r}) || T_{St}) \tag{4}$$

Note: In the above Eqn. (4), the DR_{Br} and EX_{Fr} indicate the various driver behaviors and external factors, respectively. The detailed parameters of DR_{Br} and EX_{Fr} are shown in Eqns. (5) and (6), respectively.

$$DR_{Br} = (V_{Sp} || D_{Sty} || E_{Rpm}) \tag{5}$$

$$EX_{Fr} = (R_{Sc} || T_C) \tag{6}$$

where; V_{Sp} is the vehicle speed; D_{Sty} is the driving style; E_{Rpm} is the engine revolutions per minute; R_{Sc} is the road surface condition; T_C is the traffic condition.

In public Blockchain network, each vehicle generates SMG_V for V2V communications within the cluster. Further, these SMG_V are hashed with public Blockchain hash key $PUBC_{Hk}$ and then these safety message blocks are broadcast within the network. Once these safety message SMG_V hashed blocks are received by the nearest RSU, it first records each of these blocks in private Blockchain network and then authenticates the broadcast vehicle. After successful authentication of vehicle, RSU computes vehicle's reward points based on driver behaviors and external factors by employing deep neural network model-1 as shown in Figure 6.3. Further, RSU verifies sender vehicle is fraudulent or non-fraudulent by employing deep neural network model-2 as shown in Figure 6.4. RSU records vehicle status, and if the vehicle is non-fraudulent, then SMG_V hashed with private Blockchain hash key $PRBC_{Hk}$ and forwards to ATA via private Blockchain network. ATA calculates new trust value TRT_{New} of vehicle-based on its reward points. The following Algorithms 1, 2, and 3 give a detailed step-wise description of authentication, reward point calculation, verification of vehicle, and trust computation.

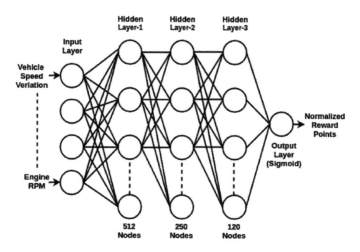

FIGURE 6.3   Deep neural network model-1.

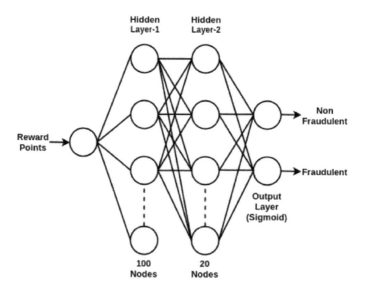

FIGURE 6.4   Deep neural network model-2.

> ➤ **Algorithm 1: Overview of Authentication, Reward Points and Trust Value Computation**

Input: SMG_V
Output: TRT_{New}

1: BEGIN
2: RSU performs sending Vehicle authentication, reward point calculation and verification of vehicle either fraudulent or non-fraudulent with the help of Algorithm-2.
3: If the sending vehicle is non-fraudulent, then ATA calculates its new trust value TRT_{New} in the Algorithm-3 by making use of reward points which are earned based on the vehicle's behavior.
4: END

> ➤ **Algorithm 2: Authentication, Reward Point Calculation, and Vehicle Verification**

Input: SMG_V
Output: V_{Rp} and fraudulent / non-fraudulent

1: BEGIN
2: After successful reception of safety message blocks SMG_V by RSU via public Blockchain network, it first verifies the message block and then authenticates vehicle by generating hash code as shown in Eqn. (7) and follows next steps.

$$H_{Code} = HMAC \ (RTO_{Srk,} \ MSG_{Alt} \ || \ TRT_{Curr}) \tag{7}$$

3: if $(H_{Code} == HMAC \ (RTO_{Srk,} \ MSG_{Alt} \ || \ TRT_{Curr}))$ then
4: RSU successfully authenticated both received message SMG_V and sending vehicle.
5: RSU records the authenticated message block and then computes the vehicle's reward points V_{Rp} using deep neural network model-1 as shown in Figure 6.3.
6: RSU then verifies sender vehicle for fraudulent or non-fraudulent using deep neural network model-2 as shown in Figure 6.4.
7: if $(V ==$ non-fraudulent) then
8: RSU forwards hashed SMG_V with private Blockchain hash key $PRBC_{Hk}$ and calculated reward points V_{Rp} forwards to ATA through private blockchain network as shown in Eqn. (8).

$$FSM_V = HASH\ (PRBC_{Hk},\ SMG_V\ ||\ V_{Rp}\ ||\ SG_{Rpr}\ (V_{Rp}\ ||\ TRT_{Cur})\ ||\ Ft_{st}) \qquad (8)$$

9:   else

10: Sending vehicle is fraudulent and hence SMG_V is drops by RSU.

11: end if

12: else

13: The SMG_V or 'V' is not integrity / authenticated, so RSU drops SMG_V.

14: end if

15: END

> **Algorithm 3: New Trust Value Calculation**

Input: $FSMG_V$ and TRT_{Cur}

Output: TRT_{New}

1:   BEGIN

2:   After successful reception of forwarded safety message block $FSMG_V$ by ATA, it first verifies the received message block using private Blockchain hash key $PRBCG_{Hk}$ and then starts calculating new trust value TRT_{New}.

3:   if $(FSMG_V == HASH'(PRBCG_{Hk,}\ SMG_V\ ||\ V_{VRp}\ ||\ SG_{Rpr}\ (V_{Rp}\ ||\ TRT_{Curr})\ ||\ Ft_{St}))$ then

4:   Received message block is successfully verified by ATA.

5:   ATA calculates the new trust value TRT_{New} by making use of reward points V_{Rp} as shown in Eqn. (9) (these reward points are forwarded by RSU in Algorithm-2).

$$TRT_{New} = \left(\frac{TRT_{Cur} + V_{Rp}}{2} \right) - \left(0.1 \times n(FA) \right) \qquad (9)$$

(where; n(FA) represents number of false alarms by vehicle).

6:   ATA updates its database of trust values of vehicles with new trust value TRT_{New} and also shares with RTO through private Blockchain network. The RTO will update these new trust value with the vehicles that make trust value update request within updation window.

7:   else

8:   Received message block $FSMG_V$ drops by ATA.

9:   end if

10: END

6.4   RESULTS AND DISCUSSION

In this section, the performance of proposed Blockchain-based authentication and trust competition scheme is analyzed. The effectiveness of the proposed scheme is measured by this performance analyzes.

6.4.1   SIMULATION ENVIRONMENT SET-UP

In order to get simulation results and perform comparative analysis with the existing scheme, this chapter simulates the proposed scheme with system environment specifications like: Linux OS, Inter processor-CORE i3-3000 series with 3.2 GHz speed. The open-source simulation tools are used, such as, traffic, and network simulators SUMO (& MOVE) and ns-3, respectively. The other open-source software library the TensorFlow 1.6.0 with python 3.5 is also used. Table 6.2 gives a detailed description of simulation parameters and their values for simulating the proposed scheme.

TABLE 6.2   Simulator ns-3 Parameters and Their Values

ns-3 Parameters	Value
Simulation area	1,000 m × 1,000 m
Total smart vehicles	100
Vehicles maximum speed	120 km/h
DSRC communication range	200 m to 400 m
RSUs in a cluster	4
Wireless channel bandwidth	6 Mbps
Safety message length	206 B
Maximum simulation time	200 s

6.4.2   EVALUATION OF PROPOSED SCHEME PERFORMANCE

The performance of the proposed scheme is evaluated by simulating and comparing with existing schemes [29, 30]. These are represented in the simulation graph as ID-MAP (identity-based message authentication using proxy vehicles) and PKI-Ceri, respectively.

6.4.2.1 TRANSMISSION OVERHEAD

The transmission overhead (OH_T) is referred to as the ratio of the total number of overhead bytes divided by the overall transmitted bytes as shown in Eqn. (10).

$$OH_T = \frac{Overhead\,bytes}{Total\,transmitted\,bytes} \tag{10}$$

Table 6.3 gives safety message format of the proposed scheme, which is compared with the existing two schemes ID-MAP and PKI-Certi. The overall size of broadcasting safety message used in the schemes proposed, ID-MAP, and PKI-Certi are 173 B, 191 B, and 247 B, respectively. These all three schemes are simulated, Figure 6.5 is drawn, which shows that the proposed scheme has achieved 7% and 16% lesser transmission overhead compared to the transmission overhead existing schemes ID-MAP and PKI-Certi, respectively.

TABLE 6.3   Safety Message Formats

Schemes	Safety Message Parameters						
	MSG	DR || EX	TRT_{Cur} || TRT_{Code}	HMAC	V_{PsID}	SG/Certi	T_{St}
Proposed	67	4	8	20	6	64	4
ID-MAP	67	X	X	X	X	120	4
PKI-Certi	67	X	X	X	X	181	X

6.4.2.2 COMPUTATION OVERHEAD

Since, the proposed scheme making use of private and public Blockchain networks during V2V communications and trust computation, respectively. The simulations of these two Blockchain networks are considered in addition to V2V and V2I communications. In Figure 6.6, the computation cost of proposed Blockchain based security scheme is compared with existing ID-MAP and PKI-Certi schemes. It shows proposed scheme computation cost is relatively less compared with the other two schemes. The schemes ID-MAP and PKI-Certi take 18.75 ms and 23 ms for signature verification, respectively. Whereas, proposed scheme takes only 4 ms as a part of message hash code verification and 3 ms for vehicle authentication. In addition to these two, the proposed scheme also verifies message block hash code, which takes an additional 4 m, hence total computation overhead of the proposed scheme is 11 ms.

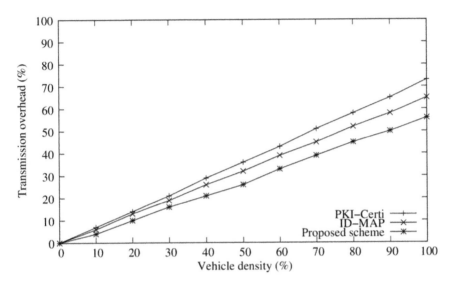

FIGURE 6.5   Transmission overhead.

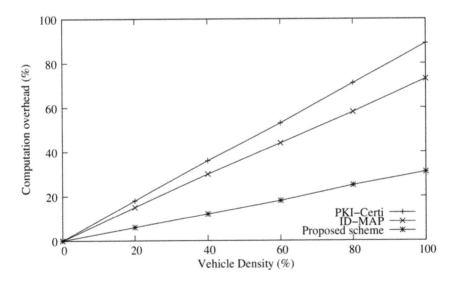

FIGURE 6.6   Computation overhead.

6.5   CONCLUSION AND FUTURE WORK

In this chapter, a Blockchain-based authentication and trust competition scheme is proposed for IoV. The proposed scheme employed private and public Blockchain networks empowered by deep neural network models for reward point and trust value computation. These Blockchain networks ensures distributed and decentralized authentication of vehicles in the IoV. Whereas, deep neural network models achieve speedy computation operations of reward points and trust value calculation at RSU and ATA, respectively. The operations of the proposed scheme are divided into two phases, phase-I and phase-II. In phase-I, IoV system is initialized by registering all entities with RTO, which is certificate authority called TA. In phase-II, firstly a vehicle is authenticated by RSU, and it further computes reward points and verifies whether the vehicle is either fraudulent or non-fraudulent by employing deep neural network model-1 and model-2, respectively. Finally, ATA computes new trust value of vehicle based on its reward points. The proposed private and public Blockchain networks are simulated along with V2V and V2I communications in IoV. The proposed simulation results are compared with existing security schemes, and the results are very impressive.

In future work, processing batch of message blocks and employing group signature will further enhance the performance of the proposed scheme. Further, speed of signature verification, reward point calculation, and trust computation by employing deep reinforcement ML at private and public Blockchain networks.

KEYWORDS

- **blockchain**
- **cryptography**
- **internet of vehicles**
- **on-board-unit**
- **trust management**
- **vehicular ad hoc networks**

REFERENCES

1. Lu, Z., Qu, G., & Liu, Z., (2019). A survey on recent advances in vehicular network security, trust, and privacy. *IEEE Trans. Intell. Transp. Syst., 20*(2), 760–776.

2. He, D., Zeadally, S., Xu, B., & Huang, X., (2015). An efficient identity-based conditional privacy-preserving authentication scheme for vehicular ad hoc networks. *IEEE Trans. Inf. Forensics Security., 10*(12), 2681–2691.

3. Mejri, M. N., Ben-Othman, J., & Hamdi, M., (2014). Survey on VANET security challenges and possible cryptographic solutions. *Veh. Commun., 2*, 53–66.

4. Ahmad, F., Franqueira, V. N. L., & Adnane, A., (2018). TEAM: A trust evaluation and management framework in context-enabled vehicular ad-hoc networks. *IEEE Access,* 28643–28660.

5. Jiang, D., & Delgrossi, L., (2008). IEEE 802.11p: Towards an international standard for wireless access in vehicular environments. *Proc. IEEE Veh. Technol. Conf. (VTC Spring)*, 2036–2040.

6. Tangade, S., Manvi, S. S., & Lorenz, P., (2020). Trust management scheme based on hybrid cryptography for secure communications in VANETs. *IEEE Transactions on Vehicular Technology.* doi: 10.1109/TVT.2020.2981127.

7. Tangade, S., Manvi, S. S., & Lorenz, P., (2018). Decentralized and scalable privacy-preserving authentication scheme in VANETs. *IEEE Transactions on Vehicular Technology, 67*(9), 8647–8655.

8. Dajun, Z., Richard, Y., & Ruizhe, Y., (2019). Blockchain-based distributed software-defined vehicular networks: A dueling deep Q-learning approach. *IEEE Transactions on Cognitive Communications and Networking, 5*(4), 1086–1100.

9. Bagga, P., Das, A. K., Wazid, M., Rodrigues, J. J. P. C., & Park, Y., (2020). Authentication protocols in internet of vehicles: Taxonomy, analysis, and challenges. *IEEE Access, 8*, 2169–3536.

10. Nagtilak, S., Sunil, R., & Rohini, K., (2018). Internet of vehicles: Motivation, layered architecture and research challenges. *Proc. IEEE Global Conf. Wireless Comput. Netw. (GCWCN)* (pp. 54–58). Lonavala, India.

11. Kaiwartya, O., Abdullah, A. H., Cao, Y., Altameem, A., Prasad, M., Lin, C. T., & Liu, X., (2016). Internet of vehicles: Motivation, layered architecture, network model, challenges, and future aspects. *IEEE Access, 4*, 5356–5373.

12. Manvi, S. S., & Tangade, S., (2017). A survey on authentication schemes in VANETs for secured communication. *Journal of Vehicular Communications, 9*, 19–30.

13. Kerrache, C., Calafate, C. T., Cano, J., Lagraa, N., & Pietro, (2016). Trust management for vehicular networks: An adversary-oriented overview. *IEEE Access, 4*, 9293–9307.

14. Zhaojun, L., Qian, W., Gang, Q., Haichun, Z., & Zhenglin, L., (2019). A Blockchain-based privacy-preserving authentication scheme for VANETs. *IEEE Transactions on Very Large Scale Integration (VLSI) Systems, 27*(12), 2792–2801.

15. Lixia, X., Ying, D., Hongyu, Y., & Xinmu, W., (2019). Blockchain-based secure and trustworthy Internet of Things in SDN-enabled 5G-VANETs. *IEEE Access, 7*, 56656–56666.

16. Huang, Z., Ruj, S., Cavenaghi, M. A., Stojmenovic, M., & Nayak, A., (2014). A social network approach to trust management in VANETs. *Peer-to-Peer Networking and Applications, 7*, 229–242.

17. Liu, Z., Ma, J., Jiang, Z., Zhu, H., & Miao, Y., (2016). LSOT: A lightweight self-organized trust model in VANETs. *Mobile Information Systems*, 1–15.

18. Ahmad, F., Hall, J., Adnane, A., & Franqueira, V., (2017). Faith in vehicles: A set of evaluation criteria for trust management in vehicular ad-hoc network. *Proc. IEEE International Conference on Internet of Things and IEEE Green Computing and Communications and IEEE Cyber-Physical and Social Computing and IEEE Smart Dat.* (pp. 44–52).

19. Li, W., & Song, H., (2016). ART: An attack-resistant trust management scheme for securing vehicular ad hoc networks. *IEEE Trans. Intell. Transp. Syst., 17*(4), 960–969.

20. Ao, Z., Jinglin, L., Qibo, S., & Cunqun, F., (2015). A security authentication method based on trust evaluation in VANETs. *EURASIP Journal on Wireless Communications and Networking, 1*, 1–8.

21. Bin, L., Xinghua, L., Jian, W., Jingjing, G., & Jianfeng, M., (2020). Blockchain-enabled trust-based location privacy protection scheme in VANET. *IEEE Transactions on Vehicular Technology, 69*(2), 2034–2048.

22. Xingchen, L., Haiping, H., Fu, X., & Ziyang, M., (2020). A Blockchain-based trust management with conditional privacy-preserving announcement scheme for VANETs. *IEEE Internet of Things Journal, 7*(5), 4101–4112.

23. Tangade, S., Manvi, S. S., & Hassan, S., (2019). A deep learning based driver classification and trust computation in VANETs. *Proc. IEEE 90th Vehicular Technology Conference: VTC2019-Fall* (pp. 1–6). Honolulu, Hawaii, USA.

24. Qi, F., Debiao, H., Sherali, Z., & Kaitai, L., (2020). BPAS: Blockchain-assisted privacy-preserving authentication system for vehicular ad hoc networks. *IEEE Transactions on Industrial Informatics, 16*(6), 4146–4155.

25. Rakesh, S., Rojeena, B., & Seung, Y., (2018). Blockchain-based message dissemination in VANET. *Proc. IEEE 3rd International Conference on Computing, Communication and Security (ICCCS)* (pp. 161–166). Kathmandu.

26. Khan, A. S., Balan, K., Javed, Y., Tarmizi, S., & Abdullah, J., (2019). Secure trust-based blockchain architecture to prevent attacks in VANET. *Sensors, 19*(4954), 1–27.

27. Canhuang, D., Xingyu, X., Yuzhen, D., Liang, X., Yuliang, T., & Sheng, Z., (2018). Learning based security for VANET with blockchain. *Proc. IEEE International Conference on Communication Systems (ICCS)* (pp. 210–215). China.

28. Azees, M., Vijayakumar, P., & Deborah, L. J., (2016). Comprehensive survey on security services in vehicular ad-hoc networks. *IET Intell. Transp. Syst., 10*(6), 379–388.

29. Bayat, M., Barmshoory, M., Rahimi, M., & Aref, M. R., (2015). A secure authentication scheme for VANETs with batch verification. *Wireless Netw., 21*(5), 1733–1743.

30. IEEE 1609 Working Group Standard (2006). *IEEE Trial-Use Standard for Wireless Access in Vehicular Environments (WAVE)-Security Services for Applications and Management Messages.* IEEE Std. 1609.2. 2006.

31. Ramakrishnan, R., & Gaur, L., (2019). *Internet of Things: Approach and Applicability in Manufacturing.* CRC Press.

32. Rajput, R., & Solanki, A., (2016). Real-time analysis of tweets using machine learning and semantic analysis. In: *International Conference on Communication and Computing Systems (ICCCS2016)* (Vol. 138, No. 25, pp. 687–692). Taylor and Francis, at Dronacharya College of Engineering, Gurgaon.

33. Ahuja, R., & Solanki, A., (2019). Movie recommender system using K-means clustering and K-nearest neighbor. In: *Accepted for Publication in Confluence-2019:*

9th International Conference on Cloud Computing, Data Science & Engineering (Vol. 1231, no. 21, pp. 25–38). Amity University, Noida.

34. Tayal, A., Kose, U., Solanki, A., Nayyar, A., & Saucedo, J. A. M., (2019). Efficiency analysis for stochastic dynamic facility layout problem using meta-heuristic, data envelopment analysis, and machine learning. *Computational Intelligence.*

35. Tayal, A., Solanki, A., & Singh, S. P. (2020). Integrated framework for identifying sustainable manufacturing layouts based on big data, machine learning, meta-heuristic and data envelopment analysis. *Sustainable Cities and Society. Vol 62.* https://doi.org/10.1016/j.scs.2020.102383.

PART II

Smart City Ecosystem Using Blockchain Technology

CHAPTER 7

Blockchain for Smart Cities: The Future of City Management

ARUN SOLANKI and TARANA SINGH

Gautam Buddha University, Greater Noida, Uttar Pradesh, India

ABSTRACT

Modern smart cities already form a large network of interconnected technologies, and this network is expected to grow rapidly, i.e., billions of IoT devices will form the typical smart city in the coming years. In 2016, India launched the Smart Cities mission, which aims to develop cities using smart solutions that provide basic infrastructure and provide citizens with a dignified quality of life, a clean and sustainable environment. These cities include smart street lights, smart waste management, smart water management, etc. As geographic networks for connecting IT devices leading to an extraordinary collection of data that needs to be protected from attacks and losses. Blockchain is the platform on which this kind of mass data, derived from smart cities, can be stored and accessible by a verified person. This chapter discusses the general architecture of smart cities using blockchain technology, applications, opportunities, and the future scope of blockchain technology in implementing smart cities.

7.1 INTRODUCTION

In recent decades, the rapid development of urbanization has made significant progress in creating capable, intelligent, and sustainable solutions. A smart city needs these solutions to provide better waste, water, energy, portability, and various e-government services. The highlights of a Smart City are characterized by government, and a significant number of them can be implemented using blockchain innovation leading to increasingly transparent, coherent, and secure solutions [1]. The combination of new advances to improve smart urban communities provides better approaches to reviewing different services. In various articles, the authors explore the uses of blockchain technology to improve smart cities. Finding new applications for Blockchain has normally associated the innovation to the rising field of smart cities [2, 3]. As these fields are extremely encouraging, researchers are studying everything from smart roads [4] to smart contracts (SCs) [5] and the maximum potential of banking and human services [6, 7]. Extensive research is being undertaken on security issues identified with blockchain [8], resolving issues identified in existing cases [9] or preventing the presence of others, long before vulnerable frameworks reappear [10]. From now on, there are big implementation of Smart Cities components. China has taken incredible steps to create Blockchain-based solutions for its smart cities [11]. Dubai is moving in the direction of Blockchain to maintain its business advantage by studying the idea of e-democracy [12]. Similarly, applications to create a recording innovation [13] and authentic support networks, especially future Smart Cities. The combination of blockchain technology with gadgets in a smart city creates a typical phase where all devices have the ability to transfer securely in a distributed domain [8].

As the technologies connect with Blockchain as a framework, the more value we can determine. By connecting regular technologies to each other, this 'Smart City' can thus begin to automate the basic services of the city. For example, IoT sensors [34, 70] can immediately detect a problem and detect the artificial intelligence (AI) of the municipal office that is competent to send a technician. Imagine that a city has an advanced record where every house or apartment has a neighborhood with all the important house data. All the information from owning and comparing the home loan to value-based information, such as using utilities, evaluating property costs, and temporary work relationships in the past. The city can link to it to facilitate the benefits and carry out authoritative commissions identified with the property [14]. The landowner will have a proven and reliable approach to trading, such as renting a room, hiring contract workers to do gardening, or selling energy generated from the solar panel on the grid. Dubai has launched a Smart City initiative, part of which is planned to transform Dubai into the first blockchain-controlled city by 2020. Dubai will apply Blockchain to the logistics and storage operations of the entire city and also move to a government and a storage system 100% paperless. The idea of smart cities has been around for some time, guided by an idealistic vision of the city structure, governments, workers, and residents working in sync for a more and more capable, safer, and more profitable state. AI is likely to be the incentive to improve the smart city. Smart urban communities will use AI, ML, DL, computer vision (CV) and NLP technologies to empower city administrations, resources, and structures to make it even more impressive, productive, and automated. These updates will provide better results for living, working, and playing [15].

In 2016, India launched the mission of smart urban communities. The aim of this mission was to develop urban communities using 'intelligent' solutions that provide the foundations of the center and provide residents with a respectful personal satisfaction, an impeccable and sustainable condition. The emphasis is on a manageable and complete turn of events. To this end, it uses innovation to incorporate and manage frameworks to provide better types of assistance and to ensure competent and ideal use of accessible resources. India has promoted its Smart Cities mission with the goal of creating 100 smart urban areas. The project includes moderate housing, coordinated multimodular vehicles, the creation and conservation of open spaces, waste, and traffic of drivers [16]. The Indian government acknowledges that many passionate urban activities will be carried out on a blockchain to enhance safety, unchanging nature, power, and transparency. There are various countries in the world that use Blockchain to provide its residents and guests with a brilliant city experience. Here are some ways in which Blockchain can be used in passionate urban areas:

1. **Smart Payments:** Inspire every single municipal payment on a Blockchain-based planning, including city programs, government assistance, finance, and many more.
2. **Identity:** Most recent decentralized Identity Management frameworks use Blockchain to give a protected system for storing and approving client identities, in this way decreasing identity theft and related cheating activities.
3. **Transportation Management:** Utilization of Blockchain to eject the rent searchers in the ridesharing economy (Uber, and many more.). This empowers a genuinely P2P platforms for transportation.
4. **Smart Energy:** Make a stronger power grid by utilizing a blockchain-based P2P power cabinet. This ejects sent looking mediators and permits people to make, purchase, sell, and exchange vitality while holding approval.
5. **Government Services:** Smart agreements can be utilized for digitizing resident rights and recognizable proof, straightforward

democratic, tax, track responsibility for, leave the paper, and digitize administrative procedures.

6. **Waste Management:** Improve efficiencies encompassed the whole waste administration process by utilizing IoT sensors and AI forecast displaying.

Blockchain can possibly make incalculable smart systems and grids, changing how we do everything from the vote and assemble credit to get energy. From several points of view, it could be a critical segment of what is expected to gap superseded frameworks and fabricate dependable answers for urban communities [17]. Smart urban communities are concentrating on the utilization of current advancements like IoT, Big Data, AI, machine learning (ML) [65–68], and so on focusing on making an increasingly livable urban condition by using assets ideally and reduce the expense. These modern technologies are the backbone of the whole smart city. The appropriate combination of these technologies results in an efficient and sustainable city for the citizens living in the smart city [18, 69].

7.1.1   ARTIFICIAL INTELLIGENCE (AI)

AI is setting an endless pattern in the public arena. How different was the world a couple of decades back and how the way of life has totally changed today is all because of man-made brainpower? Regardless of whether you are utilizing online networking Facebook or hiring an Uber taxi through your cell phone, everything is dealt with by AI. Actually, when we are searching a couple of words of this chapter on the google web search tool, it was AI who improved our hunt [19]. In spite of the fact that this is the earliest type of AI, it has a vital job to act in the advancement of urban areas and their relative domains. Brilliant urban areas are as of now utilizing AI at a more notable pace than yesterday. But in the middle of the applications offered by AI, it is significant for urban communities to have a hold over utilizing AI for a particular objective [20]. It should not be conveyed in light of the fact that it is the most current thing accessible in the innovation advertise. When discussing urban areas, the Government needs to take such choices. Thus, the initial step starts with learning-the Government can concentrate how different governments over the globe are working with AI and what kind of executions are gotten by the private segment [21]:

- With more focus on objective direction, six procedures can end up being helping devices for the Government to put the correct pace ahead with AI: Use AI as a piece of the objective based, resident driven program.
- Take commitment from the residents.
- Work towards upgrading existing assets.
- Move ahead cautiously with security and keep up-to-date with information arrangements.
- Reduce moral dangers and forestall utilizing AI for decision making.
- Augment occupations for individuals and not for AI.

AI must utilize as a device which encourages us to nail the board onto the divider where 'nailing the board' is really our choice. City authorities must furnish with this fundamental apparatus to assist them with taking care of issues effectively [22]. One more highlight be noted is, AI must put to utilize on the off chance that it is the most ideal approach to achieve an undertaking and not be constrained into the choice. Comprehensively, the focal point of each objective must be the government assistance of residents.

7.1.2   INTERNET OF THINGS (IoT)

The term IoT [48, 59] refers to the rapidly growing number of computerized devices-the quantity is currently billions-these devices can be distributed and collaborated with others on the system/web worldwide and can be remotely controlled. The IoT contains smart sensors and various devices. At the operational level of the IoT, for example, climate information is collected. The IoT opens new doors for urban communities to use information to control traffic, reduce corruption, use the foundation, and keep residents safe and clean [23]. Organizations use IoT for imaginative administration and for controlling widespread procedures. Thereafter, they can check the latter even from inaccessible points, as the data is constantly maintained in applications and in the storage of information (AoI). IoT gives a bit of scope of realizing things ahead of time. Due to the minimal effort of the IoT, it is currently possible to investigate and monitor exercises that were already inaccessible. The money related perspective is the best preferred position since this new innovation could displace people who are responsible for observing and looking after provisions. Subsequently, expenditures can be significantly reduced and progressed [24]. IoT also makes it possible to gather entirely new fragments of knowledge, for example, the climate impacts associated with modern creations. The eventual fate of IoT is boundless. It provides infrastructure in all segments, including production, design, catering, health services, training, and so on. Smart urban communities can share a typical smart urban stage, which promotes well, especially for small urban areas. The cloud-based nature of IoT responses for Smart Cities is appropriate by sharing a phase that relies on open information. Small urban areas can form a typical urban environment [25]. Along these lines, answers for small and huge smart urban areas are organized and controlled by means of the focal cloud stage. Lastly, but critically, the size of a city is not an issue if you are going to get "smart." The urban areas of each meeting can benefit from smart technologies.

7.1.3   SMART CITY

Urban communities accommodate almost 31% of India's present population and contribute 63% of GDP (Census 2011). Urban zones are relied upon to house 40% of India's population and contribute 75% of India's GDP by 2030 [26]. This requires far-reaching advancement of physical, institutional, social, and financial framework. All are significant in improving the personal satisfaction and drawing in individuals and venture, getting underway an idealistic pattern of development and advancement. Improvement of Smart Cities is a stage toward that path. The Smart Cities Mission is an inventive and new activity by the Government of India to drive financial development and improve the personal satisfaction of individuals by empowering nearby turn of events and saddling innovation as a way to make strong results for residents. Intelligent Cities center around their most squeezing needs and on the best chances to improve lives [26]. They tap a scope of approaches-advanced and data advances, urban arranging best practices, open private organizations, and strategy change-to have any kind of effect. They generally put individuals first. In the way to deal with the Smart Cities Mission, the goal is to advance urban areas that give center foundation and give a respectable personal satisfaction to its residents, a perfect and reasonable condition and utilization of 'Smart' Solutions. The attention is on reasonable and comprehensive turn of events, and the thought is to take a glance at smaller territories, make a replicable model which will act like a beacon to other hopeful urban areas [27]. The Smart Cities Mission is intended to set models that can be repeated both inside and outside the Smart City, catalyzing the formation of comparative Smart Cities in different locales and parts of the nation. The first thing is that what is the meaning of smart. Smart means 'low-cost investment and give maximum value of work.' The target of the keen city is to advance urban communities that to give core infrastructure, savvy transportation, detecting administrations and give a conventional personal satisfaction to its residents, a perfect and manageable condition and uses of "smart" arrangements [4]. A city is called smart city if it has smart people, small governance, smart mobility, smart economy, smart environment, and keen living.

There are mainly four pillars of a smart city:

- Institutional infrastructure;
- Physical infrastructure;
- Social infrastructure;
- Economics infrastructure.

7.1.4   BLOCKCHAIN TECHNOLOGY

Blockchain is an innovative technology that has had a miraculous impact on various industries and has introduced the market with the first new "bit-coin" applications. Bitcoin (BTC) is just a digital currency (cryptocurrency) that can be used to replace the legal currency of commerce. The essential novelty for achievement of cryptographic methods of currency is called Blockchain. There is a distinctive misunderstanding between entities that "bitcoin" and "Blockchain" are similar, yet there is the condition. Converting money in digital form is one of the uses of blockchain novelty, and anyways BTC, numerous applications are being created based on blockchain innovation. Simply stated, a "blockchain" can be depicted as an information structure containing information exchange records, while guaranteeing security, transparency, and decentralization [29]. It can likewise be thought of as a chain or record put away as a square that is not constrained by authorizations. Blockchain is an appropriated record that is totally open to everybody on the system. When information is put away on the Blockchain, it is hard to change or alter it. All exchanges on the Blockchain are secured by advanced marks to demonstrate their credibility. Due to the use of encoded and electronic symbols, the data placed on the Blockchain is consecrated and is not modifiable. Blockchain revolution authorize all members in the structure to approve. This is usually called agreement [30]. All data placed on the Blockchain is carefully verified and had an open past available to every member on the structure. Right now, is no requirement for outsiders to dispense with open doors for extortion and reused exchanges. To all the more likely comprehend the Blockchain, consider the case of searching for a choice to send cash to a companion who lives somewhere else. Common options that are usually available are banks or payment transfer applications such as PayPal or Paytm. With this option, additional funds are deducted from your funds as remittance fees, as a third party is involved in processing the transaction. Similarly, in this case, hackers cannot guarantee the security of funds because they can damage the network and steal funds. In both cases, customers suffer [31]. This is where Blockchain comes in. In this case, using a blockchain eliminates the need to use a bank for remittances, making the procedure easier and more secure. Since the funds are handled directly, there are no additional costs and no third parties are required. In addition, the "blockchain" database is decentralized and not limited to the only position. This means all data and accounts stored on the "blockchain" are publicly distributed. The data is not kept in one location, so hackers cannot destroy it.

7.1.5   IMPLEMENTATION OF BLOCKCHAIN TECHNOLOGY IN SMART CITY

After 10 years of close ties to BTC, immutable technology called decentralized ledgers on the Blockchain has begun to find numerous use cases in other industries. Financial services companies are exploring the use of blockchain capabilities as tamper-resistant transaction records. Blockchain can also be used in supply chain/logistics and manufacturing because it can establish reliable transaction and provenance records. He also believes that Blockchain can realize the dreams of a truly smart city, making sustainable energy (usually solar energy) more intelligent and efficient [32]. Most cities rely on traditional large power grids. However, autonomous cities are only possible with smaller, more local "micro-network" technologies, and Blockchain can be an important part of this way of generating and consuming distributed electricity. The idea is to create an ecosystem where people can generate their own energy, distribute it to other users who need it, and recharge it easily and automatically [33]. This energy ecosystem is based on generating revenue, not sharing. Installing solar panels on residential and commercial buildings and connecting them to the grid is easy. Now it is easier to create smaller microgrids and monitor consumption using private blockchains. While these blockchains are private and easy to implement, they are completely transparent in the sense that one of the benefits of the

Blockchain is that all participants can see their own consumption and where to move additional energy [18]. When a member of a microarray generates excess energy that exceeds its consumption, that energy is automatically transferred to another member of the microarray through a "smart contract" managed by the Blockchain, and the member needs the energy. Basically, the conditions and parameters of the smart contract are predefined (coded). "As long as party A needs 10 kilowatts of power, the amount that party A is willing to pay is kilowatts." If there is energy available, it is based on all the intelligence in the system [35]: contract price and consumption. SCs eliminate the need for intermediaries and match buyers and sellers directly in real-time. Blockchain networks execute peer-to-peer (P2P) transactions in real-time, debit from buyers, credit sellers, and record the descendants of the transaction. While solar energy is renewable and contributes to overall social and environmental wellbeing, blockchain approaches can further reduce energy consumption through incentives [36]. You can write a smart contract to allow a building, apartment, or office with multiple units to transfer the revenue generated by the building's solar panels in inverse proportion to the energy consumed. Low-consumption people get more, which is a great motivation, and cities become more and more sustainable. Interconnections between smart grids extend beyond homes and offices, and grids can include a variety of devices. The sensor alerts the network that a stove, water heater, manufacturing machine, etc., needs to be broken or repaired, and automatically schedules maintenance and repair times. Normally, drawers are fresh when the equipment is running. No human intervention is required unless the technician needs to repair it himself. You can imagine a scenario that charges only when needed, such as a change station with a smart contract, or only when the cost to be charged is below a threshold [37]. These use cases are just the beginning. As blockchain systems become more familiar, the applications that make use of them become easier to use, and the possibilities are endless.

7.1.5.1 RELATED WORK OF BLOCKCHAIN TECHNOLOGY IN SMART CITY

There exist lots of papers that are focused on the implementation of a blockchain-based explanation and its claims in smart urban area growth. There are lots of researchers who work on different aspects to develop a sustainable smart city. Some of the works of different authors are being discussed in the following literature review.

Maroufi et al. said in 2019 [38] that the fast development in the quantity of IoT gadgets had raised numerous hindrances that could ruin the adoption of IoT in different ventures. To start with, the market for IoT gadgets and stages is divided, with multiple models and providers. Second, interoperability is a concern, as implemented solutions often create new data silos. Data from IoT devices are typically stored securely in the cloud, but is not protected by complete device damage or tampering with sources. One important thing is that the centralized structure of most IoT explanations allows IoT gadget holders to believe these industries in order to guard information safety, regulate data, and protect data when hackers attack central servers to destroy them. On the other hand, Blockchain is a developing innovation that can help make IoT frameworks stronger. Gives a decentralized registry that avoids the challenges of centralized architecture and stores data through secure features insecure processes. Blockchain sets up trust between IoT gadgets and decreases the danger of fashioning blockchain passwords. It likewise reduces costs by disposing of middle people and the roundabout expenses of mediator parties. Automatically, while Blockchain can give a promising answer for some IoT challenges, the coordination between the two implanted advances presents new issues and hindrances. IoT gadgets have restricted force and capacity assets and cannot deal with a full capacity of the asset. In addition, Internet requirements for things such as security, data confidentiality, consensus protocols, smart contract, etc., will give require changing the characteristics of traditional Blockchain. One of the primary difficulties is diverse explanations, and we recommend integrating Blockchain with IoT technology in different types of IoT applications, depending on your needs and necessities [38]. Mathew states in 2019 that blockchain technology will continue to be developed and discover more use cases in the cutting-edge world. Network security is one of the feasible areas that have been examined and applied. Blockchain foundation is functional and addresses existing security challenges in the field of IoT gadgets, systems, and information for transport and capacity. In his article, the author had assessed the implementation of "blockchain technology" from the viewpoint of Taylor et al. as they reviewed 30 researchers work. It has been seen that most blockchain security analysts are committing a great deal of vitality to receive blockchain security in IoT gadgets. Simultaneously, the other primary regions of blockchain security are systems and information. As he saw in the conversation, blockchain innovation can be utilized to make sure about IoT gadgets through increasingly dependable validation and information transmission components. This assists in keeping the programmers from entering devices that generally have poor security settings. This expertise can also be utilized to protect network security using rigorous organization to stop unofficial associates and communications. Ultimately, the Blockchain can protect information during the broadcast and at rest via encoded chunks. Other use cases are under investigation, but these three use cases are receiving attention. Since most solutions today use different blockchains, which can hinder integration, future researchers are encouraged to investigate the usefulness of a solitary blockchain that can be used to create security arrangements. This is recommended to the upcoming researchers [39]. In 2018, Kouicem et al. [40] stated that the Internet of Things (IoT) is another standard, completely changing the world by connecting different physical items to the Internet to frame a coordinated and smart biological system. Today, another smart world is rising, with people, cell phones, PCs, and new knowledgeable objects associated with the Internet. In his article, the author reviewed the proposed security solutions for IoT applications. They classify various IoT submissions by recognizing safety needs and unique contests. Next, the author described an IoT solution based on traditional encryption solutions, including privacy, confidentiality, and availability. They also studied several new pieces of Knowledge, such as blockchains and programable networks, that are said to be effective devices for solving the IoTs scalability problem. At long last, the author talked about a few security arrangements that consider the setting related to IoT applications and the different impacts of security matters on framework security and specific countermeasures. According to some standards, a thorough comparison of the various methods is made, and technologies applicable to each type of IoTs application are also being studied. They describe all the efforts to solve the various challenges facing the IoTs, but especially because the IoT is the network of everything's, people, and data, scalability, and dynamics. There are many open issues that need to be solved, such as procedures and items that work together in a very flexible and composite system [40]. In 2018, Kouzinopoulos et al. [41] in order to improve the security of the IoT in smart home facilities, decentralized mechanisms based on blockchain technology and SCs had been proposed. IoT devices in smart homes must meet many security and confidentiality requirements. Due to the unique structure and operating characteristics of these devices, meeting these requirements is not easy. Using blockchain technology in this situation can increase security and privacy and meet many of these requirements. Possible ways are introduced and explained in his article. To continue this work, they focused on implementation details of the proposed blockchain mechanism, and validation and performance evaluation in a GHOST environment [41]. In 2018, Singh et al. [42], in his article, provided suggestions on using "blockchain technology" concluded examples to create additionally protected and reliable IoT models. Because of the top-of-the-line equipment necessities of the IoTs, they stated that the IoTs is not an official individual from the blockchain system. However, the IoT benefits from features introduced by blockchain technology via APIs provided by system hubs or special middle parties. These features can make the IoTs very secure. They discussed the cybersecurity point of emerging blockchain technology. Because they analyzed BTC as a cryptocurrency based on blockchain technology, blockchain technology is mainly used and researched in the field of finance. However, their article pursued to introduce IoT's "blockchain technology" to protect information transfer among IoT devices. To this end, they outlined blockchain technology, safety matters in the IoT environment, and discussed and proposed that Blockchain is a solution to the safety of the IoTs [42].

In 2019, Banerjee et al. [43] concluded the fact that blockchain technology has great potential had been debated, and there is great potential to more effectively shape future improved smart communities and provide higher quality technology. Life. For expertise to fulfill the promises, these needs be able to modify the status quo. Blockchain technology can deliver solutions for various problems that smart cities typically face. However, the actual implementation of this technology depends on the management and preferences of the community [43].

In 2017, Ibba et al. [44] proposed their work on a blockchain system to accomplish the environment. "City-Sense" is an arrangement which inspires people to participate in constructively monitoring the quality of urban environments to raise awareness of urban health. They want all citizens to be able to work together at all stages of software development. Because of this, they opted to use the SCRUM method. In this way, user contribution is important, and product owners need to consider the needs of citizens and their suggestions. In their approach, the role of citizens in software development is also very important. Citizens pass information to product owners, send opinions, and attend assemblies at the user's end. The conservational information gathering technique is accepted by combining the extents taken by the portable gadgets with the validation procedure performed by the particular discovery program which processes the information. Local Administration with less or no money investment can provide all citizens with digital devices that can capture environmental data and create medical cards for their resident cities to obtain real-time information and formulate for real-time countermeasures against pollution alerts [44].

7.2 BLOCKCHAIN INTEGRATED ARCHITECTURE OF SMART CITIES

Today, the IoTs is the latest technological discipline. The IoTs means that "billions of devices are connected to communications networks and generate and store large amounts of data." Today, the key issues include managing these connected devices such as fitness trackers, home energy systems, and self-driving cars. According to past research, more than 1 billion IoT devices will be installed in smart city consumer and business applications. Under certain circumstances, all of these devices have their own digital ID, and the management of user data also ensures security and confidentiality. Managing all of these devices can be tedious and even catastrophic in some cases. In previous research, there are two types of access control technology for IoT devices: centralized method and distributed method [45]. A centralized approach brings technical limitations to the world, while another decentralized approach overcomes all the limitations of previous methods. In a distributed management system, devices are associated with physically dispersed device networks. The answer is founded on "blockchain technology" that eliminates the effects of centralized systems and facilitates the connection of new IoT devices. In the next part of this chapter, we will solve the smart city architecture through the implementation of Blockchain [46, 47]. This chapter first describes the architecture of the physical components of smart cities using blockchain networks and then describes the layered architecture of smart cities using blockchain technology.

7.2.1 PHYSICAL COMPONENT ARCHITECTURE OF SMART CITY USING BLOCKCHAIN TECHNOLOGY

The architecture in Figure 7.1 shows the architecture of the physical components of a smart city using a blockchain network. With the exception of IoT devices and management centers, all components are part of blockchain technology. Blockchain networks can be very large and will continue to grow over time. Due to their limited nature, most IoT devices cannot store blockchain information. In this architecture, the Blockchain does not include

IoT devices, but it does define a new node called the Administration Center. The organization center needs a blockchain contact controller of data on for IoT devices. There are also smart city contracts that define the processes permissible by the entree regulator structure. There is a component called manager that can interact with the smart city contract to express the entree regulator strategy of the structure [18]. This segment describes the various workings of the architecture in a complete and detailed way.

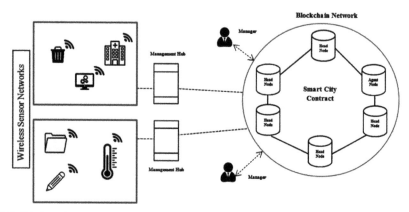

FIGURE 7.1   Physical component architecture.

7.2.1.1 *WIRELESS SENSOR NETWORK (WSN)*

This is a communication network which consents applications with limited associativity and very restricted energy and fewer necessities. The computing capability, storage space, and energy availability of these IoT devices that relates to wireless sensor networks (WSNs) are very limited. The IoT enables numerous applications around different types of sensors. Managing the growth of urban populations requires smarter, more sustainable cities and technologies that respect the environment and the economy. This architecture has a variety of WSNs that can retrieve data from a variety of data generation sources [49]. WSNs and the IoTs are key to the success of smart cities.

7.2.1.2 *MANAGER*

In a particular architecture, this component is responsible for managing access control permissions for all IoT devices. This node is considered a lightweight node in the architecture and does not store blockchain information or validate blockchain transactions as the master node does. You can register any component as an administrator node [50]. Though, gadgets enrolled as IoT devices must be registered within the control of the administrator, so that administrators cannot register to devices under their control without authentication. All devices registered in the system must be associated with anyone enrolled administrator. Else, you will not be allowed to control the device. Enrolled IoT devices can associate to various administrators at the one-point time. Later registering an IoT device within the control of the administrator who can describe precise entree regulator rights for that device [51].

7.2.1.3 PROXY NODE

A proxy node is a detailed blockchain node of this construction that deploys smart city contracts to the system. The node has a smart city contract within the life cycle of the entree regulator structure. When a smart city contract is acknowledged by the blockchain network, this node receives the address used to identify the smart city contract in the blockchain network. To establish the connection with the smart city contract, all hubs in the blockchain organization must know the location of it [52].

7.2.1.4 SMART CITY CONTRACT

The system architecture described here is regulated by processes described in a particular smart contract called smart city contract. The contract is extraordinary and irremovable from the system. Therefore, all processes defined in the smart contract are triggered by the blockchain transaction. After the transaction triggers the operation, the master node provides global access to the transaction information. Smart city contracts and their operations are also available globally [53].

7.2.1.5 BLOCKCHAIN NETWORK

In this architecture, for simplicity, the blockchain network is a remote blockchain. All fundamentals of the model have multiple dimensions and can provide reliable results during system evaluation, so you can choose a private blockchain. However, in real situations, you should use public

Blockchain. A private blockchain is a chain that anyone can read, but only private hubs are authorized for modifications. The main network node helps to retain the system protected and unchanged by allowing communications and maintaining a copy of the Blockchain [54].

7.2.1.6   *MANAGEMENT CENTER*

As declared above, IoT devices are not part of the blockchain framework. Most IoT gadgets are extremely restricted as far as CPU, storage, and battery. These limitations limit IoT gadgets from turning out to be a part of a blockchain system. To be a piece of a blockchain system, one has to keep a duplicate of the Blockchain locally and monitor network transactions. Even if you have a light solution that does not keep all the information in the Blockchain locally and relies on other nodes [55].

7.2.2   *LAYERED ARCHITECTURE OF SMART CITY USING BLOCKCHAIN TECHNOLOGY*

To designing smart city infrastructures and systems must be done in the context that is, understanding the environment in which they interact with other elements in the environment in which they exist. Figure 7.2 shows a smart city layered architecture using blockchain technology and how it describes the context of a smart city system and infrastructure. This includes six layers: goals, people, ecosystems, intangible infrastructure, urban systems, and tangible infrastructure. The following diagram is a conceptual diagram of the different stages of smart urban area structure.

7.2.2.1   *GOALS, PEOPLE, AND ECOSYSTEMS*

The smart city initiative is based on a set of goals. In general, these goals focus on creating sustainability, inclusiveness, social, and economic growth. Where strategies lead to changes in urban systems and infrastructure, thereby affecting the population, residents, workers or visitors of the city, these strategies will only be implemented through the smart city strategy. Communities that live in cities are fundamental elements of the urban ecosystem, providing support, expressing social life, representing common interests and abilities, and playing a role in communication between urban institutions and communities [56]. This includes families and social networks, communities, cultural groups, religious groups, charities, the volunteer sector, public institutions (such as municipal schools and universities), and private organizations (such as service providers, retailers, and employers). The challenge for smart city designers and designers is to create the infrastructure and services that are part of the structure and livelihood of communities and people's ecosystems.

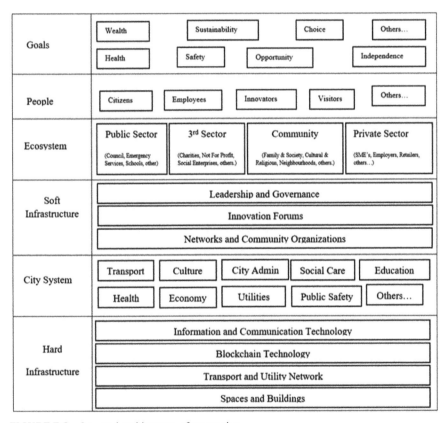

FIGURE 7.2   Layered architecture of smart city.

7.2.2.2   SOFT INFRASTRUCTURE

Intangible infrastructure elements create communication and trust in understanding how communities and individuals interact and experience smart cities. If the telecommunications process continues to grow at scale, the process and the urban community can become part of the smart city's flexible infrastructure. Some intangible infrastructure elements are very formal. For example, a standard framework that measures overall progress and performance of a single urban system according to the goals of a smart city. The governance process encourages and encourages individual purchasing decisions across the city to contribute to smart cities. Goals, norms, and principles for integration and interoperability between urban systems [57]. These are all elements of a smart city architecture that aims to implement a smart city strategy.

7.2.2.3   CITY SYSTEM

The urban system is an important part of the smart city. These literary systems provide cities with life support by providing citizens with transportation, education, healthcare, support for communities and businesses. An important element of the design process is that designer constraints must be considered [58]. All of these restrictions play a significant role in turning cities into smart cities.

7.2.2.4   HARD INFRASTRUCTURE

The field of smart cities stems from the potential of new technology platforms to transform urban systems. These platforms include 5G or 6G and broadband. These are communication tools such as phones, social media, video conferencing, IT resources (such as cloud computing), information repositories for considering open data or loading of urban observatories, and anything that can provide detailed information. Analysis and modeling tools. Overview of the operation of the city system [58]. At this layer of hard infrastructure technology, Blockchain also plays a very important role in the security and confidentiality of the data obtained at this stage or stored in the smart city environment.

7.3   APPLICATION OF BLOCKCHAIN TECHNOLOGY IN DIFFERENT COMPONENTS OF SMART CITIES

In recent past India and other emerging economics have witnessed a rapid growth in urbanization process that has resolved in the necessity of developing a trustworthy solution for his problems emerging he is responsibilities of smart cities such as waste, water, energy, education, health, and employment It is obvious that it has provided a much-needed impetus to adopt blockchain technology that focuses on providing the necessary infrastructure to ensure a perfect and feasible environment for its citizens. The technology is further used to incorporate and deal with the frameworks in an efficient manner incorporating the techniques of optimal urbanization of available resources. Blockchain technology has been described and defined by numerous researchers, and it can be considered as a specified method for recording and sharing information over various records, each having similar information records. They are kept up and controlled collectively. Blockchain is essentially a decentralized and dispersed of records and has a very wide range of applications [37]. Some important applications of Blockchain Technology in the context of smart cities are being described in subsequent sections.

7.3.1 SMART ENERGY GRID

The electric power division is exceptionally controlled, policymaker's division. Blockchain is used for understanding policymakers. They ought to effectively bolster the advancement of specialized norms and policymakers should make it workable for blockchain adventures. The most common application of Blockchain to the energy grid sector are a P2P network, grid transactions, energy financing, sustainability attribution, electric vehicles [60].

7.3.2 SECURE DATA COMMUNICATION

Although certain thread categories have been identified for smart cities such as concerning the unauthorized upholding of resources, or unauthorized manipulation and defilement of information or revelation of delicate information or increasing unapproved access to asset and touchy information or increasing unapproved access to asset and delicate information, however a Blockchain-based security framework against any type of such possibilities [60]. Moreover, a common platform that ensures the secured communication of decries in a distributed system can be achieved by integrating the blockchain technology with devices. Blockchain also ensures faster reliable and efficient operation.

7.3.3   SMART CONTRACTS (SCS)

A smart contract is a contract between buyer and seller on a written condition agreement. The agreements contained assigned decentralized blockchain network. All details of execution and transactions are trackable. Without the need for a Main power, legal system, and other procedure, SCs allow every single confided in exchange and understandings to be finished all unique and other parties. The smart contract uses for simple to complex cases. They can be used for sending money to others. SCs have the potential to disrupt many industries.

7.3.4   CITIZEN PARTICIPATION

The participation of civil society as an important aspect of the planning and development of smart cities. The main purpose of developing small cities is concerned with providing a decent and dignified life to its residents, through ensuring a clean environment and as such the participation of the citizens in managing the infrastructures, services, and resources becomes an important aspect of intensive and sustainable development of smart cities. This requires a symbiotic relationship between blockchain technology and its users. Blockchain technology can also be used to develop an efficient and transparent voting system for gathering the opinions of any professional group or the residents of specific areas or citizens at large [61]. The Blockchain can conveniently be utilized for developing secured, efficient, transparent, and decentralized voting platforms for gathering feedbacks collecting and solving grievances and improving facilities in smart cities.

7.3.5   BUDGET AND SERVICE FOR THE RESIDENTS

The blockchain technology can be gainfully utilized in the planning and development of various economic activities that have a direct impact on the employment potential of smart cities. The blockchain technology-based solutions can be adopted in almost all the business processes associated with HR, operation, supply chain, market, or financial activities at the local, regional or national level [62]. It is helpful in business registration procedures and has an equal or larger potential for planning, production, employment, and business development of local industries. Infect, there is no area of business development and management where Blockchain-based solutions cannot be gainfully employed. The Blockchain-based solutions are providing to be a boom for bringing Transparency and efficiency to various economic activities of smart cities.

7.3.6 HEALTH

For developing a public health system for a smart urban area, it is essential to maintain the health records of all the citizens. Blockchain Technology can advantageously be used in building up a solitary wellbeing record related to health for every one of its residents. Blockchain Technology is also helpful in secured maintenance of such records, which is flexible enough in the sense that it can further be improved by including new information from time to time [62]. Such records may be maintained in a decentralized pattern available to consultants as and when required. It is possible to register the medical practitioners, clinical, and consultancy psychologist, physiotherapist, and care providers by utilizing the blockchain technology. Blockchain technology can thus ensure the efficient management of medical and wellness services in smart cities. A transparent pharmaceutical supply chain based on sound ethical principles can also be created using the Blockchain. In the recent year's certain characteristics of Blockchain such as Transparency, auditability, disintermediation, industry collaboration, and new business models have resulted in creating a renewed interest for investing in the healthcare system [62]. It may, however, be mentioned that an ethical dilemma as intimately associated with availability and access to personal records of citizens that also necessitates solving the policy and privacy issues. Another sector associated with the healthcare system is insurance where distributed ledgers of Blockchain may be helpful in deceit elimination disposal, claims automation and information investigation and thus improving the efficiency of the insurance industry [62]. The Blockchain is also utilized in maintaining data recording personal historical credit information, accident environmental information and other historical information that helps the industry.

7.3.7 EDUCATION

Education is another important sector which has tremendous scope for utilization of blockchain technology, especially for supporting the academic program management and also for summative evaluation of student's performance. Blockchain technology can conveniently be used for formulating whole transcripts that include learning contents, outcomes, and students' achievements. The student's educational records, credentials, and academic certificates can also be kept in a well-secured manner. Such information can further be improved or enlarged by including information on training and skills acquired research and/or professional experience and individual interests. All such information can safely be stored on blockchain networks and accessed with the use of unique user ID. Beside Blockchain are also helpful in managing the payment of fees, scholarships, and maintaining records for training and placement of students [63].

7.3.8 PROPERTY REGISTRATION/AUCTION

Most of the registration offices in cities have been using a digitalized system for registration of property that is quite a time consuming and involves many complexities regarding and titles, approvals, investigation reports, and other related archives. Multiple numbers of authorities and offices are often involved in such a process that makes it a time consuming and costly exercise. These problems are conveniently eliminable of a blockchain bases system is adopted [63]. Such systems also facilitate the validation of the identity of citizens, thereby helping the allocation/registration of houses/properties to the rightful citizen in a transparent and verifiable manner.

7.3.9 RENEWABLE ENERGY

Blockchain-based system can also be used for efficient management of renewable energy. The renewable energy sector uses the real-time differential pricing models that can encourage neighborhood self-reliance on power and also help in containing energy uses. The use of blockchain technology can help in achieving these purposes. The Blockchain may also be helpful in connecting potential investors for raising the capital for renewable energy projects [63].

7.3.10 WASTE AND SANITATION

The waste management and sanitation offer a major challenging, smart city. The accurate assessment of tracking the production and disposal of waste involves is a complex process. However, the use of Blockchain provides an efficient solution to all such problems ensuring efficient management of waste disposal and sanitation and smart services [64].

7.3.11   SAFETY

The manipulation of records in various courts is a perpetual problem that often results in the process and procedure of the judicial system. This is more severe in property owner's cases. Such a problem puts an undesirable burden on law-reinforcement agencies also. Such records can properly be maintained and unambiguously verified using Blockchain-based system. The system may be utilized to register all evidence that ensures that there remains no further scope of altering or modifying or tempering such evidence/records [64]. The blockchain system can also be utilized in maintaining birth and death records, land records, and other necessary information in a safe and secured manner.

7.3.12   SMART ASSETS

Smarts assets are based on cloud computing. In the blockchain technology for Smart Asset model, it is especially all around adjusted to stock stock-piling and dispersion, and capacity will, in general, progressively occur outside of the cities. Smart assets allow to apply constraints on all operations and offer for autonomy, anonymity, and low-cost of transactions. By the Blockchain required less space for inventory management and storage [62]. Thus, we can say that more space is accessible by upgrading dissemination and Smart Cities to profit by a greater decent variety of products.

7.4   OPPORTUNITIES AND CHALLENGES OF BLOCKCHAIN TECHNOLOGY IN SMART CITY IMPLEMENTATION

Blockchain can perform a big role in smart cities. Blockchain also gives important prospects for a smart city's secluded division like fraud-fighting, to start small businesses, and many other ways [56]. Following are the main roles:

1. **Increased Transparency and Connectivity:** Using blockchain technology, smart cities/cities can be interconnected for various services such as mobility, energy, security through one system [56].
2. **Direct Communication:** Blockchain can provide direct communication between government bodies/dept. And public without any intermediaries [57].
3. **Integrity Over Information:** Using Blockchain technology, any data file can be encrypting partially or fully. With this, there will not ant manipulations by a third party [57].
4. **Security:** Data can be secure or protect by the Blockchain. Nobody can change/update this data [57].

There are various other applications of Blockchain in smart cities which are being discussed as in the following points:

1. **Educations:** Blockchain Technology can be applied in academic credentials for students, faculties, and learners. It can be used for student educational platforms to secure and share their records. By this, the third parties cast can be reduced and increase the security and reliability of records/documents. Blockchain system can work in the easiest process to verification and attestation of any certificate of any student [58].
2. **Healthcare:** Blockchain networks makes trust decentralized. Present situations, healthcare is playing the most important role in life, and patients is a basic part of the medical organization [58]. We can say that by the blockchain technology, it can improve mainly following majors uses in the field of healthcare:

    i.   Drugs traceability;
    ii. Patient data management;
    iii. Clinical trials.

    Healthcare institutions can store medical records containing sensitive personal information on Blockchain-based systems. The advantage of storing data in Blockchain is that it is impossible to get hacked as the ledger is not maintained on just a repository. It is spread out different databases in duplicate copies [58]. Blueprint of genetic information, patient's medical data and other valuable results and information from the clinical trial remains safe and secured in Blockchain.

3. **Banking:** Using Blockchain technology, each and every transaction can be encrypted to protect the integrity and identity of all parties involved. The main question now is that how future bank will adopt to Blockchain. Blockchain technology can help in the creation of a decentralized client identification systems. All payment company/bank required 'KYC' (know your client) measures before accepting any applications/task [57]. Using the blockchain systems, users have to identify once, and the information will be stored in a secure location where all payment companies/banks in the system can access it. In many more services/applications of banking, Blockchain can perform an easy task. The blockchain technology can be used in a multinational way in a smart city field [58]. We can use in following applications too for smart city:

    i.   Identity services;
    ii.   Government services;
    iii.   Publishing services;
    iv.   Aid and development;
    v.   Supply chain management;
    vi.   IoT;
    vii.   Cloud computing;
    viii.   Constructions industry;
    ix.   Digital advertising;
    x.   Emerging;
    xi.   Human resource/recruitment.

7.5 CONCLUSION

In near future, billions of IoT devices will be the backbone of the smart city in the world. India already launched the Smart Cities mission to develop smart cities using. The main features of these cities include smart street lights, smart waste management, smart water management, smart hospitals, smart transportation, smart governance, etc. But to maintain the security in these domains, the Blockchain is used as a security measure. This chapter is started with the general discussion of smart city, IoT, and AI technologies. After that a critical review is done about the implementation of blockchain technology in the smart city domain. This chapter proposed a blockchain architecture for the smart cities. In the last section, opportunities and challenges of blockchain technologies in smart city are discussed.

KEYWORDS

- **artificial intelligence**
- **blockchain**
- **computer vision**
- **Internet of Things**
- **smart city**
- **wireless sensor network**

REFERENCES

1. Sanjaya Baru & Neel Ratan (2018). *Blockchain: The Next Innovation to Make Our Cities Smarter*. PricewaterhouseCoopers Private Limited India. doi: http://ficci.in/spdocument/22934/Blockchain.pdf.
2. Sun, J., Yan, J., & Zhang, K. Z., (2016). Blockchain-based sharing services: What blockchain technology can contribute to smart cities? *Financial Innovation, 2*(1), 26.
3. Ibba, S., Pinna, A., Seu, M., & Pani, F. E., (2017). City Sense: Blockchain-oriented smart cities. In: *Proc. of the XP 2017 Scientific Workshops* (p. 12). ACM.
4. Sharma, P. K., Moon, S. Y., & Park, J. H., (2017). Block-VN: A distributed blockchain based vehicular network architecture in smart city. *Journal of Information Processing Systems, 13*(1), 184–195.
5. Watanabe, H., Fujimura, S., Nakadaira, A., Miyazaki, Y., Akutsu, A., & Kishigami, J. (2016). Blockchain contract: Securing a blockchain applied to smart contracts. In: *2016 IEEE International Conference in Consumer Electronics (ICCE)* (pp. 467, 468).
6. Peters, G. W., & Panayi, E. (2015). Understanding modern banking ledgers through blockchain technologies: Future of transaction processing and smart contracts on the internet of money. *Banking Beyond Banks and Money*, 239–278.
7. Laure A. Linn & Martha B. Koo (2016). *Blockchain for Health Data and its Potential use in Health it and Healthcare Related Research*. Use of Blockchain for healthcare and research workshop in 2016.
8. Biswas, K., & Muthukkumarasamy, V. (2016). Securing smart cities using blockchain technology. In: *2016 IEEE 14th International Conference on Smart Cities* (pp. 1392, 1393).
9. Luu, L., Chu, D. H., Olickel, H., Saxena, P., & Hobor, A., (2016). Making smart contracts smarter. In: *Proc. ACM SIGSAC Conference on Computer and Communications Security* (pp. 254–269).
10. Kshetri, N., (2017). Can Blockchain strengthen the Internet of Things? In: *IEEE IT Professional* (Vol. 19, No. 4, pp. 68–72).
11. Guo, M., Liu, Y., Yu, H., Hu, B., & Sang, Z., (2016). An overview of smart city in China. *China Communications, 13*(5).

12. Karmakar, A., & Sahib, U., (2017). *SMART DUBAI: Accelerating Innovation and Leapfrogging E-Democracy* (pp. 197–257). "E-Democracy for Smart Cities" Singapore: Springer.
13. Davidson, S., De Filippi, P., & Potts, J., (2016). *Economics of Blockchain.* Available: https://www.nber.org/system/files/working_papers/w25407/w25407.pdf (accessed on 28th July 2021).
14. Christidis, K., & Devetsikiotis, M., (2016). Blockchains and smart contracts for the IoTs. *IEEE Access, Special Section on the Plethora of Research in IoT,* 2292–2303.
15. Shuling, L., (2018). Application of blockchain technology in smart city infrastructure. *IEEE International Conference on Smart Internet of Things (SmartIoT), 1,* 276–2766.
16. Christian, N., & Stelian-Mihai, M., (2018). Using Blockchain as a platform for smart cities. *Journal of E – Technology, 9*(2).
17. Barzilay, O., (2017). *3 Ways Blockchain Is Revolutionizing Cybersecurity.* Forbes. Available: https://www.forbes.com/sites/omribarzilay/2017/08/21/3-ways-blockchain-is-revolutionizing-cybersecurity/ (accessed on 28th July 2021).
18. Biswas, K., & Muthukkumarasamy, V., (2016). Securing smart cities using blockchain technology. In: *2016 IEEE 18th International Conference on High Performance Computing and Communications; IEEE 14th International Conference on Smart City; IEEE 2nd International Conference on Data Science and Systems (HPCC/SmartCity/DSS)* (pp. 1392–1393). doi: http://dx.doi.org/10.1109/ HPCC-SmartCity-DSS.2016.0198.
19. Christidis, K., & Devetsikiotis, M., (2016). Blockchains and smart contracts for the Internet of Things. *IEEE Access 4,* 2292–2303. doi: http://dx.doi.org/ 10.1109/ ACCESS.2016.2566339.
20. Giulio, C., Giuseppe, D., Michele, M., Marco, O., & Roberto, T., (2013). Micropatterns in agile software. In: *International Conference on Agile Software Development* (pp. 210–222). Springer.
21. Steve, C., Giuseppe, D., Xiaohui, L., Sigrid, E., Andreas, E., & Kenneth, A., (2016). Comparing test and production code quality in a large commercial multicore system. In: *Software Engineering and Advanced Applications (SEAA), 2016 42th Euro Micro-Conference* (pp. 86–91). IEEE.
22. Eduardo, C. F., (2016). *The Blockchain: A New Framework for Robotic Swarm Systems.* CoRR abs/1608.00695 (2016). http://arxiv.org/abs/1608.006956.
23. Michal, L., Ondrej, P., & Tomas, Z., (2016). *Hybrid-Agile Approach in Smart Cities Procurement. 1839,* 182–196.
24. Marco, O., Giuseppe, D., Stephen, S., & Michele, M., (2016). Measuring high and low priority defects on traditional and mobile open-source software. In: *Proceedings of the 7th International Workshop on Emerging Trends in Software Metrics* (pp. 1–7.). ACM.
25. Aafaf, O., Anas, A. E., & Abdellah, A. O., (2017). FairAccess: A new blockchain-based access control framework for the Internet of Things. *Security and Communication Networks.* doi: http://dx.doi.org/10.1002/sec. (1748). SCN-16-0184.
26. Andrea, P., (2016). A Petri net-based model for investigating disposable addresses in bitcoin system. In: *2nd International Workshop on Knowledge Discovering on the Web (KDWEB), CEUR Workshop Proceedings* (Vol. 1748). ceur-ws.org. http://ceur-ws.org/ Vol-1748/paper-14.pdf (accessed on 28th July 2021).
27. Porru, S., Pinna, A., Marchesi, M., & Tonelli, R., (2017). *Blockchain-oriented Software Engineering: Challenges and New Directions. 2203,* 203–223.

28. Jiong, J., Jayavardhana, G., Slaven, M., & Marimuthu, P., (2014). An information framework for creating a smart city through the Internet of Things. *IEEE Internet of Things Journal, 1, 2,* 112–121.

29. Stocchi, M., Lunesu, I., Ibba, S., Baralla, G., & Marchesi, M., (2016). The future of bitcoin: A synchro squeezing Wavelet Transform to predict search engine query trends. In: *2nd International Workshop on Knowledge Discovering on the Web (KDWEB), CEUR Workshop Proceedings* (Vol. 1748). ceur-ws.org. http://ceur-ws.org/Vol-1748/paper-14. pdf (accessed on 28th July 2021).

30. Feng, T., (2016). An agri-food supply chain traceability system for China based on RFID blockchain technology. In: *2016 13th International Conference on Service Systems and Service Management (ICSSSM)* (pp. 1–6). doi: http://dx.doi.org/10.1109/ICSSSM.2016.7538424.

31. Andrea, Z., Nicola, B., Angelo, C., Lorenzo, V., & Michele, Z., (2014). Internet of things for smart cities. *IEEE Internet of Things Journal, 1*(1), 22–32.

32. Yu, Z., & Jiangtao, W., (2016). The IoT electric business model: Using blockchain technology for the Internet of things. *Peer-to-Peer Networking and Applications.* doi: http://dx.doi.org/10.1007/s12083-016-0456-1.

33. Pieroni, A., Scarpato, N., Di Nunzio, L., Fallucchi, F., & Raso, M., (2018). Smarter city: Smart energy grid based on blockchain technology. *Int. J. Adv. Sci. Eng. Inf. Technol., 8*(1), 298–306.

34. Solanki, A., & Nayyar, A., (2019). Green Internet of Things (G-IoT): ICT technologies, principles, applications, projects, and challenges. In: *Handbook of Research on Big Data and the IoT* (pp. 379–405). IGI Global: Hershey, PA, USA.

35. Sun, J., Yan, J., & Zhang, K. Z., (2016). Blockchain-based sharing services: What blockchain technology can contribute to smart cities. *Financial Innovation, 2*(1), 1–9.

36. Rivera, R., Robledo, J. G., Larios, V. M., & Avalos, J. M., (2017). How digital identity on Blockchain can contribute in a smart city environment. In: *2017 International Smart Cities Conference (ISC2)* (pp. 1–4). IEEE.

37. Li, S., (2018). Application of blockchain technology in smart city infrastructure. In: *2018 IEEE International Conference on Smart Internet of Things (SmartIoT)* (pp. 276–2766). IEEE.

38. Maroufi, M., Abdolee, R., & Tazekand, B. M., (2019). *On the Convergence of Blockchain and Internet of Things (IoT) Technologies.* arXiv preprint arXiv:1904.01936.

39. Mathew, A. R., (2019). Cybersecurity through blockchain technology. *International Journal of Engineering and Advanced Technology.*

40. Kouicem, D. E., Bouabdallah, A., & Lakhlef, H., (2018). Internet of things security: A top-down survey. *Computer Networks, 141,* 199–221.

41. Kouzinopoulos, C. S., Spathoulas, G., Giannoutakis, K. M., Votis, K., Pandey, P., Tzovaras, D., & Nijdam, N. A., (2018). Using blockchains to strengthen the security of the Internet of Things. In: *International ISCIS Security Workshop* (pp. 90–100). Springer, Cham.

42. Singh, M., Singh, A., & Kim, S., (2018). Blockchain: A game-changer for securing IoT data. In *2018 IEEE 4th World Forum on Internet of Things (WF-IoT)* (pp. 51–55). IEEE.

43. Banerjee, M., Lee, J., & Choo, K. K. R., (2017). A blockchain future to Internet of Things security: A position paper. *Digital Communications and Networks.* URL: https://www.sciencedirect.com/science/article/pii/S2352864817302900 (accessed on 28th July 2021).

44. Ibba, S., Pinna, A., Seu, M., & Pani, F. E., (2017). CitySense: Blockchain-oriented smart cities. In: *Proceedings of the XP2017 Scientific Workshops* (pp. 1–5).
45. Jesus, E. F., Chicarino, V. R., De Albuquerque, C. V., & Rocha, A. A. D. A., (2018). A survey of how to use Blockchain to secure Internet of things and the stalker attack. *Security and Communication Networks, 2018.*
46. Kshetri, N., (2017). Can Blockchain Strengthen the Internet of things? *IT Professional, 19*(4), 68–72.
47. Eddine, D., Bouabdallah, A., & Lakhlef, H., (2018). *Internet of Things Security: A Top-Down Survey, 1905,* 1–24.
48. Krishnamurthi, R., Nayyar, A., & Solanki, A., (2019). Innovation opportunities through the Internet of Things (IoT) for smart cities. In: *Green and Smart Technologies for Smart Cities* (pp. 261–292). CRC Press: Boca Raton, FL, USA.
49. Ahmed, S., Rahman, M. S., & Rahaman, M. S., (2019). A blockchain-based architecture for integrated smart parking systems. In: *2019 IEEE International Conference on Pervasive Computing and Communications Workshops (PerCom Workshops)* (pp. 177–182). IEEE.
50. Rathore, S., Kwon, B. W., & Park, J. H., (2019). BlockSecIoTNet: Blockchain-based decentralized security architecture for IoT network. *Journal of Network and Computer Applications, 143,* 167–177.
51. Mendiboure, L., Chalouf, M. A., & Krief, F., (2018). Towards a blockchain-based SD-IoV for applications authentication and trust management. In: *International Conference on Internet of Vehicles* (pp. 265–277). Springer, Cham.
52. Nagothu, D., Xu, R., Nikouei, S. Y., & Chen, Y., (2018). A microservice-enabled architecture for smart surveillance using blockchain technology. In: *2018 IEEE International Smart Cities Conference (ISC2)* (pp. 1–4). IEEE.
53. Hakak, S., Khan, W. Z., Gilkar, G. A., Imran, M., & Guizani, N., (2020). Securing smart cities through blockchain technology: Architecture, requirements, and challenges. *IEEE Network, 34*(1), 8–14.
54. Cali, U., & Fifield, A., (2019). Towards the decentralized revolution in energy systems using blockchain technology. *International Journal of Smart Grid and Clean Energy, 8*(3), 245–256.
55. Wei, X., Duan, Q., & Zhou, L., (2019). A QoE-driven tactile internet architecture for smart city. *IEEE Network, 34*(1), 130–136.
56. Ramachandran, G. S., & Krishnamachari, B., (2018). *Blockchain for the IoT: Opportunities and Challenges.* arXiv preprint arXiv:1805.02818.
57. Khan, M. A., & Salah, K., (2018). IoT security: Review, blockchain solutions, and open challenges. *Future Generation Computer Systems, 82,* 395–411.
58. Xie, J., Tang, H., Huang, T., Yu, F. R., Xie, R., Liu, J., & Liu, Y., (2019). A survey of blockchain technology applied to smart cities: Research issues and challenges. *IEEE Communications Surveys & Tutorials, 21*(3), 2794–2830.
59. Rameshwar, R., Solanki, A., Nayyar, A., & Mahapatra, B., (2020). Green and smart buildings: A key to sustainable global solutions. In: *Green Building Management and Smart Automation* (pp. 146–163). IGI Global: Hershey, PA, USA.
60. Al Nuaimi, E., Al Neyadi, H., Mohamed, N., & Al-Jaroodi, J., (2015). Applications of big data to smart cities. *Journal of Internet Services and Applications, 6*(1), 25.

61. Hao, L., Lei, X., Yan, Z., & ChunLi, Y., (2012). The application and implementation research of smart city in China. In: *2012 International Conference on System Science and Engineering (ICSSE)* (pp. 288–292). IEEE.

62. Lee, J. H., Phaal, R., & Lee, S. H., (2013). An integrated service-device-technology roadmap for smart city development. *Technological Forecasting and Social Change, 80*(2), 286–306.

63. Mohammed, F., Idries, A., Mohamed, N., Al-Jaroodi, J., & Jawhar, I., (2014). UAVs for smart cities: Opportunities and challenges. In: *2014 International Conference on Unmanned Aircraft Systems (ICUAS)* (pp. 267–273). IEEE.

64. Cimmino, A., Pecorella, T., Fantacci, R., Granelli, F., Rahman, T. F., Sacchi, C., & Harsh, P., (2014). The role of small cell technology in future smart city applications. *Transactions on Emerging Telecommunications Technologies, 25*(1), 11–20.

65. Pandey, S., & Solanki, A., (2019). Music instrument recognition using deep convolutional neural networks. *Int. J. Inf. Technol., 13*(3), 129–149.

66. Rajput, R., & Solanki, A., (2016). Real-time analysis of tweets using machine learning and semantic analysis. In: *International Conference on Communication and Computing Systems (ICCCS2016)* (Vol. 138 No. 25, pp. 687–692). Taylor and Francis, at Dronacharya College of Engineering, Gurgaon.

67. Ahuja, R., & Solanki, A., (2019). Movie recommender system using K-means clustering and K-nearest neighbor. In: *Accepted for Publication in Confluence-2019: 9th International Conference on Cloud Computing, Data Science & Engineering (Vol. 1231, No. 21, pp. 25–38). Amity University, Noida.

68. Tayal, A., Kose, U., Solanki, A., Nayyar, A., & Saucedo, J. A. M., (2019). Efficiency analysis for stochastic dynamic facility layout problem using meta-heuristic, data envelopment analysis, and machine learning. *Computational Intelligence*.

69. Tayal, A., Solanki, A., & Singh, S. P. (2020). Integrated framework for identifying sustainable manufacturing layouts based on big data, machine learning, meta-heuristic and data envelopment analysis. *Sustainable Cities and Society*. https://doi.org/10.1016/j.scs.2020.102383.

70. Pramanik, P. K. D., Solanki, A., Debnath, A., Nayyar, A., El-Sappagh, S., & Kwak, K. S., (2020). Advancing modern healthcare with nanotechnology, nanobiosensors, and internet of nano things: Taxonomies, applications, architecture, and challenges. In *IEEE Access* (Vol. 8, pp. 65230–65266). doi: 10.1109/ACCESS.2020.2984269.

CHAPTER 8

Application of Blockchain Technology to Make Smart Cities Smarter

YOGITA BORSE,[1] PURNIMA AHIRAO,[1] KUNAL BOHRA,[2]
NIDHI DEDHIA,[2] YASH JAIN,[2] ROHIT KASALE,[2] and UNMESH MADKE[2]

[1]Faculty of Department of Information Technology,
K.J. Somaiya College of Engineering, Mumbai, Maharashtra, India,
E-mail: yogitaborse@somaiya.edu (Y. Borse)

[2]UG Student of Department of Information Technology,
K.J. Somaiya College of Engineering, Mumbai, Maharashtra, India

ABSTRACT

Today, most of the population has started migrating to urban areas. Quick urbanization centers around circumstances introduced by this urban change to make rising urban areas that go about as amazing and comprehensive advancement devices. A smart city utilizes data innovation to coordinate and oversee physical, social, and business frameworks so as to offer better types of assistance to its occupants while guaranteeing productive and ideal use of accessible assets. The reason for this report is to comprehend one of the new innovation achievements to be specific blockchain and break down how it tends to be used for making urban areas smarter. It will consider different regions of a smart city and distinguish those zones where block-chain advancement can be utilized to improve our urban communities and give better livability and financial improvement.

8.1 INTRODUCTION

Urban communities on the planet are confronting tremendous pressures from difficulties related to fast urbanization. The United Nations assesses that 4.334 billion or 58% of the world's people live in urban domains in 2020 and a 2.5 billion, or total, 68% will be incorporated by the year 2050 [20]. City development increases population blast, yet also, extreme issues, for example, traffic clog, contamination, non-inexhaustible asset exhaustion, and expanding social imbalance. These urban issues do not regard the outskirts of the countries or borders between new areas. Incredible obligations on comprehending them lie at the city level, where clashes concerning financial, social, and genetic improvement remain neglected [5].

For as far back as decades, various urban maintainability and intelligent metropolitan area systems have been proposed. They provide devices to help urban decision-makers plan, conduct activities, and survey the urban communities' advancement towards an increasingly economical future. As of late, scientists began to advocate the idea of "Blockchain Cities" as the successive wave in changing the urban setting to address the urbanization difficulties. Many accept that blockchain is ready to assume a significant job in the practical advancement of the worldwide economy, improving individuals' satisfaction and eventually carrying essential changes to the real world. A World Economic Forum report appraises that 10% of global GDP will be put away on blockchain innovation till 2027 [21]. Blockchain highlights a decentralized, shared database that gives transparency and immutability of transaction records.

A smart city utilizes data innovation to coordinate and oversee physical, social, and business foundations to offer better types of assistance to its tenants while guaranteeing proficient and ideal use of accessible assets. Nowadays, due to rapid urbanization, there is a lot of strain on basic amenities and other facilities like healthcare, government services, power distribution, etc. In the healthcare sector, data is handled in a very haphazard and inconvenient manner by the facility providers as well as the insurance providers. Digitization has helped to some extent, but centralization has created many doubts. The government sector also has faced criticism for the mishandling of identities of citizens and lack of security of essential data (AADHAR). Social welfare schemes like MGNREGA, old-age pension scheme implementation is also affected due to these fake identities and lack of transparency over the current centralized approach. The shortage of power is also now an issue due to rapid urbanization.

This chapter talks about the recent implementation of a few major sectors in a city: healthcare, governance, energy, and social benefits. The chapter also proposes blockchain-based solutions for the problems in these sectors. Lastly, the chapter concludes with results and observations and talks about the future work needed.

8.2   METHODOLOGY

Below mentioned are some of the detailed use-cases, benefits, and advancements of blockchain in turning a city into a smart city. This section covers how the process goes on in each sector, including all the concerned factors. It also discusses blockchain-based solutions and their different use cases in each field.

8.2.1   HEALTHCARE

The healthcare industry is a prominent sector in every city of the world, with branches ranging from clinical care to drugs to surgical and other medical instruments. It opens up a broad scope for innovations and welfare of the citizens with an increase in the quality of living, leading to a better and smarter city of tomorrow.

8.2.1.1   CURRENT SITUATION

Nowadays, medical data is handled in a very haphazard and inconvenient manner by various medical institutions. This medical data not only comprises illness history, drug history, but also medical insurance records. This medical data can be visible to more than one authority, including the medical facility providers as well as the insurance providers. The data is processed and managed by institutional policy. It may lead to loss of accuracy and integrity. Therefore, it becomes difficult to gather a patient's medical records in case of any medical emergency. Such a problem could mean the difference between life and death for any patient. In the current scenario, there are no appropriate standards for handling or accessing these medical data. Blockchain would provide a solution to this problem and thereby improve the quality of handling and accessing the medical data.

8.2.1.2   BLOCKCHAIN IMPLEMENTATION

8.2.1.2.1   Medical Records

Digital medical records can be stored on the blockchain, which will help in binding all the aspects of the medical data together. Therefore, using

Blockchain technology, the work of the medical facilities would get simplified as the healthcare providers would get a complete idea of the patient's medical data in one place. Medical records can use blockchain in a below-mentioned way:

i.   A patient visits a doctor without any previous medical records;
ii. The doctor checks on the blockchain for his/her past medical history;
iii. The doctor is unable to find the records;
iv. The doctor checks the patient and treats him;
v.   A block gets created with the latest records on the public blockchain where they reside forever, securely;
vi. Once the block goes through processing, the patient can visit any other doctor for a check-up;
vii. This time, the other doctor can access the patient's previous medical history securely and efficiently without any hassle of paperwork;
viii. The patient's medical history will be a collection of all the records that have occurred throughout his life, thereby making them immutable.

As stated in Ref. [18], Figure 8.1 gives an idea of how a person's health information generated from various providers and devices can reside on a decentralized blockchain. After collecting the data, it will be encrypted and digitally signed for maintaining authenticity upon which this encrypted data will be stored in a data lake. At this point, a health record of that respective user will get registered on the blockchain. The patient will finally be informed about his health data being added successfully on the blockchain, and he can now easily access it and insert data with the help of digital signature and encryption.

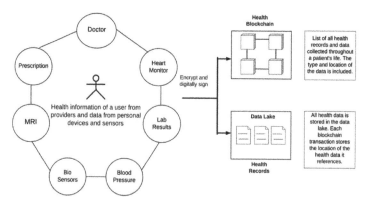

FIGURE 8.1   Collection, encryption, and storage of medical records on the blockchain. *Source:* Adapted from Ref. [18].

8.2.1.2.2   DOCUMENT RECOVERY

Blockchain technology is a decentralized technology. It means that data will not remain at a single location. Therefore, in the case of disaster recovery, Blockchain technology would prove helpful. Due to the decentralized manner of storing the data, it would always be safe. It would particularly prove beneficial in the scenario when the paper documents are lost, damaged, or missing as they now reside on a permanent architecture powered by blockchain.

8.2.1.2.3   FRAUD PREVENTION

Fraud prevention is also a requirement for better healthcare. The frauds can occur in prescriptions or billing. Traceability is one of the features of blockchain. This feature would help trace the prescription history for a particular patient. This prescription history will always contain a timestamp. Also, since Blockchains are immutable, no alterations would be allowed in the data. It will decrease the possibility of fake prescriptions or prescription mishandling. Billing fraud is a severe issue. Currently, the billing procedures are finding it difficult to detect these frauds. Blockchain can help in this situation. Blockchain would provide for the authenticity of such bills and also provide security for payment processing. It can be implemented in this scenario as follows:

- Whenever a patient undergoes any sort of treatment, he will get a receipt which will have the prescription details and a unique transaction number generated by the blockchain, after the block is mined.
- This transaction number can be checked in the blockchain to prove its authenticity.
- It would give us an insight into the receipt, whether it is authentic or not.
- Similarly, doctors will have to prove their authenticity by storing their medical certificates on the blockchain in a secure manner.
- Once the block is mined, the doctor will be verified and vice versa.
- Similarly, for drugs to be proved authentic, Blockchain smart contracts (SCs) can be used. It will securely store and check whether a drug is original or fake.

8.2.1.2.4   CLINICAL TRIALS

Clinical trials are an essential aspect of the health sector. They are used for determining the effect of a particular remedy, be it a drug or treatment for an illness. The clinical trials include multiple tests and lots of data. It becomes challenging to follow all the data and the results. This data and results would be stored on the blockchain. As stated above, the information on a blockchain is immutable. Therefore, there are no chances of data being manipulated. Thus, the researchers could trust the data and be confident about data accuracy. The blockchain implementation for this part will be similar as in medical records shown above.

8.2.1.2.5   PATIENT INCENTIVE PROGRAMS

Patients upon completing a specific task can be incentivizing. This idea would help in gaining customers. If the patient follows the given plan or keeps the doctor updated with his health conditions, he would be rewarded. Blockchain technology would help in managing the incentives for the patients.

8.2.2   GOVERNANCE

Identity is an essential factor in our lives. Nowadays, everyone lives in such a space where all do spin around identity. Furthermore, with simple access to it, modern viewpoints can sparkle. Hence, managing identities of citizens become the prime motive of the government sector as these identities then have their use-cases in other defined sectors like banking, e-commerce, healthcare, insurance, payments, travel, governance itself, and so on.

8.2.2.1   CURRENT SITUATION

Government divisions have functional interdependence yet work in silos, which impacts the accessibility of administrations and breaks down resident experience. So, what happens currently is that a lot of corruption going on at the government officials' side, mishandling of the identities of citizens, no proper communication about the citizen's information with them, and lack of security of their valuable data. Conservation of our own identity and having control over it is of utmost importance as, without it, one cannot even prove who they say they are. In most of the areas, still paper-based identity management systems work, which themselves can have specific human errors. Some of the prime disadvantages of the current system are:

- No regulations for access restriction;
- The browser and device security concerns;
- KYC repetition and registration process;
- Identity risk from online platforms;
- Persistent identity hack/threat;
- Weak authentication protocols.

Current structures to approve, guarantee, and oversee identity depend on centralized, top-down methodologies that rely on authorities and other administrators. Also, identity not being available on a single secured platform compels the user to enter his details and perform the KYC process repetitively each time and everywhere, which is tedious, as shown in Figure 8.2. Further, the verification is performed by every platform individually on the same uploaded identity documents, which waste resources, time as well as adds inconsistencies with no proper security and privacy of the data. The strength of authentication protocols used for this cannot be assured, and hence this can cause data leaks, which can necessarily prove useful for hackers. After the verification process is done, this data gets stored into individual data-bases of respective platforms, which causes duplication of the original data and hence leading to any hacks/threats by the hackers. The current process of how the user identity travels through different platforms and how is it been verified and stored can be represented through the following flowchart shown in Figure 8.2.

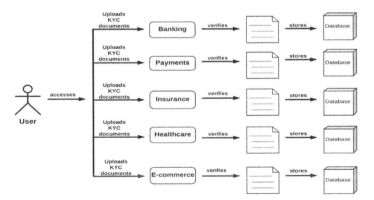

FIGURE 8.2   Current situation of handling records.

Hence it is now vital that traditional identity management systems see the transformation onto digital identity schemes using blockchain, which will let users manage their identity easily.

8.2.2.2 BLOCKCHAIN IMPLEMENTATION

Intelligent government services in a smart city enable a simple association with its residents, eliminating delays and dissatisfaction in collaboration with the government. Implementing blockchain-based solutions in the governance sector might eliminate the discrepancies or problems caused by the current implementation. Blockchain innovation can be used to break the silos, check government debasement, increment effectiveness, and straightforwardness.

8.2.2.2.1 Governance Use Cases Using Blockchain

Land registry, electronic voting, income tax service is the vital area where blockchain has found its application [31]. Land ownership record is always an issue while purchasing land. Digital registry is the proposed solution for secured and transparent land trading. Information about the land can get stored on the blockchain, and immutable records related to subsequent changes in the ownership can get recorded. It will prevent all types of fraudulent activity in the land trading arising out of irregular update in the current land registry system. Electronic voting is a trivial blockchain application which can provide security and transparency to the system. A voter has to register to the digital system, and complete the voting process on the election day. Votes are directly recorded and stored as a transaction on the blockchain, including the voter identity. Only the voter will know who he has voted for, votes stored on the blockchain can undergo decoding and counting. It leads to a more transparent and secured way of election procedure. Income tax return, including GST, which is the backbone of the country's economy faces much of chaos even though it has been digitalized. Blockchain technology-based implementation easily find its importance in this process. The overall procedure of taxation can become much simpler if it can get recorded on the blockchain transaction system. The stakeholders can get benefitted from this system and eventually add more revenue to the country's economic gain.

8.2.2.2.2 *Identity Management Using Blockchain*

Putting up a person's identity on a blockchain will help him access it across all other sectors. Such utilization may take out desk work and further promote the Digital India push. Genuinely speaking, doing this will improve trust during issuance, and minimal misuse or fraud will take place while utilizing these credentials. Publication of these accreditations on a blockchain can be attached to the identity of an individual. Once the individual identity block is in the blockchain, it can be produced further for any authorization and verification across different entities. The user will have control to give consent about revealing related identifiable information to the concerned authorities. It will ensure that only specific identity information becomes visible according to the need of the verification procedure [4, 17, 24]. Authors Bayu et al. [27] focus on identity management in the educational domain. During college admission many times, the students have to produce their identity, leading to a repetition of the authentication and verification process at different institutes. The personal details of the student along with biometric information, are hashed and stored on the blockchain. The student will get a digital certificate issued by the college once the information furnished by the student gets authenticated and validated. This certificate can be produced by the student in the future admission process and verified by the data stored on the blockchain. It will eradicate the need for authentication and verification of identity at numerous institutes. As discussed in Ref. [19], Figure 8.3 shows the flow of how blockchain can be implemented in managing the digital identities of the citizens.

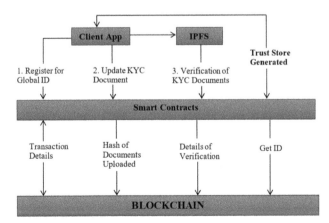

FIGURE 8.3 Identity management through blockchain.
Source: Adapted from Ref. [19].

Comparing the approach shown in Figure 8.3 with the traditional method, the first thing that can be noted is that there is only one single point secure platform, i.e., blockchain. It can be used for storing all the identities instead of multiple databases on multiple platforms. It will eliminate the duplication of identities and inconsistencies. So, the user will initially register and get a global ID through the app where in the next step, he will upload all his KYC documents just once. The hash of these documents will be uploaded and stored on the blockchain. Since the blockchain is highly secure and immutable, the privacy aspect can be achieved easily. At each step, the SCs will be executed and interact with the blockchain. The transaction details stored will be unique for every user. Documents uploaded by the user will be verified every time. IPFS here assumes a significant job in letting us address a lot of information and spot the immutable, perpetual connections into the blockchain exchanges. These timestamps and ties down the data without putting the information itself on the blockchain. Finally, a trust score is generated based on the verification, and then the final ID gets stored on the blockchain. This procedure is successful in preventing threats from the attackers as much as possible. Hence, document, and information transfer between divisions through a blockchain would expand visibility into the procedure and guarantee that the information pushes ahead progressively. This will help people connect virtually with the government and build a trust factor among them.

Blockchain is a decentralized, distributed, and immutable ledger. These characteristics make it suitable for storing data over it. One such field where these features of blockchain would be most suitable is the Identity Management. Financial as well as healthcare institutions where there is a need to store a vast amount of personally identifiable information (PII), a blockchain solution would be most suitable. However, blockchain is mainly of two types which is the Public and Private or permissioned one. The Public blockchain is a better option for storing identity in databases as it is decentralized and has no redundancy. Comparatively, the private blockchain has a centralized node which can be vulnerable to attacks by stealing the private keys. In such cases, the public blockchain, with its advantages for storing the identity of an individual comes with a shortcoming. The decentralized nature takes the blockchain ledger open to be viewed by any new node that is added to the blockchain. In healthcare or finance, where certain identity attributes like health-related issues and bank balance should be kept private. If such data is stored over the public blockchain, any node in the blockchain can easily access this data. Thus, keeping certain attributes of the identity anonymous becomes a challenge for Identity Management over the public blockchain.

A solution for this problem was developed by combing Zero-Knowledge Proof with the Pedersen Commitment scheme and a protocol for Identity Management was generated. The solution focuses on the issue of keeping certain confidential attributes of an identity secret while storing it over the public blockchain [36].

Blockchain is a technology where it is assumed that fraudulent activities can be part of the system. In case of PII, Blockchain usage has advantages as well as disadvantages. The blockchain makes use of distributed ledger which is updated in every node periodically to maintain the integrity of the ledger. This property of blockchain has positives as well as negatives when it comes to PII management using blockchain. But when it is compared to the way in which such information is handled, the positives of blockchain make the choice of blockchain necessary. In conventional methods, the identity documents are required to be presented for official works. When an individual hands over the Personal Identity document, whether copy or original, He /She does not have control over the usage of that document. If such important Personal Documents find their way to the hands of criminals or fraudsters, the loss is irrevocable. Thus, use of blockchain can play a vital role in the management of personally Identifiable documents [37]. In the proposed system [37], the complete process is divided into three crucial stages Creation of Document, Storing of Document, and Sharing of Document. The Creation is performed by government agencies, by validating the applicants' details. Also, it is necessary for the applicant or any other verifier of document, that the node should be able to verify the integrity of the document by using PKI. They have suggested the partial storing using IPFS. The addition of every document created can be recorded and included in blockchain by using proof of work (PoW). According to the authors, the sharing is the most crucial part of this management, where the purpose is to produce a proof of identity without revealing personally identifiable sensitive information. They have proposed zero-knowledge proof to achieved this. Figure 8.4 shows an overview of how the management of Personal Identity can be done over blockchain.

8.2.3 ENERGY

In the making of a smart city, energy plays a vital role. Energy acts as a source of income for some and works as a way of survival for others. Energy has not only revolutionized the industries during the industrial revolution but brought about a significant rise in the urban populace. Irregular inexhaustible power generation is additionally on the ascent, and framework steadiness on the local and national level is the crucial goal of power grid management. Maintaining energy supplies and trade is a complex process involving several stakeholders and intermediaries who help in the efficient management of energy. To sustainably develop the cities into smart cities, it is of utter importance to transform the energy sector using blockchain.

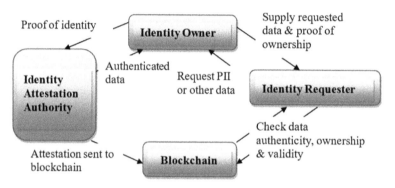

FIGURE 8.4   Personal identity on blockchain.
Source: Reprinted with permission from Ref. [37], ©Springer Nature; https://doi.org/10.1007/978-981-15-0222-4_41.

8.2.3.1   CURRENT SITUATION

Energy is a critical player in modernizing the cities, but due to improper planning and execution, the prices of electricity have risen significantly. Due to the decrease in supply, a massive increase in the number of intermediaries and a rise in industrial demand takes place. The energy supply is owned by private companies who have all the data stored on centralized servers, which has led to a strong security concern. The meters that gage and measure the electricity units are outdated and are not smart enough to measure inactivity and cannot comply with the current need for electricity [11, 23, 35]. The demand for renewable energy has risen in the past few years, but the market for trading these sources of energy is still premature. It involves a lot of intermediaries, making it difficult for the common man to afford renewable sources.

Current strategies include manual post-handling and expanded interchanges to solidify data held independently by each piece of the transaction.

Accordingly, existing techniques are moderate and tedious, as transactions should be confirmed and accommodated on numerous occasions from initialization to final settlement. Due to the lower speed of trade, it prompts frictional costs that are prohibitive to a certain extent. Figure 8.5 shows the diagrammatic representation of the current wholesale energy trading and supply market, which largely relies on the bank unit for the money, highlighting its centralized nature.

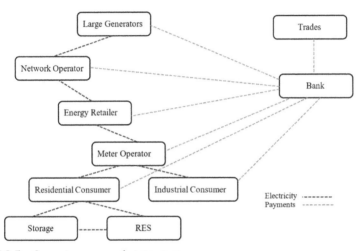

FIGURE 8.5   Current energy market structure.
Source: Adapted with permission from Ref. [11].

To overcome this, blockchain can play an essential role in decreasing costs by streamlining energy procedures and enhancing its security. The blockchain-based solution can also help in advancing the sensibility by empowering practical knowledge and low-carbon game plans, thereby overcoming the shortcomings of the energy sector. Figure 8.5 shows the diagrammatic representation of the current wholesale energy trading and supply market, which largely relies on the bank unit for the money, highlighting its centralized nature.

8.2.3.2   BLOCKCHAIN IMPLEMENTATION

Blockchain innovations can be useful to various use cases associated with the tasks and business procedures of energy organizations. Existing work directs potential applications and parts of plans of action that may get influenced. In the energy sector, blockchain can see its implementation in energy supply, source, and efficiency as summarized in subsections.

8.2.3.2.1   ENERGY TRADING AND SUPPLY

Blockchain has pulled in gigantic consideration and is presently being effectively sought after in the energy area too. A potential application is using Distributed ledger innovations in wholesale energy trading markets. The energy markets like the wholesale ones have more complicated processes, and hence they need mediators and traders to carry out trade. Distributed ledger methodology can make it possible to remove these middlemen during the transaction by producing units that will be solely executed by the energy providers and purchasers. The operator would look for the best arrangement in the commercial center that fulfills a purchaser's figure of interest for a given timeframe. The understanding would be securely recorded in the blockchain and naturally executed at the predetermined time of conveyance. Instalments would happen consequently at the hour of transportation as determined in the concurred agreement. All this data would be accessible by the primary system administrators through a single point of access. Relative use cases would require significant changes in the authoritative framework with perhaps certifiable effects at work of go-betweens, for instance, intermediaries, intermediaries, and exchanging organizations. The below-mentioned steps show how this can be implemented:

- A customer wants to buy a few units of energy.
- He will be directed to the retailer.
- There will be no middle-men in this process, and an autonomous trading agent will look for the best suitable figure that fits the needs of the customer.
- Once a consensus is reached, the transaction would be securely recorded on the blockchain permanently.
- Once the transaction is mined, it will be available in the distributed ledger for reference without revealing the anonymity of the buyers.
- This process will be completely transparent and secure.

Figure 8.6 represents the energy market implemented through blockchain, which eliminates all the intermediaries and has a shared trading platform for everyone to trade.

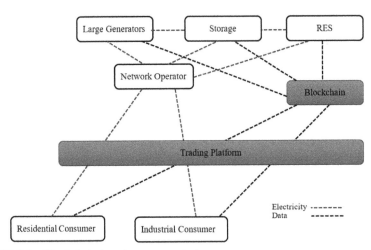

FIGURE 8.6   Blockchain-based energy market.
Source: Reprinted with permission from Ref. [11]; Open access creative commons CC-BY
license, Elsevier.

8.2.3.2.2   Billing

SCs, smart meters, and blockchains can comprehend automated charging for
the buyers. Administration associations may benefit from the potential for
energy micropayments, pay-more only as costs arise, or installment stages for
prepaid meters. Also, concerning irregularity settlement, blockchain-based
SCs define a set of rules to charge accordingly and diminish any chance of
irregularities arising out of the network. On any occasion, digitalization offers
an opportunity for sufficient storage, automation, examination, charging
stations, and pricing of new sources of energy. At the point when coordinated
with the metering framework, blockchains give a chance to computerized
charging in energy administrations for customers and distributed generators,
which accompanies the capability of regulatory cost decrease. Blockchain
increases transparency in the energy market by recording the units generated
and consumed as well as educating the consumers about the origin and costs
associated with the energy supply. This opens up the door for boosting social
change and requesting a response. What is more, upgraded secure power of
blockchains might be utilized to defend information protection, character
the executives, and strengthen towards digital dangers. Below mentioned is
a way how one can implement smart billing:

- Computerized billing will be incorporated with smart meters itself.
- Once the activity is recorded in the meter, a bill will be generated for that particular day or period and will be priced appropriately.
- This bill will be recorded in the blockchain in a tamper-proof environment.
- This gives the power to prove that the energy consumed has not been altered, and the price matched the current energy price in the area.
- Once the transaction is verified, the block will be mined, and at the end of the month, a total bill will be generated based on the monthly consumption reading the values from the blockchain.
- This will create a conflict-free, transparent, and tamper-proof environment in the billing process.

8.2.3.2.3 Blockchain-Based Smart Meters

They can help quicken the authorization of constant differential valuing models to contain energy utilization. Such models can advance neighborhood independence on energy. Blockchain-based sustainable power source microgrids associated with the fundamental matrix can enable nearby networks to deal with the force supply and the burden they need. A blockchain-based energy charging framework can productively uphold continuous differential valuing for utilization. These meters will provide an efficient redressal mechanism system to avoid conflicts among the producers and consumers. They will eliminate the need for third-party consent and verification as they will work autonomously and securely without the intervention of anybody. This meter works in the following manner:

- The producer installs smart meters.
- Consumers consume electricity.
- The meter records the activity.
- It initiates a transaction independently every time energy is consumed.
- The transaction, once verified, gets stored in a public/private blockchain permanently.
- The meter generates a monthly bill based on the electricity consumed during that period.
- In case of any fraud, the meter detects it and reports the error to the blockchain. Here security is imposed, indicating that a fraud has occurred at that instance.

8.2.3.2.4   Renewable Energy on the Blockchain

By putting resources into a smaller scale, the buyers can become prosumers and choose from sustainable power sources like sun or wind. They can pick the measure of the energy required for everyday use and sell the abundance back to the framework for which they will be rewarded. A blockchain-based sustainable power source charging framework will boost residents to move to inexhaustible sources because of the trust in the framework. This framework will work in the following way:

- Consumers invest in renewable sources.
- These renewable sources produce energy.
- Consumers choose how much energy they need on a day-to-day basis.
- The energy used is recorded and stored on a blockchain in the form of transactions.
- The excess energy is calculated.
- This excess energy is sold to other consumers, thereby earning profit for the consumers.
- SCs can record the transactions happening and provide immutability to them.

8.2.3.2.5   Electric Vehicle Charging

With blockchain intervention in electric vehicle charging, one can assure that the bill will be generated on the owner's name regardless of the place where they charge to pace up the process and speed up the payments. At the end of the requirement for an intermediary EV charging framework, adaptation to internal failure, additionally an end of significant value setting and middle-men between charging stations and electric vehicle owners [25]. Nonetheless, additionally, right now, it needs to attain absolute protection and defeat security concerns. Blockchain arrangements mean to give impetuses to privately created EV charging architecture. With the advent of the blockchain in energy, the electric vehicle owners can choose their mode of electrical supply, thereby increasing the transparency associated with it. Besides, what blockchains offer as a bit of leeway to different arrangements, is a remarkable check and correspondence stage that would work in various areas, including cross-fringe voyaging. It is a step-by-step process on how this can be implemented:

Customers with electric vehicles can charge their vehicles at any charging point.

- The bill will be metered accordingly and recorded in the form of a block on the blockchain.
- This transaction will be verified with other miners that will check the energy prices elsewhere.
- If found that the charging station is measuring inappropriately or asking money that is more than other stations, the consumer will be informed.
- It will create transparency in the system, as no private proprietor of the stations can perform fraud of any kind.
- The prices of energy will be stored on a decentralized architecture, and this will increase competition among private companies, thus benefiting the consumers.

8.2.4   SOCIAL BENEFITS

Governments across the world spend a lot of money from their budget and significant resources for the welfare of the poor, sick, underprivileged, elderly, and the marginalized citizens, which helps these citizens survive in this fast-paced world at times. This includes a wide range of facilities provided by the government to keep its citizens happy. Keeping this sector aligned with modern technologies is an integral part of the development of the city, thus making it better and smarter in terms of efficiency, trust, and transparency.

8.2.4.1   CURRENT SITUATION

The money from the government's annual budget has to be disbursed in the form of funds to all the beneficiaries, ensuring that there are no leakages, else it shall be a loss of the valuable taxpayers' money. Monetary consideration of residents is one of the vital empowering agents for the disbursal of sponsorships and funds directly to recipient accounts, in this way diminishing any kind of fraud.

However, an expected 1.7 billion adults overall are barred from the formal financial inclusion framework, which means that they do not have a bank account. In India, roughly 600 million individuals need access to banking administrations, and near 300 million individuals live beneath the official destitution line. Various schemes like the Mahatma Gandhi National Rural Employment Guarantee Act, direct bank transfer (DBT), Kisan Bima Yojana, medical care funds, or inadequate funds help several individuals sustain in the rapidly developing cities.

The fund disbursal process occurs today with the help of the DBT Scheme, which involves the payments to the citizens where the money gets directly transferred to the beneficiary's bank accounts. Be that as it may, a couple of difficulties despite everything exist, including budgetary misfortunes through extortion and mistake, an enormous number of unbanked government assistance petitioners, cost of unapproved exchanges, high exchange costs, and prioritizing the most vulnerable residents.

Blockchain will increase disintermediation in the system leading to a state where it will not require to rely on any third-party agents for processing in the network. Utilizing blockchain innovation for social advantage plans will bolster the administration's more extensive approach targets of maintainability, accordingly decreasing poverty and creating an incentive for cash in public expenditure.

8.2.4.2   BLOCKCHAIN IMPLEMENTATION

A blockchain implementation of the above fund disbursal process in the government can be generalized for other fund disbursal platforms also. This platform will incorporate two aspects into picture *viz.* the top-down approach and a bottom-up approach.

8.2.4.2.1   The Top-Down Approach

For fund transfer, as shown in Figure 8.7(a), the top-down approach is what most governments already follow but without blockchain. This process involves the transfer of funds after a fixed budget for the scheme is approved. The top ministry reviews and passes a budget for a particular scheme. These funds are transferred on to the lower-level ministries for the scheme-specific department. Once this disbursal occurs, the funds are transferred further to the state department's handling that particular scheme disbursal process. The state departments then transfer the funds to the lower authorities who have to handle the disbursal process further. The lower authorities can follow the task of verification of the citizens whether they need the funds or have carried out malpractices to gain an advantage. Once verified, the money would be transferred directly to the bank account of the beneficiary.

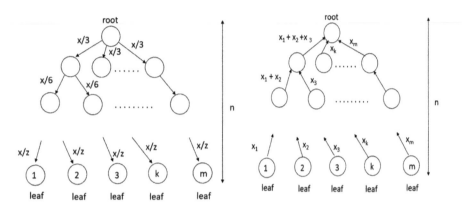

FIGURE 8.7   (a) Top-down implementation; (b) bottom-up implementation.

Followings are the observations:

- At each level, Sum of funds = x (where; x is the total funds transferred by the root node).
- The funds are equally distributed among its children ($x = x/3 + x/3 + x/3$).

8.2.4.2.2   THE BOTTOM-UP APPROACH

The bottom-up approach, as shown in Figure 8.7(b) is the reverse of the process discussed above. It would start with the fund request process. A needy beneficiary would request an amount of money from the government for a particular scheme after being adequately verified. These requests will be combined and will be sent to the district level authorities. Each district-level authority would request the total amount of money needed for his respective district to the state department ministry. This ministry would collect all the requests from all the district officers and sum up the total funds required. This ministry would pass on the requests to the higher authority, and this will go on until the parent node is reached, which will take the necessary action to disburse the asked funds. This process would continue until each requirement is satisfied at the lowest levels, and the money is transferred completely.

This approach has its advantages, one of which being only the amount required by the beneficiaries will be spared by the government. Also, not a single extra rupee needs to be disbursed, unlike in the top-down approach where there can arise situations where an unequal amount of funds remains within the state governments. This process also eliminates fund stagnation at an intermediary level and ensures a smooth flow of funds through each level.

Followings are the observations:

- At each level, Sum of funds = x (where x = total funds transferred by the root node);
- The funds passed on by the children nodes to their parent node sum up to become the amount that needs to be disbursed at every level ($X = x_1 + x_2 + x_3 + ... + x_m$).

Step by step process, as shown in Figure 8.7 is discussed below:

- For a beneficiary claiming funds for a scheme, he will first prove his identity to the upper-level official.
- Once verified, the request will be mined as a block in the blockchain by the miners and will hold the necessary timestamp, hash, nonce, and other necessary encrypted details.
- The officials will have to disburse the funds in the stipulated time to all the lower-level entities.
- This process can get monitored at every step by the beneficiaries as well as the other stakeholders in the chain.
- Once the funds are transferred to the bank account of the beneficiary, the transaction will be completed. After this step block will be mined, stating that the funds are received.
- A verification process will be performed by the miners to check if the beneficiary who received the funds is the right person to receive them.
- At each level, there will be verification checks using the summation method shown above to catch any leakages in the system.

The fund tracking process will thus be transparent as it will be recorded at each step, and the entire process can be tracked and viewed at any time by all the citizens. The process would execute on a public blockchain specific for the country and would involve miners in the form of citizens of the country only. These miners would mine transactions and would be given a reward in the form of a native cryptocurrency specific to the Indian market. This will, in turn, make sure that there is no fraud occurring, and the transactions are mined faster so that the entire process is smoother. Every transaction would be adequately timestamped so that no user or official can claim otherwise. Once the funds are transferred to the respective beneficiary, a verification check will be done to check whether the said citizen has received the entire amount on time. This process will be completed, considering the process in the form of a tree, as shown in Figure 8.8.

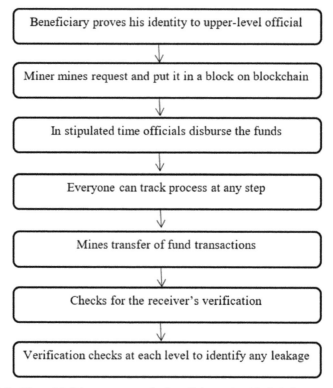

FIGURE 8.8 Financial disbursements to the beneficiary using blockchain.

Every beneficiary fall at the leaf or the end node of the process from where the funds cannot proceed further. At the root, the central government ministry will lie. The process involves 'n' levels of intermediaries to ease up the process for the parent node so that the distribution increases in scale and speed, keeping in mind that the process is entirely transparent. Every official or ministry is treated as a node in this system. The fund verification process occurs at each level of the tree to ensure that each node or intermediary has performed the task with complete honesty, and no fraud has happened in the system. The funds transferred by each node is matched with the total funds that the node received for the disbursal process. In this way, if at all a fraud is detected wherein a malicious official does not pass the entire money to its child nodes, then the fraud can be detected immediately, and the necessary action can be taken. This verification check will take place at each level and will be matched with the leaf nodes, i.e., the beneficiaries, and if the numbers do not add up, then it will be proved that a fraud has occurred in the system and also where it has occurred.

8.3 STATISTICS ABOUT BLOCKCHAIN

Figure 8.9 predicts the blockchain technology paradigm shift over the period majorly from 2019 to 2025. It shows the initial challenges of lack of best practices shifting towards leading technology and finally predicted to become the mainstream technology.

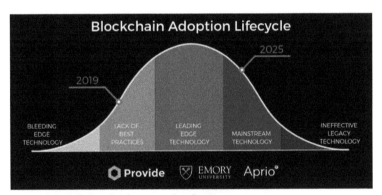

FIGURE 8.9 Blockchain adaption over timeframe.
Source: Reprinted with permission from Ref. [33]; Downloaded from: https://provide. services/state-of-enterprise-blockchain-study-report/.

As shown in Figure 8.10, currently many states of India are working on implementing blockchain technology for different use cases. The three main use cases being land registry, Farm Insurance, and Digital certificates. Hence it can be foreseen that many blockchain applications will be available in the market along with its major benefits in the near future.

As shown in Figure 8.11, the blockchain adaptation is growing in many countries around the world. Here it shows how energy utility applications are experimented in more countries for better functionality and results.

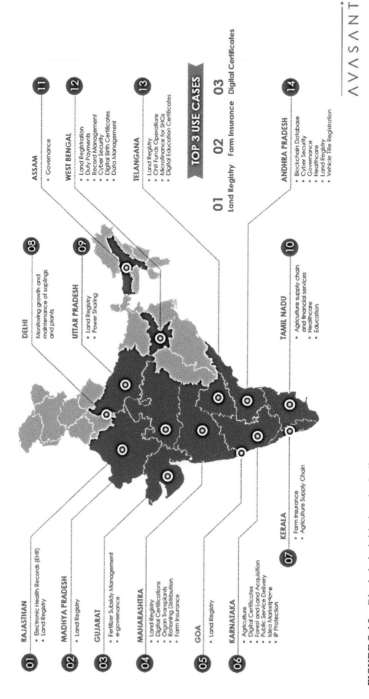

FIGURE 8.10   Blockchain initiatives by Indian states.

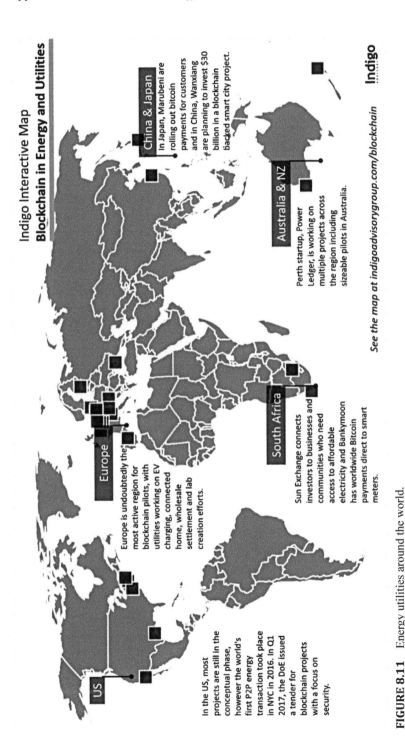

FIGURE 8.11 Energy utilities around the world.

In the various blockchain-based healthcare system are compared with respect to different applicability features. Hence the healthcare applications can gain momentum and popularity due to the numerous characteristics of the blockchain technology.

8.4   ADVANTAGES OF BLOCKCHAIN

There are many advantages of blockchain technology, which makes it useful in the HealthCare industry. Doctors can focus on the medical data provided at one place from the records in the blockchain, rather than tracking the patient's data from various sources. Also, frauds would be eliminated in the industry because of the authenticity of the data provided by the technology. Various incentive programs based on the blockchain network would increase the market share of the industry, increasing innovation, transparency, and trust, and decreasing conflicts. It would open a door for a decentralized way to store important health records, documents, and data.

The intervention of blockchain in the government sector would lead to more transparent government operations and an increase in the trust of the citizens in the government. The governments that use blockchains in their schemes, policies, and operations can claim about the transparency and immutability they provide in their country. Citizen identity management becomes easier and much more efficient using blockchain. It paves a path for removing fraudulent identities in the system, removing redundancy, removing duplicate IDs, decrease in cost and efforts required for verification over and over again by providing a mode to process data and store it permanently on a distributed ledger safely and securely.

Blockchain enables the complete exploration of the potential of the energy sector by opening a way for wholesale trading markets. The different modes to distribute renewable energy, improving transparency in the system by providing smart meters which record activity according to the usage of energy, smart billing and electric vehicle charging stations over a decentralized architecture. It disintermediates the energy sector, thus making it robust, tamper-proof, and conflict-free for the upcoming smart world by reducing the hefty costs involved in the sector, currently.

In the sector of social benefits, blockchain proves to be a viable option to disburse funds to the needy beneficiaries, as it stores the records immutably, purges the fraudulent and corrupt officers, and provides a way to transact completely cost-free. The hierarchical systems, as shown in Figure 8.8, represent the fund disbursal architecture as a tree and offer a simple yet efficient solution to calculate the required entities and resolve any arising conflicts. Use of Blockchain can guarantee immutability, and this is how a system can be developed with no fraud or malpractices by any person in the network.

8.5   CRITICAL ACCLAMATIONS OF BLOCKCHAIN

Though blockchain has its plus points in each sector, surveys still show the lack of acceptance of blockchain technology as a solution to existing industrial problems in the country or even worldwide. According to the 'PwC Global Blockchain Survey' in 2018 [14], when a question came up as which of the following will be the most significant barriers to blockchain technology adoption in the industry in next 3–5 years the stats observed were as follows:

- 31% of companies told PwC that the cost was the primary reason their organization had not progressed further with blockchain;
- 47% said lack of trust amongst potential users;
- 39% showed regulatory uncertainty;
- 37% gave the reasonability to bring the network together;
- 36% said blockchain interoperability as a reason;
- 33% admitted to the option' inability to scale.'

The industry is said to be going through a cultural paradigm shift when using blockchain technology, and the promoters will encounter a hard time in making this technology sink in the veins of the economic market for revenue generation. The stakeholders of the Indian environment are presently at different phases of arranging, structuring, executing, and receiving a blockchain-based solution to relieve a portion of their key difficulties. Still, there are some companies and projects which have taken the initiative to try and implement blockchain into their system. The acceptance of blockchain is gradual as of now, but it might evolve as a significant technology in the near future.

8.6   CONCLUSION

The apparent effect of blockchain-based solutions and their integration in the system in various sectors can be listed as follows:

- Reduced transaction cost;
- Immutability and easy accessibility of identity;
- Security and privacy of the data;
- Trust-building through the use of simpler, quicker, and effective means of interaction between the government and citizens.

The significance of blockchain is verifiable as a rearrangement system for new inclusive chances. As seen from the above-mentioned statistics, blockchain is beneficial and helps boost up the modern world technologies in various sectors. A proper and in-depth study is important to decide and select segments where blockchain technology can make a positive change in the existing system. Blockchain is more secure than the present world technologies, and it is not confined to just one particular use-case. This chapter has tried to discuss some of the positive real-life use-cases of Blockchain application. The results and discussion try to promote it becoming a superior technology for the future of society.

Though there are multiple benefits associated with blockchain technology, this technology is not widely accepted by the mainstream industries. One important reason behind this is the absence of standardization. For data to be stored in the technology, the data needs to be standardized. For this purpose, universal standards need to be prepared and accepted by organizations. Another reason behind non-acceptance is that there is no central control over the technology. This will require a network of devices for verification of data before inputting in the blockchain. This is very expensive. Also, since blockchain is a much newer technology, the industry is quite hesitant towards this technology.

All in all, it could be concluded that blockchain is beneficial for the industry. Still, for this technology to get accepted in the mainstream industries, it is highly essential to overcome the initial barriers. Any system can be developed consisting of trust, transparency, and security for the benefit of all using blockchain, unlike the traditional solutions. Such a system will help us achieve a new and efficient approach to evolve modern technologies over time with the power and potential of blockchain.

KEYWORDS

- **blockchain**
- **digitization**
- **direct bank transfer**
- **personally identifiable information**
- **transaction cost**
- **urbanization**

REFERENCES

1. Mohite, A., & Acharya, A., (2018). Blockchain for government fund tracking using Hyperledger. In: *2018 International Conference on Computational Techniques, Electronics and Mechanical Systems (CTEMS)* (pp. 231–234). Belgaum, India.
2. Joshi, P., Kumar, S., Kumar, D., & Singh, A. K., (2019). A blockchain-based framework for fraud detection. In: *2019 Conference on Next Generation Computing Applications (NextComp)* (pp. 1–5). Mauritius.
3. Qiu, J., Liang, X., Shetty, S., & Bowden, D., (2018). Towards secure and smart healthcare in smart cities using blockchain. In: 2018 *IEEE International Smart Cities Conference (ISC2)* (pp. 1–4). Kansas City, MO, USA.
4. Rivera, R., Robledo, J. G., Larios, V. M., & Avalos, J. M., (2017). How digital identity on a blockchain can contribute in a smart city environment. In: *2017 International Smart Cities Conference (ISC2)* (pp. 1–4). Wuxi.
5. Li, S., (2018). Application of blockchain technology in smart city infrastructure. In: *2018 IEEE International Conference on Smart Internet of Things (SmartIoT)* (pp. 276–2766). Xi'an.
6. Vipul, H. N., Ajinkya, N., & Rahul, W., (2018). Overview of blockchain technology in government/public sectors. *International Research Journal of Engineering and Technology, 5*(6).
7. Felix, H., Axel, V. P., Thomas, H., Erwin, S., Lena, L., & Maximilian, C., (2016). *Blockchain – an Opportunity for Energy Producers and Consumers?* PwC global power & utilities.
8. Rizal, B. F., Jolien, U., & Marijn, J. (2018). Challenges of blockchain technology adoption for e-government: A systematic literature review. In: *Proceedings of the 19th Annual International Conference on Digital Government Research: Governance in the Data Age (dg.o '18)* (pp. 1–9). Association for computing machinery, New York, NY, USA, Article 76.
9. Ølnes, S., Ubacht, J., & Janssen, M., (2017). Blockchain in Government: Benefits and implications of distributed ledger technology for information sharing. *Government Information Quarterly, 34*. 10.1016/j.giq.2017.09.007.

10. FICCI-PwC Report on *"Blockchain: The Next Innovation to Make Our Cities Smarter,"* (2018). Retrieved from: http://ficci.in/spdocument/22934/Blockchain.pdf (accessed on 28th July 2021).
11. Merlinda, A., Valentin, R., David, F., Simone, A., Dale, G., David, J., Peter, M., & Andrew, P., (2019). Blockchain technology in the energy sector: A systematic review of challenges and opportunities. *Renewable and Sustainable Energy Reviews, 100,* 143–174.
12. PWC Report on *"Reimagining Health Information Exchange in India Using Blockchain"* (2019). Retrieved from: https://www.pwc.in/assets/pdfs/consulting/technology/it-function-transformation/insights/reimagining-health-information-exchange-in-india-using-blockchain.pdf (accessed on 28th July 2021).
13. Bell, L., Buchanan, W. J., Cameron, J., & Lo, O., (2018). *Applications of Blockchain Within Healthcare* (Vol. 1, p. 8). Blockchain Healthc today. https://doi.org/10.30953/bhty.v1.8.
14. PWC Report on *"A Prescription for Blockchain and Healthcare-Reinvent or be Reinvented."* Retrieved from: https://www.pwc.com/us/en/health-industries/health-research-institute/assets/pdf/pwc-hri-a-prescription-for-blockchain-in-healthcare_27sept2018.pdf (accessed on 28th July 2021).
15. Maria, R., (2018). *Blockchain and Healthcare: Use Cases Today and Opportunities for the Future.* Retrieved from: https://mlsdev.com/blog/blockchain-and-healthcare-use-cases-today-and-in-the-future (accessed on 28th July 2021).
16. Randy, B., CIO Network, (2018). *Will Blockchain Transform Healthcare?* Retrieved from: https://www.forbes.com/sites/ciocentral/2018/08/05/will-blockchain-transform-healthcare/#732b5ed95 53d (accessed on 28th July 2021).
17. Hasib, A., (2019). *Blockchain for Digital Identity: The Decentralized and Self Sovereign Identity (SSI).* Retrieved from: https://101blockchains.com/digital-identity/#5 (accessed on 28th July 2021).
18. Linn, L. A., & Koo, M. B., (2016). *Blockchain for Health Data and its Potential Use in Health IT and Healthcare-Related Research.* ONC/NIST Use of Blockchain for Healthcare and Research Workshop. Gaithersburg, Maryland, United States: ONC/NIST.
19. Akash, T. (AkashTakyar is the founder and CEO at LeewayHertz). *Blockchain Identity Management: Enabling Control Over Identity.* https://www.leewayhertz.com/blockchain-identity-management (accessed on 28th July 2021).
20. *Department of Economic and Social Affairs, News.* Retrieved from: https://www.un.org/development/desa/en/news/population/2018-revision-of-world-urbanization-prospects.html (accessed on 28th July 2021).
21. *Cisco Reports, News and Updates.* Retrieved from: https://www.etorox.com/news/10-of-global-gdp-will-be-on-blockchains-by-2027-says-cisco-report/ (accessed on 28th July 2021).
22. Murugan, A., Chechare, T., Muruganantham, B., & Ganesh, K. S., (2020). Healthcare information exchange using blockchain technology. *International Journal of Electrical & Computer Engineering (2088–8708), 10*(No. 1, Part 1), 6, 421–426.
23. Li, Z., Kang, J., Yu, R., Ye, D., Deng, Q., & Zhang, Y., (2018). Consortium blockchain for secure energy trading in industrial Internet of Things. In: *IEEE Transactions on Industrial Informatics* (Vol. 14, No. 8, pp. 3690–3700). doi: 10.1109/TII.2017.2786307.

24. PwC Global Power and Utilities. (https://www.pwc.ru/en/publications/blockchain. html). *Blockchain: An Opportunity for Energy Producers and Consumers?* Retrieved from: https://www.pwc.com/gx/en/industries/assets/pwc-blockchain-opportunity-for-energy-producers-and-consumers.pdf (accessed on 28th July 2021).

25. Su, Z., Wang, Y., Xu, Q., Fei, M., Tian, Y., & Zhang, N., (2019). A secure charging scheme for electric vehicles with smart communities in energy blockchain. In: *IEEE Internet of Things Journal* (Vol. 6, No. 3, pp. 4601–4613). doi: 10.1109/JIOT.2018.2869297.

26. Kumaresan, M., Harshal, P., & Prasenjit, B., (2018). A comprehensive integration of national identity with blockchain technology. *International Conference on Communication, Information & Computing Technology (ICCICT).*

27. Bayu, A. T., Bruno, J. K., Youngho, P., & Kyung-Hyune, R., (2017). A critical review of blockchain and its current applications. *International Conference on Electrical Engineering and Computer Science(ICECOS).*

28. Jignasha, D., Meenal, C., Himani, G., & Sanjana, T., (2020). Verification of identity and educational certificates of students using biometric and blockchain. *International Conference on Advances in Science and Technology, ICAST.*

29. Karnika, P., Vennis, S., Devi, D., & Purnima, K., (2020). Trade finance using blockchain. *International Conference on Advances in Science and Technology, ICAST.*

30. Bharambe, A., Motwani, R., Rathi, A., & Kothari, P., (2020). Smart contract for real estate using blockchain. *International Conference on Advances in Science and Technology, ICAST.*

31. Ashutosh, R., & Zaheed, S., (2020). A survey on blockchain technology with use-cases in governance. *International Conference on Advances in Science and Technology, ICAST.*

32. Retrieved from: https://info.avasant.com/blockchain-adoption-indian-states# (accessed on 28th July 2021).

33. Retrieved from: https://provide.services/state-of-enterprise-blockchain-study-report/ (accessed on 28th July 2021).

34. Retrieved from: https://energycentral.com/c/ec/energy-and-blockchain-go-global-utilities-startups-and-use%C2%A0cases (accessed on 28th July 2021).

35. Vidhya, R., Tanesh, K., Braeken, A., Madhusanka, L., & Mika, Y., (2018). Secure and efficient data accessibility in blockchain based healthcare systems. *IEEE Global Communications Conference (GLOBECOM).*

36. Borse, Y., Chawathe, A., Patole, D., & Ahirao, P., (2019). Anonymity: A secure identity management using smart contracts. *Proceedings of International Conference on Sustainable Computing in Science, Technology and Management (SUSCOM).* Amity University Rajasthan, Jaipur – India. Available at SSRN: https://ssrn.com/abstract=3352370 or http://dx.doi.org/10.2139/ssrn.3352370 (accessed on 28th July 2021).

37. Patole, D., Borse, Y., Jain, J., & Maher, S., (2020). Personal identity on blockchain. In: Sharma, H., Govindan, K., Poonia, R., Kumar, S., & El-Medany, W., (eds.), *Advances in Computing and Intelligent Systems: Algorithms for Intelligent Systems.* Springer, Singapore.

38. Zhang, P., Walker, M. A., White, J., Schmidt, D. C., & Lenz, G., (2017). Metrics for assessing blockchain-based healthcare decentralized apps. In: *2017 IEEE 19th International Conference on e-Health Networking, Applications and Services (Healthcom)* (pp. 1–4).

39. Azaria, A., Ekblaw, A., Vieira, T., & Lippman, A., (2016). Medrec: Using blockchain for medical data access and permission management. In: *2016 2nd International Conference on Open and Big Data (OBD)* (pp. 25–30).
40. Xia, Q., Sifah, E. B., Asamoah, K. O., Gao, J., Du, X., & Guizani, M., (2017). Medshare: Trust-less medical data sharing among cloud service providers via blockchain. *IEEE Access, 5*, 14 757–14 767.
41. Xia, Q., Sifah, E. B., Smahi, A., Amofa, S., & Zhang, X., (2017). BBDS: Blockchain-based data sharing for electronic medical records in cloud environments. *Information, 8*(2), 44.
42. Yang, H., & Yang, B., (2017). *A Blockchain-Based Approach to the Secure Sharing of Healthcare Data.* Retrieved from: https://www.semanticscholar.org/paper/A-Blockchain-based-Approach-to-the-Secure-Sharing-Yang-Yang/af404093a0fe066a43e5021b7d0f2a1c7b9105e6 (accessed on 28th July 2021).

PART III

Blockchain Technologies: Opportunities for Solving Real-World Problems in Healthcare and Biomedical Science

CHAPTER 9

Blockchain Technology for Biomedical Engineering Applications

DINESH BHATIA,[1] ANIMESH MISHRA,[2] AND ANOOP KUMAR PRASAD[3]

[1]*Department of Biomedical Engineering, North Eastern Hill University, Shillong – 793022, Meghalaya, India, E-mail: bhatiadinesh@rediffmail.com*

[2]*Department of Cardiology, North Eastern Indira Gandhi Regional Institute of Health and Medical Sciences, Shillong, Meghalaya, India, E-mail: animesh.shillong@gmail.com*

[3]*Royal School of Engineering and Technology, Assam Science and Technology University, Guwahati, Assam, India, E-mail: anoopkprasad@rgi.edu.in*

ABSTRACT

The healthcare industry is one of the largest and fastest-growing industries involving more than 10% of the gross domestic product (GDP) of several countries globally. The industry is under strict regulatory norms, involves a complex system of different interconnected entities leading to fragmented patient data and enhanced costs due to system inefficiencies leading to lack of complete transparency in data transactions, data traceability issues, and security concerns for the patient. To overcome this problem, blockchain technology has gained considerable attention in several domains, including the healthcare industry, to ensure secure and reliable management of real-time clinical healthcare patient data. This would help in reducing healthcare costs, accurately report all clinical trials to overcome substandard and fake drug usage, which are still major areas of concern in the healthcare industry. The technology could be employed to support patient data management, hospital supply chain management (SCM), pharmaceutical research, online patient monitoring through hospital telemedicine set-up, and hospital billing system. The blockchain follows peer to peer integrated network where any transaction request made is broadcasted to all nodes and needs to be approved by all nodes to be added to the blockchain network. The blockchain system comprises of a decentralized database with open access, transparency, and autonomy to all users who are logged into the network to access information and regularly update after approval when required. Once the records are updated in the blockchain network, it cannot be deleted or removed easily without authentication and approval from more than half (51%) the number of users on the network.

Soon, the blockchain technology would immensely benefit and revolutionize the healthcare sector by ensuring data transparency, reducing costs of service delivery, central server for management of all patient data, and hospital services. The clinicians can monitor their patient's condition in real-time without worrying regarding the level of honest information being provided by their patients. Similarly, the patients need not worry about getting a second referral for their medical condition due to the transparency of data and services being provided by the healthcare service provider. Further, they can willingly share their data to enable them to learn regarding similar cases being treated or reported in other countries, which would help them in improving their present medical condition and nature of services being provided by their clinician. Overall, the practical utilization of blockchain technology can benefit patients, clinicians, healthcare service providers, researchers, and biomedical specialists to improve service efficiency, quality of healthcare services, accurate diagnosis, and treatment of diseases, and share patient data with enhanced security and privacy. It would prevent loss or modification of patient data being decentralized and transparent in nature. The major concerns for acceptance of this technology are the misuse of data by sharing it with third parties for profit or indirect identification of users through pseudonymous identifiers.

9.1 INTRODUCTION

The healthcare industry is one of the largest and fastest-growing industries worldwide. The healthcare industry which is under strict regulatory norms involves the integration of multiple complex entities, scattered patient data with huge costs involved in patient care. This leads to lack of transparency in data transactions, complex security issues and difficulty in easy retrieval of patient data [1]. Tandon A. wrote to overcome this problem, blockchain technology has gained considerable attention in several domains, including the healthcare industry, to ensure secure and reliable management of real-time clinical healthcare patient data [1, 11]. This would help in reducing healthcare costs, accurately report all clinical trials, and overcome substandard and fake drug usage, which are still major areas of concern in the healthcare industry. Qamar et al. referred that the technology has widespread scope for use in pharmaceutical drugs discovery, supply chain management in hospitals, hospital billing and online monitoring of patients over a distance. The blockchain follows peer to peer integrated network where any transaction request made is broadcasted to all nodes and needs to be approved by all nodes to be added to the blockchain network [2].

Das et al. affirmed that the blockchain concept was first introduced as an integral component of the cryptocurrency bitcoin (BTC). The chapter that led to the creation of BTCs was published in 2008, and the first genesis block of the blockchain was mined on 3rd January 2009 [31]. The blockchain system comprises the decentralized database with open access, transparency, and autonomy to all users who are logged into the network to access information and regularly update after approval when required. Once the records are updated in the blockchain network, it cannot be deleted or removed easily without authentication and approval from more than half (51%) the number of users on the network. Hasselgren et al. inducted that Blockchain would help to overcome several problems associated with the complex integrated healthcare network, including the lack of traceability during different transactions, non-reporting of clinical trials, and issues related to sub-standard or fake medicines being supplied to patients [18]. This would avoid the lack of confidentiality in handling the patient data and non-sharing with for-profit organizations for commercial benefits or misuse of the information in any form. It would also remove the presence of middlemen in the healthcare system by cutting down on costs, thereby enhancing trust and transparency in the system. Drosatos et al. referred that it would also enable healthcare systems to remove the difficulty of depending on a single vendor or "vendor lock-in" problem by not being dependent on single cloud technology providers and easily moving to different vendors as per requirement without incurring substantial costs or legal constraints. Presently, most of the blockchain systems are at the proof-of-concept stage, however, their potential is being acknowledged and realized with several projects involving blockchain technology increasing rapidly [19, 20].

With a decentralized database, blockchain can help solve several underlying problems within the healthcare system, giving more autonomy and control to patients over their data with selected access rights to researchers relating to the medical records. It is a more patient-centric healthcare model where trust is paramount, and the records cannot be altered or deleted as per convenience. Further, integrated solutions can be built to avoid medical fraud and drug counterfeiting. Priyadharsan et al. referred that using IoT and machine learning (ML), the technology can help in overcoming the deletion of clinical trials record or hiding sources of funding for carrying out medical research. It can also help in saving huge costs in the pharmaceutical industry by having a defined network of supply chains [14]. The contracts incorporated during implementing the blockchain system can be legally binding and enforceable as per law of the land. Thus, they could help in automating the healthcare system such as billing, insurance, diet planning, and pharmacy management, thereby significantly reducing costs involved. Shakeel et al. and Padmavathi et al. took up that blockchain technology research in healthcare focused on integrity, integration, and access control of patient-related health records. In future work is ongoing on other aspects such as clinical trials, medical research, SCM, and medical insurance [15, 19].

9.2   UNDERSTANDING A BLOCKCHAIN SYSTEM

The blockchain technology has been growing rapidly promising applications in several technological fields. It may be defined as a decentralized system where the data entries are immutable and cannot be altered at convenience. The time-stamped chain of blocks is connected in a blockchain framework using cryptographic hashes affecting all industrial domains exponentially, even influencing the healthcare sectors through its processes, services, management, education, etc. The blockchain has largely affected the functioning of electronic and personal health records (EHR and PHR) in the healthcare set-up. Cocosila et al. and Lafourcade et al. cited that the blockchain allows better data integrity in all aspects, access control to medical records, interoperability, and smooth transactions. The term blockchain originated from "bitcoins" or cryptocurrency in the year 2008 [17, 44]. This slowly expanded and got acceptance with several million users by 2019, wherein it was felt that its potential could be employed in different applications including healthcare. Cachin talked about blockchain, cryptography, and consensus. As per the IBM report in 2019, several healthcare leaders predicted that blockchain technology will have a major impact in the healthcare sector due to its decentralized framework for sharing medical records safely, better management of clinical trials, and regulatory compliance enforceability. Further, it is expected that the global market for blockchain technology in the healthcare sector may soon cross $500 million by 2022 [11, 33, 38]. Still, technology has miles to go and adapt to the existing systems and lend useful value to improve the further efficiency of the present healthcare industry.

The blockchain allows decentralization of records with no central authority controlling the addition of blockchain content. The immutable entries are added to the blockchain network based on consensus through peer-to-peer (P2P) network. Balandina et al. stated that the blocks are linked to each other through cryptographic protocols which are self-executable smart contracts (SCs). Persistence is also a key feature of blockchain as accepted entries cannot be deleted due to a distributed ledger stored across multiple nodes [7]. It also allows anonymity to be maintained in the blockchain with easy traceability as the blocks are interconnected with each other through the hash. The transactions in each block are arranged in the form of a Merkel tree with each leaf value verified to its established root for traceability. Although the blockchain has been most used in the Finance sector as it allows security and privacy of data, it is being applied in several computational fields where the transaction is voluminous. Kumar et al. and Liang et al. told that it is being used in the Internet of Things (IoT) for recording, analyzing, and transfer of real-time data from nodes or objects, humans, and sensors to automate different assignments or tasks [3, 5].

The different types of blockchain are a public, consortium, and private. The public blockchain is also known as permissionless as anyone can join it to view and contribute to the blockchain and allow changes to its core software. The two largest cryptocurrencies that widely employ public blockchain are BTC and Ethereum (ETH). The efficiency is low, but data immutability is high. The consortium of public permission allows a restricted number of identified group(s) entities to access, view, and participate in the consensus protocol. In consortium blockchain, the efficiency is high, but data could easily tamper. Priyadharsan et al. and Drostos et al. stated that in the private blockchain, only selected nodes can participate in the distributed network controlled by a single organization having centralized authority overall transactions in the blockchain. In this blockchain, although the efficiency is high, the data could be tampered with [14, 20].

All users connected in the block can view ongoing transactions in the network. The record of previous blocks and transactions is available with each block, thereby making it a secure and immutable ledger record. The chain increases as new blocks are added and arranged in a P2P network comprising of several nodes and each node maintaining a copy of the same. The nodes may comprise of "users" or "miners" who check and validate every transaction occurring on the network is appropriate or not. With the "consensus" of users and miners, the data is added to each block. Three majorly employed consensus protocols are Proof-of-Work (PoW), Proof of Stake (PoS), and practical Byzantine fault tolerance (PBFT) protocols. The PoW is the consensus protocol strongly associated and integrated with BTC. The miners are involved in solving some complex computational problems by finding lesser value than the predetermined one in the proposed block. The miner who finds the solution first is rewarded and validates the transaction. However, Kuo et al. claimed that it is limited due to its high computational energy requirements when applied on a large blockchain which may be equivalent to the electricity required for a small country [22]. The stake of each node in the blockchain determines the approval node selection process in the PoS. This may give an unfair advantage to richer nodes and influence the decision-making. Hence, the randomization process is employed to select the approving node. Tandon A. and Padmavathi et al. referred that ETH, the second-largest cryptocurrency is in the process to shift from PoW to PoS. The PBFT is based on the Byzantine agreement protocol limiting its usage in the public domain as all nodes must be known in the network. It comprises three phases, namely as pre-prepared, prepared, and commit [11, 19]. Each node requires a majority two-thirds vote share approval from all nodes in the network to move to the next phase. The PBFT is currently employed in the Hyperledger Fabric. Saraf et al. stated that till the year 2018, the number of frameworks rolled out by Hyperledger Fabric is equal to five, namely [36]:

- Burrow;
- Fabric;
- Indy;
- Iroha; and
- Sawtooth.

The Blockchain 3.0 version is based on the direct acyclic graph (ADG) and has several advantages over previously employed versions 1.0 and 2.0. Blockchain 1.0 also known as BTC came into existence in 2008 that uses proof of work (PoW) whereas Blockchain 2.0 which came into existence in 2014 with ETH and SCs employs Proof of Stake concept. Both of these are vertical schemes that have several limitations such as privacy, self-sustainability, scalability, interoperability, and governance. Hasselgren et al. stated that Blockchain 3.0 is a horizontal scheme that addresses several limitations of the previous two versions [18].

9.3 THE FUNCTIONING OF THE BLOCKCHAIN

A blockchain comprises of a sequential chain of blocks which contains all transaction records. The first block is the genesis block, with successive blocks connected through hash values. ETH is one such blockchain system that has become popular since 2013. The block comprises the block header and a set of validation rules that needs to be followed. Hasselgren et al. talked about The Merkel Tree Root Hash comprises of the relevant transactions and corresponding hash values in the block. The Timestamp records the actual time of transactions as Universal time during block creation. The Nonce a 4-byte field has 0 value initially which gets incremented after the calculation of every hash function. The parent block hash comprises of 256-bit values which link to previous blocks in the blockchain [18]. After each transaction in the block, the transaction counter is incremented. The size of the block and transactions decides the maximum number of transactions a block can hold. Asymmetric cryptography techniques are utilized to authorize and validate the transactions using a digital signature technique. Each user in the network uses two keys, namely private and public keys. The private key is confidential, not shared, and employed for digital signatures. The signed transactions are then shared over the entire network [11, 19]. Blockchains are a new technology built on a relatively simple premise. The premise is that a continuously generated stream of information can be commonly shared amongst all members of a group.

Hashing is a key cryptographic tool which uses the concepts of Mathematics for code making and code breaking aspect. Using Hashing, the message of any length can be converted to an output of a fixed length of alphanumeric characters. Therefore, when transactions happen, large pieces of information are converted into manageable small outputs which represent the original information. When a large number of transactions happen, then need of verification and compilation may arise. It is collectively called 'validation' and requires the knowledge of mining as the validation of a block leads to the creation of new units of cryptos in the environment. Once the problem is solved by any miner and a new block has been created, everyone on that network must agree that the problem is solved and accept it as the next block. The validation of the work is performed by solving the problem by oneself and matching with the work done by the other successful miner(s). This is a tedious task where PoW comes into play. The miner(s) who solves the problem also broadcasts the PoW to other members, so that they could accept it and move on to solve the next problem and create the block. Niya et al. stated that the Proof of Stake (PoS) concept involves the validation of block that depends on the number of cryptocurrency tokens that a person already owns [58]. It is said that a number of cryptos is directly proportional to the mining power by miner(s). Instead of utilizing energy to answer PoW puzzles, a PoS miner is limited to mining a percentage of transactions that reflects the ownership stake.

9.4 BLOCKCHAIN FOR HEALTHCARE AND BIOMEDICAL APPLICATIONS

Healthcare comprises sensitive patient data, and with digitization, it is easy to transfer and move manual records in the digitized form which helps in easy storage and retrieval. Talpur cited that The Electronic Patient Record (EHR) is a digitized that allows easy sharing and analysis of patient data by healthcare providers [4]. Catarinucci et al. and Tandon A. brought into light that the digitized record is always under threat of cybercrime to manipulate the confidential patient information. The amount of security, privacy, and confidentiality of the medical data and its interoperability between different healthcare agencies are crucial, and blockchain technology can help to provide robust structure against attacks and system failures. The blockchain also prevents the manipulation and tampering of healthcare data. Besides, it reduces costs, enhances efficiency, and ensures data safety with regulatory compliance with the creation of a secure environment and immutable records [6, 11].

Blockchain technology has various applications in the healthcare field. It can be employed for patient-based requirements such as seeking consent, allows privacy, and ensures secrecy with continuous monitoring and interaction with the system. The issues that may be associated with its implementation include scalability, interoperability, and patient management. Blockchain technology can also be employed for effective management of the electronic health record (HER) by employing deep learning (DL) algorithms. Balandina et al. showed their concern over the advantages of Blockchain as it reduces paperwork, storage costs, improves efficiency, automates processing, and lower chances of fraud or mismanagement of the secure patient data [7]. The blockchain technology allows the removal of intermediaries' giving more control over the information to the patient, which is a huge advantage. The technology also aids in biomedical research with the help of smart digital contracts to regulate the functioning of the biomedical databases. It ensures data integrity and non-repudiation. Gong et al. and Amin et al. unveiled different pathways in the healthcare system for privacy protection as it allows for remote patient monitoring with the secure management of medical sensors. The monitoring of patients may be done in real-time with continuous updates regarding the health condition to the patient and the doctors [8, 10]. It also provides medical interventions with automated notifications to the concerned parties. The only disadvantage being data transfer through open channels from smart devices to blockchain nodes and due to certain delays in block verification, it cannot be used in emergency response situations as the response time would increase comprising patient health. The blockchain technology would also help in overcoming fake drug prescription and labeling due to prudent check mechanism. The use of counterfeit drugs leads to adverse drug reactions with over 10 lakh deaths annually as per the World Health Organization (WHO) data. Padmavathi et al. introduced to avoid the inclusion of such fake drugs in the supply chain transparency must exist to avoid tampering, modifications, or stealing of drugs. The pharmacists also need to ensure that drugs of correct quantity and strength are available. The blockchain also helps in the medical insurance storage system that allows transactions to be verified before being added to the Blockchain network, thereby saving time, costs, and the transaction verification process for the insurance company [18, 19].

Before blockchain can be employed in the healthcare sector, the need to understand the application areas where it has been employed for present applications in biomedical domains such as clinical medical records, clinical trials, clinical education, and evidence-based databases, etc. Further, Alhussein et al. and Cocosila et al. clarified that the types of data such as personal clinical records, medical health records, willingness forms, drug supply chain, patient location, health insurance transactions have been addressed with blockchain technology that need to be elaborated with the level of present expertise available such as architectural design, system components availability of technical tools, etc. [12, 17] The need for employing blockchain technology in the biomedical healthcare along with its pros and cons need to be understood. It needs to understand the level of usage of blockchain technology in access control, data integrity, data auditing, and provenance needs to be studied. Drosatos G. referred to the type of blockchain framework such as ETH, Hyperledger Fabric, BTC, etc., that would be employed to address the problems encountered in the biomedical domain needs to be understood [11, 20].

Gong et al. cited that the healthcare data can be broadly classified into two main categories, mainly as medical records focused on medical care produced from healthcare departments or hospitals [8]. The other being personal health records comprising information entered usually by patients willingly. Blockchain technology has shown immense potential in providing privacy, security, and integrity of healthcare records. Alhussein et al. referred that it is being nowadays, used in creating patient record ledgers residing at different locations, integration of medical records with healthcare enterprises, and sharing of these resources across different regions and users [12]. The technology also allows patients to have control over their information and can be employed to seek a second opinion or referral by sharing if required. Blockchain can lead to integration and unification of personal health records with medical records and allows easy management of one unified healthcare record. The insurance-based health record is encrypted and stored without tampering in the blockchain network for future use. Another area of application for blockchain is to safeguard the transactions of researchers with reference biomedical databases such as clinical trials, pharmaceuticals, etc., that are updated regularly. Hasselgren et al. stated that blockchain can also be employed to track the drug supply chain from the manufacturer to the consumer to avoid counterfeit drugs entering the supply chain. Most of the blockchain technologies for medical applications are at the initial stage of maturity or being tested for implementation shortly. However, any technology depicting real-time solutions is not presently available, and more time is required for the blockchain technology to mature in the future (Figure 9.1) [18, 20].

The blockchain technology allows independence to patients to assign rules permitting specific use of their recorded data for research purposes. It also gives freedom to patients to connect to hospitals directly and procure their medical data automatically. Tandon et al. told that with the help of ledger technology in blockchain, it is possible to securely transfer patient medical records without consideration of any boundaries, SCM of medicines administered to patients, improved coordination among healthcare organizations, fewer transaction costs and risks involved and help in genetic research or genomics. This is can help in streamlining healthcare processes and prevent costly and fatal medical errors [11]. Genomics has been considered to improve human health, and several companies are exploring the possibility of employing this to understand human health for scientific advancement in patient treatment and quality healthcare. Bhatia et al.; Engelhardt et al.; and Randall et al. concluded that the Blockchain technology is found to be perfectly suitable to this area by helping patients to share their encrypted genetic information which allows for wider database creation thereby helping the scientific community in their research endeavors [21, 23, 26].

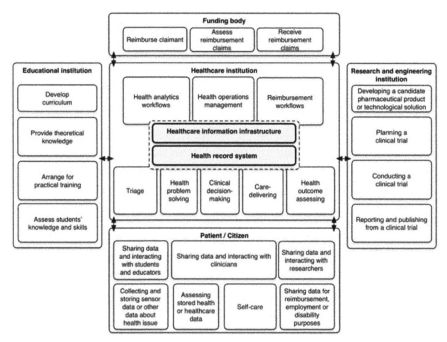

FIGURE 9.1 Blockchain and healthcare sector.

Source: Reprinted with permission from the authors of Ref. [18]. © 2019 The Authors.

Blockchain can help another area such as Telemedicine to thrive by developing mutual trust between patients and healthcare professionals. It helps in data integrity, authentication, and identification of parties ensures transparency by providing incentives in the form of crypto tokens for players to act fairly. Shakeel et al. cited that with the use of regulated artificial intelligence (AI) tools, it creates a seamless platform for global healthcare by effective use of healthcare resources and needs [15]. Mendes et al.; Bahga et al.; and Reyna et al. showed that the major benefit would be providing remote qualitative diagnostic services, especially when a present situation such as COVID-19 lockdowns are prominent across the globe with a large number of people deprived of quality healthcare facilities. A large number of start-ups are coming up in this field, and several are found in the dermatology field due to its close association with the beauty industry [16, 28, 30].

9.5 IoT FOR HEALTHCARE

Qamar et al. claimed that the healthcare industry is in a state of great despair with rising healthcare costs, an aging global population, and an increase in new and chronic diseases [3]. This could lead to a state of healthcare services being out of reach of the common man making them unproductive and prone to chronic diseases. As mentioned by Catarinucci et al. that the IoTs has tremendously transformed and revolutionized the way present healthcare services are being provided to patients [6]. The technology allows remote monitoring in the healthcare sector, helping in continuous and safe monitoring of the patient's health condition, empower physicians to deliver quality healthcare at reasonable costs. Fonseca et al. acknowledged that the level of patient-doctor interaction has increased tremendously, leading to widespread satisfaction with improved healthcare efficiency among patients due to ease of interaction with the healthcare providers [13]. This decreased the hospital duration stay for the patients and their families, showing a tremendous impact on diminishing healthcare costs and improving treatment outcomes. Several applications in the healthcare sector employing IoT are presently available to benefit the healthcare providers, hospitals, insurance companies, and patients at large. Mendes et al. disclosed that IoT has largely benefitted critically ill patients and the elderly population living alone who require continuous monitoring due to their diseased condition by tracking their health condition on a regular basis and allowing any disturbances or change in routine daily living activities to trigger an alarm to the family member or concerned healthcare providers for an immediate check on their condition and provide instant care and treatment to avoid any fatalities or long term damage [15, 16]. However, Priyadharsan et al. projected that several potential challenges exist while designing any IoT based healthcare system that needs to be addressed such as security and privacy, user identification and authentication, and regular communication and exchange of healthcare data before the adoption of the technology on a mass scale [14].

Cocosila et al. stated that IoT is being employed in several organizations to enhance operational efficiency, deliver better services by acknowledging customers' concerns, aid in decision-making, thereby enhancing the business outcome [17]. It is believed that the forecasted global revenue growth of IoT products would be around $6.2 trillion with major income from the health-care sector by 2025. Drosatos et al. showed the concern over potential to transform healthcare and global public health. The IoT business running on a cloud server (CS) model would generate global revenues of approximately $ 490 billion soon. The major business tycoons such as Google and Samsung having Net Labs and SmartThings solutions respectively are keen on providing innovative IoT technology to their users [19, 20]. The IoT devices are having greater penetration in manufacturing, industrial automation, busi-nesses, wearable devices, and the healthcare sector at large. The wearable or carrying IoT devices employed in the healthcare sector are smart electronic devices that can be worn as clothing or body implants or accessories. These devices in the form of the smartwatch, fitness trackers, or Google applica-tions can collect human physiological data that could be transmitted over the network employing Bluetooth or Wi-Fi setups. Fuqaha referred that these devices help in reducing costs and improving efficiency as they can monitor remotely vital patient's signs and health status [29]. The wearable devices are playing a vital role in the healthcare industry by monitoring health fitness levels, wellness factor regularly, motivate users for healthy living, and sharing of medical data with healthcare providers. Prasad established 'SMART Asthma Alert using IoT.' This device collects the data from the asthmatic patient's environment and understand the pattern of allergen that triggers the attack. Here, the device predicts the threshold value and alerts the patient [32]. To maintain integrity and security, blockchain-IoTs concept can be added for the smooth functioning of the device and connectivity with patient's family and healthcare providers. Bhatia et al. considered IoT as a boon in the healthcare industry, several difficulties exist in their acceptance such as privacy, security, and since data is being transferred to the CS it may easily fall into the wrong hands and get misused thereby comprising safety and privacy of the patient [21]. Kuo et al. embarked on data authenticity, integrity, security, and confidentiality are major vulnerabilities of an IoT network (Figure 9.2) [22].

9.6 IoT AND BLOCKCHAIN

Engelhardt et al. and Randall et al. mentioned about the smart devices employed in IoT cannot be directly used in blockchain as they are light-weight with energy and memory constraints. By using the traditional client-server model in IoT, problems related to data synchronization, security, and data privacy may exist. This can be overcome by employing blockchain incorporated within the IoT systems [23, 24]. Jing et al. powered that a blockchain with hypergraphs used in smart homes can maintain security and privacy protection by reducing cyber-attacks possibilities. Rohers et al. considered that the system must be robust enough to allow data privacy and integrity, which could be managed with a unique identity of devices with little scope of tampering. The system should allow scalability and solve problems of data exchange and trading. With the help of data fusion techniques, the analysis could be done on the cloud, thereby improving the healthcare delivery and reducing operational expenditures [25, 26]. Reyna et al. affirmed that the healthcare data privacy is maintained between the cloud nodes and fog with security through blockchain. The advantages include smooth data aggregation, accuracy in the management of data, low-cost IoT software, less consumption of resources. Sultan et al. cited that for smooth IoT functioning integration with 5G network is required, and the blockchain transactions may be visible to all if secure communication protocols on members are not maintained in the blockchain system [27, 30].

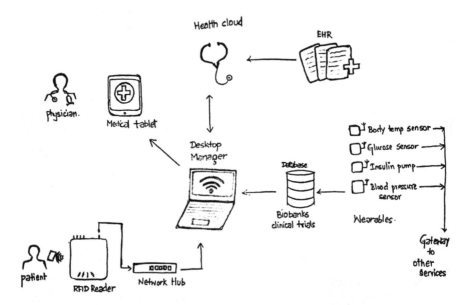

FIGURE 9.2   IoT for the healthcare sector [21].

Tandon stated that by employing blockchain in IoT has several distinct advantages. These are *adaptability* as blockchain and IoT integration can be easily done without the requirement of intermediate blocks [11, 13]. Bagha et al. referred that the cross-communication can be implemented easily with middleware technologies among blockchain devices. The next is *scalability*, it is the ability of the blockchain network to accommodate more IoT devices or sensors on the network without any problem. Fuqaha et al. added that the third advantage is *security*, which is of prime importance between the fusion of both blockchain and IoT technologies [28, 29]. Shakeel et al. showed that the security issues in IoT such as denial of services (DoS), Authorization, Identity Management, and data protection can be overcome by using blockchain. The last one is *consistency*, which implies that the data in the system is consistent with the combination of blockchain and IoT. This ensures data integrity and allows the autonomous administration of IoT systems or domains. It allows every domain in the blockchain network to act independently for managing the IoT devices [15, 16].

9.7   IMPLEMENTATION OF BLOCKCHAIN AND IoT IN HEALTHCARE AND BIOMEDICAL APPLICATIONS

Several researchers, Dorri et al. mentioned in their research works that they are working on possible solutions to devise blockchain-based IoT (BIoT) architectures that are lightweight to reduce the communication induced network overhead and are concentrated mostly on smart homes. They cited about distributed time-based consensus algorithm (DTC) which reduces the mining processing overhead and delay [76]. The BIoT architecture is divided into three layers, namely Home, Network, and Cloud. The home layer comprises interconnected sensing devices that store locally collected data with blockchain features. The network layer connects the devices with the network or the internet. The cloud is employed for the storage of recorded data. Hasselgren et al. cited that blockchain can be employed with several IoT applications such as healthcare, smart home, smart city, supply chain, e-governance, etc. This fusion technology finds several applications in industrial automation. If Big Data is combined with BIoT, it could be used to effectively control and monitor the huge amounts of data collected in the IoT ecosystem [18, 20]. Presently, the BIoT consortium is quite less in number and proper regulations are required to control different factors that presently affect the implementation of BIoT. Padmavathi et al. showed that the issues of prime importance are security, scalability, and cross-platform applications. The commercial products on BIoT presently available include Walton chain, Ambrosus, Power Ledger, Block Mesh, etc. [19].

The different challenges in implementing BIoT include technical integration, interoperability, legal hurdles, and government regulations. Tandon and Fuqaha et al. paved that the technical integration due to scalability, standardized development of products is an important challenge being faced in its smooth implementation. Interoperability can be achieved by having mutual agreement among all stakeholders involved in the implementation BIoT framework [11, 29]. The government regulations with legal frameworks that need to be addressed are real-time product testing to avoid system malfunctioning and seamless integration of different aspects of the combined technology. Drosatos et al. and Bhatia et al. stated that due to the high demand for real-time data by industry and research organizations, it puts them at a larger risk of theft, break-ins, unauthorized sharing. Further, due to malpractices such as counterfeit drugs, skills of medical staff, etc., in the healthcare system, it may erode public trust in the system. Hence, it is important to implement a system that can overcome such disadvantages and improve the overall management of the system [20, 21]. Talpur lead that the Blockchain technology leads the way due to its decentralized approach and data integrity it provides a potential solution to the potent problems affecting the present-day healthcare system [4]. Catarinucci et al. claimed that Blockchain allows greater interoperability, data integrity, sharing of information, and provenance among the stakeholders, thereby enhancing efficiency, system performance, and most importantly, mutual trust among parties involved [6]. Xia et al. introduced the indicator centric schema (ICS), a data model used to simplify the various medical data (like scans, reports, x-rays, etc.), and store them efficiently. Separation of frequently used data from the infrequently used data can be done using this blockchain model [34].

9.8 DISCUSSION AND CONCLUSION

The blockchain technology allows less dependence on huge servers and reducing redundant work. By providing high security and data privacy for users, the system can become cost-effective with fast computing capabilities. Soon, the blockchain technology would immensely benefit and revolutionize the healthcare sector by ensuring data transparency, reducing costs of service delivery, central server for management of all patient data, and hospital services [14, 19]. The clinicians can monitor their patient's condition in real-time without worrying regarding the level of honest information being provided by their patients. Similarly, Tandon A.

emphasized that the patients need not worry about getting a second referral for their medical condition due to the transparency of data and services being provided by the healthcare service provider [11]. Further, Kuo et al. pressed that they can willingly share their data to enable them to learn regarding similar cases being treated or reported in other countries, which would help them in improving their present medical condition and nature of services being provided by their clinician. The blockchain technology could assist multiple people involved in patient welfare that could improve overall efficiency and quality of patient care required with robust security set-up to prevent loss of patient data in anyway due to its decentralized network [20,22]. The major concerns for acceptance of this technology are the misuse of data by sharing it with third parties for profit or indirect identification of users through pseudonymous identifiers. Since research is ongoing in this area and the number of publications is available presently, the future for the growth and expansion of this technology looks promising. In the healthcare sector, it is expected that the blockchain technology market may cross $500 million by 2022. Hasselgren et al. informed that due to its data integrity and enhancing mutual trust among users, its demand would be ever-increasing and required by healthcare institutions to perform their tasks with seamless efficiency and precision [18].

KEYWORDS

- **blockchain**
- **clinicians**
- **denial of services**
- **direct acyclic graph**
- **electronic patient record**
- **gross domestic product**
- **Internet of Things**
- **practical Byzantine fault tolerance**
- **proof of stake**
- **pseudonymous**
- **transparency**

REFERENCES

1. Kumar, P. M., Lokesh, S., Varatharajan, R., Babu, G. C., & Parthasarathy, P., (2018). Cloud and IoT-based disease prediction and diagnosis system for healthcare using fuzzy neural classifier. *Future Generation Computer Systems, 86*(1), 527–534. Elsevier.
2. Qamar, S., Abdelrehman, A. M., Elshafie, H. E. A., & Mohiuddin, K., (2018). Sensor-based IoT industrial healthcare systems. *International Journal of Scientific Engineering and Science, 11*(2), 29–34.
3. Kumar, S. M., & Majumder, D., (2018). Healthcare solution based on machine learning applications in IoT and edge computing. *International Journal of Pure and Applied Mathematics, 119*(16), 1473–1484.
4. Talpur, M. S. H., (2013). The appliance pervasive of Internet of Things in healthcare systems. *International Journal of Computer Science Issues (IJCSI), 10*(1), 1–6.
5. Liang, Z., Zhang, G., Huang, J. X., & Hu, Q. V., (2014). Deep learning for healthcare decision making with EMRs. *IEEE International Conference on Bioinformatics and Biomedicine*, 556–559.
6. Catarinucci, L., De Donno, D., Mainetti, L., Palano, L., Patrono, L., Stefanizzi, M. L., & Tarricone, L., (2015). An IoT-aware architecture for smart healthcare systems. *IEEE Internet of Things, 2*(6), 515–526.
7. Balandina, E., Balandin, S., Balandina, E., Koucheryavy, Y., Balandin, S., & Mouromtsev, D., (2015). IoT use cases in healthcare and tourism. *IEEE 17th Conference on Business Informatics*, 37–44.
8. Gong, T., Huang, H., Li, P., Zhang, K., & Jiang, H., (2015). A medical healthcare system for privacy protection based on IoT. *Seventh International Symposium on Parallel Architectures, Algorithms and Programming*, 217–222.
9. Alam, M. M., Malik, H., Khan, M. I., Pardy, T., Kuusik, A., & Moullec, Y. L., (2018). a survey on the roles of communication technologies in IoT-based personalized healthcare applications. *IEEE Access, 6*(1), 36611–36631.
10. Amin, S. U., Hossain, M. S., Muhammad, G., Alhussein, M., & MRahman, D. A., (2019). Cognitive smart healthcare for pathology detection and monitoring. *IEEE Access, 7*(1), 10745–10753.
11. Tandon, A., (2019). An empirical analysis of using blockchain technology with the Internet of Things and its application. *International Journal of Innovative Technology and Exploring Engineering, 8*(9S3), 1469–1474.
12. Alhussein, M., Muhammad, G., Hossain, M. S., & Amin, S. U., (2018). Cognitive IoT-cloud integration for smart healthcare: Case study for epileptic seizure detection and monitoring. *Mobile Networks and Applications, 23*(1), 1624–1635.
13. Fonseca, C., Mendes, D., Lopes, M., Romão, A., & Parreira, P., (2017). Deep learning and IoT to assist multimorbidity home based healthcare. *J. Health and Medical Informatics, 8*(3), 1–4.
14. Priyadharsan, M. J. D, Sanjay, K. K., Kathiresan, S., Karthik, K. K., & Prasath, K. S., (2019). Patient health monitoring using IoT with machine learning. *International Research Journal of Engineering and Technology (IRJET), 6*(3), 7514–7520.
15. Shakeel, P. M., Baskar, S., Dhulipala, V. R. S., Mishra, S., & Jaber, M. M., (2018). Maintaining security and privacy in healthcare system using learning-based deep-Q-networks. *Journal of Medical Systems, 42*(186), 1–10.

16. Mendes, D. J. M., Rodrigues, I. P., Baeta, C. F., & Solano-Rodriguez, C., (2015). Extended clinical discourse representation structure for controlled natural language clinical decision support systems. *Int. J. Rel. Qua. E-Health, 4*(1), 1–11.

17. Singh, G., Gaur, L., & Ramakrishnan, R., (2017). *Internet of Things-Technology Adoption Model in India, 25,* 835–846.

18. Hasselgren, A., Kralevska, K., Gligoroski, D., Perdersen, S. A., & Faxvaag, A., (2020). Blockchain in healthcare and health sciences-A scoping review. *International Journal of Medical Informatics, 13*(1), 1–10.

19. Padmavathi, U., & Rajagopalan, N., (2019). A research on the impact of blockchain in healthcare. *International Journal of Innovative Technology and Exploring Engineering, 8*(9S2), 35–40.

20. Drosatos, G., & Kaldoudi, E., (2019). *Blockchain applications in the Biomedical Domain: A Scoping Review, 17*(1), 229–240.

21. Bhatia, D., Bagyaraj, S., Arun, K. S., Mishra, A., & Malviya, A., (2020). Role of the Internet of Things (IoT) and deep learning for the growth of healthcare technology. *Trends in Deep Learning Methodologies Series: Hybrid Computational Intelligence for Pattern Analysis and Understanding.* Elsevier (in press).

22. Kuo, T., Kim, H. E., & Machado, L. O., (2017). Blockchain distributed ledger technologies for biomedical and healthcare applications. *Journal of the American Medical Informatics Association, 24*(6), 1211–1220.

23. Engelhardt, M. A., & Espinosa, D., (2017). Hitching healthcare to the chain: An introduction to blockchain technology in the healthcare sector. *Technology Innovation Management Review, 7*(10), 22–35.

24. Randall, D., Goel, P., & Abujamra, R., (2017). Blockchain applications and use cases in health information technology. *Journal of Health and Medical Informatics, 8*(3), 8–11.

25. Rohers, A., André, C., & Righi, R., (2017). Omni PHR: A distributed architecture model to integrate personal health records. *Journal of Biomedical Informatics, 71,* 70–81.

26. Jing, Q., Vasilakos, A. V., Wan, J., Lu, J., & Qiu, D., (2014). Security of the Internet of Things: Perspectives and challenges. *Wireless Networks, 20*(8), 2481–2501.

27. Sultan, A., Sheraz, M. A. M., & Mushtaq, A., (2018). Internet of things security issues and their solutions with blockchain technology characteristics: A systematic literature review. *Am. J. Compt. Sci. Inform. Technol., 6*(3)1–5.

28. Bahga, A., & Madisetti, V., (2014). *Internet of Things: A Hands-On Approach* (1st edn.). Universities Press, Delhi.

29. Fuqaha, A. A., Guizani, M., Mohammadi, M., Aledhari, M., & Ayyash, M., (2015). Internet of things: A survey on enabling technologies, protocols, and applications. *IEEE Communications Surveys and Tutorials, 17*(4), 2347–2376.

30. Reyna, A., Cristian, M., Chen, J., Soler, E., & Díaz, M., (2018). BOn blockchain and its integration with IoT- challenges and opportunities. *Future Generation Computer Systems, 88*(1), 173–190.

31. Susan, D., Debbarma, M., & Deka, A., (2019). A brief review on blockchain and distributed ledger technology. *International Conference on Computing for Sustainable Global Development,* 2–5.

32. Prasad, A. K., (2020). SMART Asthma alert using IoT and predicting threshold values using decision tree classifier. *International Conference on Computer Communication and IoT,* pp. 2–4. Springer Nature.

33. Casino, F., Dasaklis, T. K., & Patsakis, C., (2018). A systematic literature review of blockchain-based applications: Current status, classification and open issues. *Telematics and Informatics 36*(1), 55–81.

34. Xia, Q., Sifah, E. B., Asamoah, K. O., Gao, J., Du, X., & Guizani, M., (2017). MeDShare: Trust-less medical data sharing among cloud service providers via blockchain. In: *IEEE Access* (Vol. 5, No. 1, pp. 14757–14767).

35. Ameer, R., (2016). *What is Blockchain Technology? A Step-by-Step Guide for Beginners.* https://blockgeeks.com/guides/what-is-blockchain-technology/ (accessed on 29th July 2021).

36. Saraf, C., & Sabadra, S., (2018). Blockchain platforms: A compendium. *IEEE International Conference on Innovative Research and Development (ICIRD)* (pp. 1–6). Bangkok.

37. Satoshi, N., (2008). *Bitcoin: A Peer-to-Peer Electronic Cash System [Online].*

38. Cachin, C., & Vukolić, M., (2017). *Blockchain Consensus Protocols in the Wild.* arXiv preprint arXiv:1707.01873 [cs. Dc].

39. Kurt, F., & David, P. C., (2016). Blockchain and its coming impact on financial services. The *Journal of Corporate Accounting and Finance, 27*(5), 53–57. Wiley Periodicals.

40. Supriya, T. A., & Vrushali, K., (2017). Blockchain and its applications – a detailed survey. *International Journal of Computer Applications (0975–8887), 180*(3), 29–35.

41. Brett, S., (2016). *How Can Cryptocurrency and Blockchain Technology Play a Role in Building Social and Solidarity Finance? 1*(1), 1–25. UNRISD Working Paper.

42. Joshua, B., Angela, O., David, M., & Dion – Schwarz, C., (2015). *National Security Implications of Virtual Currency* (pp. 1–102). ISBN: 978-0-8330-9183-3.

43. Roehrs, A., Da Costa, C. A., Da Rosa, R. R., Da Silva, V. F., Goldim, J. R., & Schmidt, D. C., (2019). Analyzing the performance of a blockchain-based personal health record implementation. *Journal of Biomedical Informatics, 92*(1), 103140.

44. Lafourcade, P., & Lombard-Platet, M., (2020). About blockchain interoperability. *Information Processing Letters, 161*(1), 105976.

45. Singh, G., Gaur, L., & Ramakrishnan, R., (2017). Internet of things-technology adoption model in India. *Pertanika Journal of Science & Technology, 25*, 835–846.

46. Liu, S., & He, S., (2019). Application of block chaining technology in finance and accounting field. *International Conference on Intelligent Transportation, Big Data and Smart City (ICITBS)*, 342–344.

47. Tapscott, D., & Tapscott, A., (2016). *Blockchain Revolution: How the Technology Behind Bitcoin is Changing Money, Business, and the World.* Penguin, ISBN-13: 978-1101980149.

48. Korpela, K., Hallikas, J., & Dahlberg, T., (2017). Digital supply chain transformation toward blockchain integration. *Proceedings of the 50th Hawaii International Conference on System Sciences, 50*, 1–10.

49. Cong, L. W., & He, Z., (2019). Blockchain disruption and smart contracts. *The Review of Financial Studies, 32*(5), 1754–1797.

50. Hackius, N., & Petersen, M., (2017). Blockchain in logistics and supply chain: Trick or treat? In digitalization in supply chain management and logistics, smart and digital solutions for an industry 4.0 environment. *Proceedings of the Hamburg International Conference of Logistics (HICL), 23*(1), 3–18.

51. Milutinovic, M., He, W., Wu, H., & Kanwal, M., (2016). Proof of luck: An efficient blockchain consensus protocol. *Proceedings of the 1ˢᵗ Workshop on System Software for Trusted Execution*, pp. 1–6.
52. Conoscenti, M., Vetro, A., & Martin De, J. C., (2016). Blockchain for the Internet of Things: A systematic literature review. *IEEE/ACS 13ᵗʰ International Conference of Computer Systems and Applications (AICCSA)*, 1–6.
53. Tian, F., (2017). A supply chain traceability system for food safety based on HACCP, blockchain and Internet of things. *International Conference on Service Systems and Service Management*, 1–6.
54. Atlam, H. F., Alenezi, A., Alassafi, M. O., & Wills, G., (2018). Blockchain with Internet of Things: Benefits, challenges, and future directions. *International Journal of Intelligent Systems and Applications, 10*(6), 40–48.
55. Jo, B. W., Khan, R. M. A., & Lee, Y. S., (2018). Hybrid blockchain and internet-of-things network for underground structure health monitoring. *Sensors, 18*(12), 4268.
56. Kouicem, D. E., Bouabdallah, A., & Lakhlef, H., (2018). Internet of things security: A top-down survey. *Computer Networks. 141*(1), 199–221.
57. Siyal, A. A., Junejo, A. Z., Zawish, M., Ahmed, K., Khalil, A., & Soursou, G., (2019). Applications of blockchain technology in medicine and healthcare: Challenges and future perspectives. *Cryptography, 3*(1), 3–6.
58. Niya, S. R., Eryk, S., Ile, C., Fabio, M., Kürsat, A., Timo, S., & Thomas, B., (2019). Adaptation of proof-of-stake-based blockchains for IoT data streams. In: *2019 IEEE International Conference on Blockchain and Cryptocurrency (ICBC)* (pp. 15, 16). Seoul, Korea (South). doi: 10.1109/BLOC.2019.8751260.
59. Brodersen, C., Kalis, B., Leong, C., Mitchell, E., Pupo, E., & Truscott, A., (2016). *Blockchain: Securing a New Health Interoperability Experience*. ONC/NIST use of blockchain for healthcare and research workshop.
60. Crosby, M., Pattanayak, P., Verma, S., & Kalyanaraman, V., (2016). Blockchain technology: Beyond bitcoin. *Applied Innovation, 2*(6–10), 71.
61. Antonopoulos, A. M., (2017). *Mastering Bitcoin: Programming the Open Blockchain*. O'Reilly Media, Inc.
62. Eyal, I., (2017). Blockchain technology: Transforming libertarian cryptocurrency dreams to finance and banking realities. *Computer, 50*(9), 38–49.
63. Corbet, S., Larkin, C., Lucey, B., Meegan, A., & Yarovaya, L., (2020). Cryptocurrency reaction to fomc announcements: Evidence of heterogeneity based on blockchain stack position. *Journal of Financial Stability, 46*(1), 100706.
64. Chuen, L. D. K., & Linda, L., (2018). *Inclusive Fintech: Blockchain, Cryptocurrency and ICO*. World Scientific.
65. Miraz, M. H., & Ali, M., (2018). *Applications of Blockchain Technology Beyond Cryptocurrency*. arXiv preprint arXiv:1801.03528.
66. Lee, J. Y., (2019). A decentralized token economy: How blockchain and cryptocurrency can revolutionize business. *Business Horizons, 62*(6), 773–784.
67. Crandall, J., (2019). Blockchains and the "Chains of Empire": Contextualizing blockchain, cryptocurrency, and neoliberalism in Puerto Rico. *Design and Culture, 11*(3), 279–300.
68. Martino, P., Wang, K. J., Bellavitis, C., & DaSilva, C. M., (2019). An introduction to blockchain, cryptocurrency and initial coin offerings. *New Frontiers in Entrepreneurial Finance Research* (pp. 181–206). Chapter 7. World Scientific Publishing Co. Pte. Ltd.

69. Valdeolmillos, D., Mezquita, Y., González-Briones, A., Prieto, J., & Corchado, J. M., (2019). Blockchain technology: A review of the current challenges of cryptocurrency. *International Congress on Blockchain and Applications* (pp. 153–160). Springer, Cham.

70. Manzoor, A., Liyanage, M., Braeke, A., Kanhere, S. S., & Ylianttila, M., (2019). Blockchain based proxy re-encryption scheme for secure IoT data sharing. *IEEE International Conference on Blockchain and Cryptocurrency (ICBC)*, 99–103.

71. Wang, H., & Zhang, J., (2019). Blockchain based data integrity verification for large-scale IoT data. *IEEE Access, 7*(1), 164996–165006.

72. Novo, O., (2018). Blockchain meets IoT: An architecture for scalable access management in IoT. *IEEE Internet of Things Journal, 5*(2), 1184–1195.

73. Sagirlar, G., Carminati, B., Ferrari, E., Sheehan, J. D., & Ragnoli, E., (2018). Hybrid-IoT: Hybrid blockchain architecture for Internet of Things-pow sub-blockchains. *IEEE International Conference on Internet of Things (iThings) and IEEE Green Computing and Communications (GreenCom) and IEEE Cyber, Physical and Social Computing (CPSCom) and IEEE Smart Data (SmartData)*, pp. 1007–1016. IEEE.

74. Singh, S., & Singh, N., (2016). Blockchain: Future of financial and cybersecurity. In: *2nd International Conference on Contemporary Computing and Informatics (IC3I)* (pp. 463–467). IEEE.

75. Ramakrishnan, R., & Gaur, L., (2019). *Internet of Things: Approach and Applicability in Manufacturing*. CRC Press.

76. Dorri, A., Kanhere, S. S., Jurdak, R., & Gauravaram, P., (2019). LSB: A lightweight scalable blockchain for IoT security and anonymity. *Journal of Parallel and Distributed Computing, 134*(1), 180–197.

CHAPTER 10

Decentralized and Secured Applications of Blockchain in the Biomedical Domain

MEET KUMARI,[1] MEENU GUPTA,[2] and CHETANYA VED[3]

[1]*Department of Electronics and Communication Engineering, Chandigarh University, Punjab, India, E-mail: meetkumari08@yahoo.in*

[2]*Department of Computer Science and Engineering, Chandigarh University, Punjab, India, E-mail: gupta.meenu5@gmail.com*

[3]*Department of Information Technology, Bharati Vidyapeeth's College of Engineering, Maharashtra, India, E-mail: chetanyaved@gmail.com*

ABSTRACT

In this book chapter, the decentralized and secured applications of blockchain in the biomedical domain have been discussed to offer visibility and privacy to the biomedical domain. After extensive review of recent work, it has been explained how the blockchain features can be used in biomedical filed. The results show the proposed decentralized electronic-medical blockchain (DEMB) based system and its feature. Also, its future applications, challenges and solutions are presented.

10.1 INTRODUCTION

Blockchain has emerged as new technology and gained significant attention, with numerous applications, ranging from financial services, data management, food science, IoT, and financial services to the biomedical research domain. There has been a significant interest in employing blockchain applications for the safe as well as secure medical data managing in the biomedical field [1, 2]. In the era of the Internet, it has become simple to obtain medical information by patients through smart modules and consulting multiple doctors, thus enhancing the secrecy of shared knowledge. Moreover, the medical history can be obtained by patients easily through request permission from providers and an extensive accessible network. It requires interoperability, authentication, considerations, and sharing biomedical data regarding e-Health [3–5]. The digital rising in healthcare produced a standard shift in the biomedical domain. The significant advantage of the current electronic medical record and digital healthcare system is to enhance the medical history both for patients and health professionals. This successful initiative provides access to patients for their medical electronic medical records to maximize the efficiency, quality, as well as scalability of the biomedical system. The biomedical data consists of omics data, laboratory testing, and clinical records [6, 7]. For a particular single disease patient, the interpretation accuracy and reliability are based on health information issued to the doctor by the patient. In the context of doctor and patient, this method would not work significantly for diagnosis because sometimes patients do not remember the details of their medical history [4]. Even the event occurred long previously and limited medical knowledge by most of the patients to discuss the treatments which they are receiving professionally. Limited medical knowledge may affect the present doctor's medical judgment [8–10]. Hence, the current doctor may stop to conclude the precise data while observing out the disease diagnosis. For solving doctors and patient problem, an innovative scheme for private and secure blockchain technology within the medical care unit has been put forward [11, 12]. The basic design of the blockchain is shown in Figure 10.1.

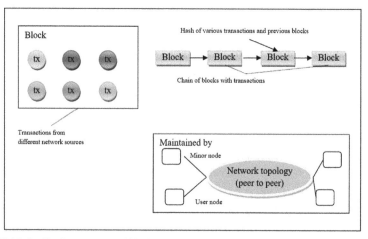

FIGURE 10.1   Basic structure of blockchain [7].

Blockchain technology introduces a distributed database that maintains connected objects or devices recordset. Blockchain has several vital features, such as decentralized, immutable, transparent, open-source, and autonomous [13, 14]. Here, each device has equal command and an overall chain copy. A blockchain database can be executed as a public as well as private blockchain. The public blockchain allows every candidate for reading, writing, and verifying the recorded database, for example, bitcoin (BTC), and Ethereum (ETH) [15, 16]. Private blockchain permits the only candidate of the trusted group for reading, writing, and testing the registered database. With the help of blockchain, the biomedical data collected in the private blockchain, which provides better security, fast transaction, privacy, and cost effeteness [17, 18]. Besides, doctors can visit the hospital's private blockchain easily to observe the associated blockchain for index data about access to original medical records.

Biomedical is a clinical data-intensive domain where a large amount of data is produced, accessed, and distributed regularly. In the biomedical field, biomaterials have conventionally been generated from natural sources to replace lost tissues to disease. However, early in the current century, these materials began to be put back by metal alloys, ceramics, and synthetic polymers, due to better reproducible properties and performance than natural materials. Storing and distributing the massive amount of data is critical, remarkably challenging, because of the sensitive medical data, security, and privacy [19, 20]. The biomedical system performance can be improved and make protective against a cyber-attack more vigorous by engaging the adaptive techniques. These techniques can help to diagnosis various forms of biomedical data [21] like images, bio-signals, records, etc. A distinct, emerging application of blockchain technology is the biomedical domain. It includes three paradigms' applications. A first approach consists of blockchain to maintain participant data surety and integrity for evaluation of patient's trial inclusion and provide access to medical data after trial subscription [22]. A second approach consists of versioning and non-repudiation of trial forms. A third approach consists of ensuring clinical trial protocol storage considering private blockchain. In these applications, the blockchain technology is utilized as a distributed ledger of activities such as research for ensuring immutability, integrity, and non-repudiation [23, 24].

The primary blockchain application in the biomedical field is electronic health records (EHR). It includes patient clinical data fragments managed by healthcare suppliers. EHR is used for integration, controlling, sharing, preserving, and overall managing. Thus, to improve the EHR security, a biomedical EHR security structure is shown in Figure 10.2. Here, the patient writes in the blockchain database using an authentication key, which is known as facts creations. The doctors and other members, after authorization, can decrypt and see the file with transparency. Also, the users' accounts cannot be changed as it is encrypted by respective users' private keys, also known as authentication. The whole process is described in Figure 10.2. The underlying blockchain features used in biomedical are shown in Table 10.1.

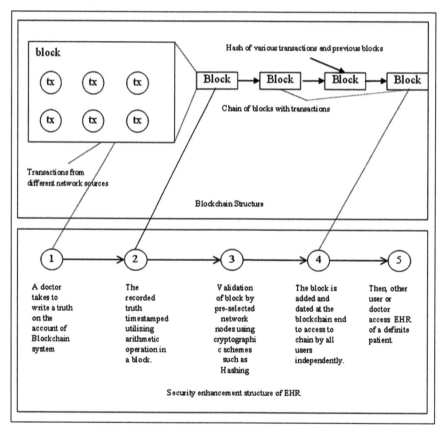

FIGURE 10.2   Biomedical EHR security enhancement structure in blockchain [6, 25].

This book chapter presents an overview of the background studies that depict the latest research field of the blockchain application in biomedical. The main aim is to recognize the biomedical issues concerned with blockchain, the maturity levels of respective approaches, considered biomedical security in terms of its secure structure, features, and functionalities, used in blockchain for the biomedical domain. This chapter is significantly focused on safety, integrity, integration, and medical records access control, along with patient data in EHR. The exciting and diverse applications are medical research addressing clinical trials, medical insurance, and medicines supply chain. Further, this chapter summarized the background of the blockchain (its actual progress), and the latest developments in the biomedical domain have been discussed. Then the various potential applications of the blockchain in the biomedical field are described.

Further, the open issues and challenges while utilizing blockchain in the biomedical domain are also presented. Furthermore, the future perspectives of the biomedical area using blockchain technology have been highlighted.

In this book chapter, after the introduction, Section 10.2 presents the summarized "Literature Review." Sections 10.3–10.5 present the "Results Analysis;" "Applications;" and "Open Challenges." Section 10.6 describes the future scope, along with the conclusion.

TABLE 10.1 Blockchain Features Used in Biomedical [26, 27]

SL. No.	Blockchain Feature	Examples
1.	Digital access	On-chain, clinical medical data, or collected off-chain is joined to the patient's public key. By using blockchain technology such as smart contracts in biomedical, assign the rules for data access, e.g., authorizing release for some time to a patient registry.
2.	Data accumulation	A connected patient to the various institutional restrictions through institution-defined patient portal suppliers public blockchain key having security permission. It is done across multiple clinical data and institutions using blockchain technology.
3.	Data liquidity	The highly sensitive medical data such as advanced supervision planning or the medication allergies on the public blockchain can be declared, keeping the information as correct.
4.	Patient identity	The public keys of respective patients can be managed by a mobile device or multi-sig wallet along with the public-key infrastructure (PKI) for establishing the uniqueness for clinical data retrieval.

10.2   LITERATURE REVIEW

Blockchain provides a secure and reliable platform for sharing data (in a secured form) in different domains such as the financial sector, energy sector, supply chain management (SCM), food industry, and healthcare, etc. Various researchers have given their views about the role of blockchain in these fields. The different researcher's opinions in the biomedical domain using the application of blockchain technology are discussed below.

Drosatos, G. and Kaldoudi, E. reviewed the recent research of various blockchain applications in scoping and demanding the biomedical field. The primary aim of the work was to find the biomedical problems used with blockchain, the maturity level of approaches, biomedical data types, features, frameworks, and functionalities exploited. Here, the preferred reporting items for systematic reviews and meta-analysis (PRISMA-SGR) methodology has been used. It has shown that the blockchain application, i.e., biomedical, is still in its initial stage, having various studies in architectural design and conceptual phase. Thus, researchers need to be focused on integrity, integration, and access control of the biomedical domain along with patient healthcare data. The other emerging applications of blockchain technology are medical addressing research, the supply chain of medicines, medical insurance, etc. [23]. Kleinaki et al. presents a notarization resource based on blockchain technology that utilizes digital contracts to take biomedical stored database results and queries. The primary aim of the chapter was to guarantee that the retrieved data do not be validated deny and modify. The proposed approach was implemented using an actual blockchain infrastructure along with testing of biomedical evidence of different databases, PubMed database, literature references, abstracts, and healthcare risk factors [25].

Service, C., and Via, P. MeDShare proposed a MeDShare system that inscribes the medical information issue sharing with healthcare big data guardians in a faithless environment. The proposed method was based on blockchain technology and gives data control, auditing, and provenance for medical shared data among entities of big data in cloud repositories. MeDShare helps in monitoring access information entities for malicious use in a data guardian system. It helps in data sharing and transitions in substances having tasks performed on the proposed method. It shows that the proposed system, along with cloud service and data guardians, provides data provenance, control, and auditing, which helps in medical institutions and various research institutions with less risk factor of data privacy [28]. Hussein et al. presents a blockchain technology-based managing technique to provide interpretation enhancement concerning biomedical fields. Here, two blockchain methods have been used for constructing base and blockchain algorithm utilized to produce a secure series for a hash key. It shows that it helps in handling various kinds of data types, as well as biomedical images and text. Again, it shows latency up to 750 ms indicating its use in hospitals and clinics at 400 requests per second [29].

Workman, J. P. presented the various merits and demerits of blockchain technology in the biomedical domain. It was explored how multiple challenges in biomedical such as data ownership, accessibility, and cost can be inscribed by blockchain, equally important or not. The clear merits and demerits of blockchain technology identify the open area's innovation for biomedical entrepreneurs, academics, and executives [30]. Jovic et al. analyzed the biomedical signal has been analyzed based on web systems, and they describe the own design of software architecture. Here, the applications, medical education, home care, visualization capabilities, signal repositories, signal analysis circumstances, and data mining have been examined of biomedical signals. Also, the various open challenges such as frontend workflow, frontend interactions, backend interactions, data analysis integration, reporting libraries, programming language challenges, and data privacy have been discussed [31].

Mamoshina et al. presented a review of the blockchain technologies and artificial intelligence (AI)-based on next-generation innovative solutions to accelerate the research in the biomedical domain with the latest tools for controlling, profiting from data used personally has been provided for continuous health monitoring. Here, the new concepts have been presented to evaluate and appraise personal records with relationship-value and time combinations of data. It is also shown that how blockchain-based decentralized personal data enables new techniques for biomarker development, discover of the drug as well as preventative medical care. The combined deep learning (DL) and blockchain technologies help in resolving the open challenges accepted by the personal data such as medical records and regulators [32].

10.3 RESULT ANALYSIS

10.3.1 DECENTRALIZE ELECTRONIC-MEDICAL BLOCKCHAIN-BASED SYSTEM (DEMB)

It is an interconnected decentralized system that consists of a decentralized HER and Medication diagnostics blockchain record. HER consists of a patient's medical history. Medication health record (MHR) system consists of a patient's medication history, and it is interconnected with an HER. MHR is used for the estimation of what type of medication has improved the patient's health condition or to track what kind of medical salts person had taken earlier. Figures 10.3 and 10.4 show the detailed architecture of the DEM blockchain-based system.

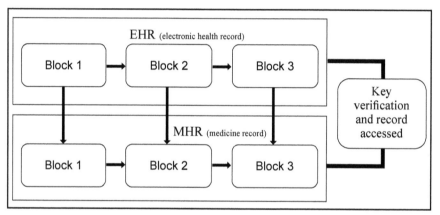

FIGURE 10.3   Decentralize electronic-medical (DEM) blockchain-based system.

Medical Health Record Blockchain consists of data related to medicines given to the patients during the treatment and response of the patient's body towards the medication. The doctor's corresponding will give improvement labeling to the reaction of the patient's body.

Both types of blockchain can be accessed separately by different organizations for treatment or medication purposes. Medical Records of patients can be obtained by a critical verification system only (i.e., public, and private key concept). The essential verification process will enhance the security and interoperability of this system. Separate keys will be generated corresponding to each type of record to maintain the trustability and integrity of EHR's. This system can be used for predicting a patient's body behavior towards a particular kind of medication. Private medical practitioners just have read access (i.e., they cannot add/write block) in the DEMB until they consist of authorization to access its private key. The patient will have only read access. Approved medical institutions/ hospitals have the right to read/write both types of records for treatment and diagnosis purposes.

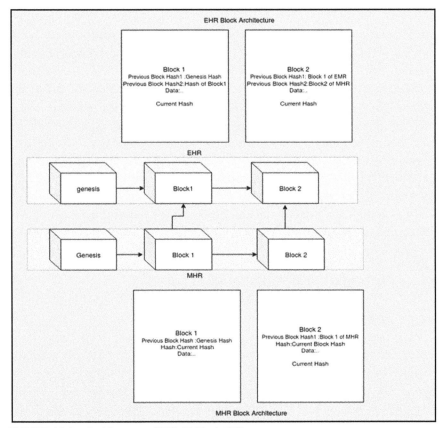

FIGURE 10.4   Architecture of DEMB block [32].

10.3.2   FEATURES OF THIS SYSTEM

1. It will improve the diagnostic capabilities of the current healthcare system.
2. It is enhancing the predictability of medication treatment over the body of a patient.
3. System model will possess ACID properties.
4. It has limited access rights to private medical/clinical organization, which enhances its security and trustability feature.

10.3.3   DEMB BASED ANALYSIS OF BIOMEDICAL DOMAIN

Blockchain comes up with decentralized and security solutions. In the domain of biomedical applications security, correct diagnosis, proper medication treatment plays a significant role in curing any health-related issue or problem. The motive of decentralize electronic-medical blockchain-based system (DEMB) is to evolve innovations, advance plans of action, and business models in the biomedical domain. This objective can be realized by binding up to two different record systems together to enhance diagnosis facility as well as productive treat purpose. It is clear to observe the upside of these networks, yet there are some immense challenges also. It can manifest a big hit in increasing transparency, scalability, immutability, enhancing security by maintaining the patient's electronic record history with medication records. As stated in this proposed method, it provides the feature of security by giving limited access to decentralized nodes of reading and writing data so that patient's medical history cannot be tempered or invade easily [33].

10.4   APPLICATIONS AND OPEN CHALLENGES

10.4.1   APPLICATIONS

Blockchain technology is considered the best technology for making systemic, lower costs, equal access services, enhance longevity, etc. It plays an essential role in the biomedical field with various application areas such as biomedical healthcare management, automated biomedical adjustment, medical data exchanging, drug counterfeiting, precision medicine, and longitudinal biomedical healthcare records. Mainly, blockchain in the biomedical domain resolves the problems of reliability of looking at losing data, data dredging, endpoint switching, etc., in the clinical trials. Besides, it solves the issues of patients' consent [34, 35]. To manage the electronic healthcare records (EHRs) data of patients is highly relevant. An EHR consists of patients' medical history, predictions, clinical progress of all treatment. Blockchain technology for HERs can be seen as a protocol by which patients may access as well as maintain health information simultaneously guarantees privacy and security. It helps in storing the medical data in a distributed way, having no centralized controller are an owner to breach. Thus, data is always available and updated while data from distributed sources combined in the single data repository. The various potential applications of blockchain technology in the biomedical field are shown in Figure 10.5 as follows [30, 36, 37].

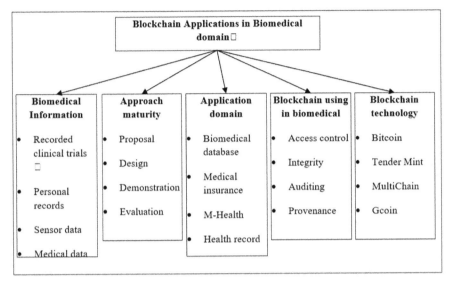

FIGURE 10.5   Potential blockchain applications in the biomedical domain [2, 3].

10.5   OPEN CHALLENGES

Nowadays, blockchain is adopted in various business and research fields, providing unlimited opportunities. However, this emerging technology has some significant challenges and issues. Thus, in this section, the significant open problems in the biomedical domain are discussed along with avenues for further research opportunities in future [3, 38, 39]:

- **Blockchain Suitability:** Various sectors Companies are attracted to blockchain and its capability to drive its digital transformation for real-life issues solution. However, lots of IT specialists predict the utilization of blockchain in every project. Still, they are not able to recognize the fundamental reason for blockchain technology usage, especially in data management. As if no data is added, blockchain shall not be capable of storing any data to already existing technical solutions. On the contrary, blockchain technology is suitable for permanent-past records or trustless sources transaction. Thus, before embracing blockchain-enabled opportunities, the blockchain suitability should be examined against usage requirements. Moreover, in the form of trust, blockchain technology restricts the usage of third-party trust, hence improve the verifiability and reliability of the content. It is also appropriate for operations and transactions to require to be discovered or if the processes need secure privacy and security.

- **Scalability and Latency:** As the various cryptocurrencies have a minimum transaction rate. Such as, a BTC transaction which cannot compare to VISA's card system that regularly manages the number of transactions. Undoubtedly, the extensive crypto-currencies need adoption to overcome this latency issue. Thus, the blockchain design faces severe latency issues. On the other side, a private blockchain is more efficient as they do not require standards.

- **Data Management Security and Privacy:** Instead of a large number of advantages of blockchain in terms of confidentiality and security of data storage and management, still blockchain has various issues and weaknesses. Generally, confidentiality and privacy are the major blockchain problems as the information is collected in terms of a public ledger. Thus, various encryption-based and anonymization schemes can be utilized to save the security of the data. However, these schemes are not a magic bullet. These depend on the context and implementation of the system.

- **Artificial Intelligence and Big Data:** The increasing speed of data acquisition, enhances the accuracy and effectiveness of information across AI and Big data domain. It paves the path for more accurate and scalable AI models as well as solutions within various contexts, improving data analytics possibilities by using the public blockchain interoperability and standard systems. Also, the blockchain architecture using data analytics implies overhead problems [40]. Thus, efficient, and intermediate auxiliary architectures are implemented, thus improving efficiency. Some other challenges are also given in Table 10.2.

TABLE 10.2   Barriers to Blockchain-Enabled Biomedical Applications [26, 41]

S. No.	Challenge	Solutions
1.	Transaction amount of clinical data	• Data exchange on clinical summarized data. • Blockchain having permission to handle large transaction amount except validation of time-Intensive for regional geographies. • Latest technologies along with blockchain-based research methodologies.

TABLE 10.2   (Continued)

S. No.	Challenge	Solutions
2.	Security and privacy	• Permissioned blockchain consortium member to reduce public exposure.
		• On-chain focusing with data storage around meta-data or permissions.
3.	Patient engagement	• It is an "app" based on a patient-friendly ecosystem to handle permissions and public keys.
4.	Incentives	• They are expanding federal continued incentives for API coverage, such as VA data pledge.
		• Open data association is used for reimbursement of data.
		• API-enabled to non-enabled systems for API.
		• Infrastructure invests.

10.6   CONCLUSION AND FUTURE SCOPE

The blockchain applications in the biomedical domain are widely used with lots of issues that have to be addressed. The combination of various blockchain features offers uniqueness to biomedical applications to justify the interest of several users. It provides decentralized security in the biomedical domain. The results analysis shows that blockchain technology is a mature technology having lots of applications that can overcome open challenges. As blockchain technology in the biomedical field has been widely deployed, lots of problems have yet to be addressed. Thus, blockchain technology will become efficient and scalable as well as durable. It makes the blockchain technology ideal for lots of other applications instead of biomedical. In the coming years, blockchain technology will become mature with claims that are expected to infiltrate more biomedical domain as other applications. It has been described that using blockchain in the biomedical domain will benefit a vast amount of patients, healthcare providers, healthcare entities, medical practitioners, biomedical researchers, and research and development specialists to dispense a large amount of information significantly. Blockchain recommendations in the biomedical field have guaranteed security excellent privacy protection. It will surely open innovative research avenues for the next future advanced biomedical research. This would help in the sharing of safe, scalable acquisition, secure, etc., clinical data in developing potential methods for the treatment of diseases. The transparent and decentralized feature of blockchain would restrict medical data from being stolen or changed. Consequently, the blockchain-based biomedical domain system will grab patients more in their excellent healthcare to ultimately enhance the quality of human life.

KEYWORDS

- **blockchain**
- **data-sharing**
- **electronic health record**
- **healthcare**
- **medication health record**
- **public-key infrastructure**

REFERENCES

1. Biswas, K., & Muthukkumarasamy, V., (2017). Securing smart cities using blockchain technology. *Proc. – 18th IEEE Int. Conf. High Perform. Comput. Commun. 14th IEEE Int. Conf. Smart City 2nd IEEE Int. Conf. Data Sci. Syst. HPCC/SmartCity/DSS 2016* (pp. 1392, 1393). https://doi.org/10.1109/HPCC-SmartCity-DSS.2016.0198.

2. Al-Jaroodi, J., & Mohamed, N., (2019). Blockchain in industries: A survey. *IEEE Access, 7*, 36500–36515. https://doi.org/10.1109/ACCESS.2019.2903554.

3. Casino, F., Dasaklis, T. K., & Patsakis, C., (2019). A systematic literature review of blockchain-based applications: Current status, classification and open issues. *Telemat. Informatics, 36*, 55–81. https://doi.org/10.1016/j.tele.2018.11.006.

4. Siyal, A. A., Junejo, A. Z., Zawish, M., Ahmed, K., Khalil, A., & Soursou, G., (2019). Applications of blockchain technology in medicine and healthcare: Challenges and future perspectives. *Cryptography, 3*(1), 3. https://doi.org/10.3390/cryptography3010003.

5. Warkentin, M., & Orgeron, C., (2020). Using the security triad to assess blockchain technology in public sector applications. *Int. J. Inf. Manage*, 102090. https://doi.org/10.1016/j.ijinfomgt.2020.102090.

6. Tamazirt, L., Alilat, F., & Agoulmine, N., (2018). Blockchain technology: A new secured electronic health record system. In: *2018 Int. Work. Adv. ICT Infrastructures Serv.* (p. 134).

7. Nofer, M., Gomber, P., Hinz, O., & Schiereck, D., (2017). Blockchain. *Bus. Inf. Syst. Eng., 59*(3), 183–187. https://doi.org/10.1007/s12599-017-0467-3.

8. Eklund, J. M., (2019). *Blockchain Technology in Healthcare: A Systematic Review*. https://doi.org/10.3390/healthcare7020056.

9. Crosby, M., Pattanayak, P., Verma, S., & Kalyanaraman, V., (2016). Blockchain technology: Beyond bitcoin. *Appl. Innov. Rev.*, (2), 5–20.

10. Jamil, F., Hang, L., Kim, K. H., & Kim, D. H., (2019). A novel medical blockchain model for drug supply chain integrity management in a smart hospital. *Electron., 8*(5), 1–32. https://doi.org/10.3390/electronics8050505.

11. Chen, H., & Huang, X., (2018). Will blockchain technology transform healthcare and biomedical sciences? *EC Pharmacol. Toxicol., 6*(11), 910–911.

12. Wu, X., & Lin, Y., (2019). Blockchain recall management in pharmaceutical industry. *Procedia CIRP, 83*, 590–595. https://doi.org/10.1016/j.procir.2019.04.094.

13. Rantos, K., Macedonia, E., Drosatos, G., Ilioudis, C. A., & Papanikolaou, A., (2018). ADvoCATE: A consent management platform for personal data processing in the IoT using blockchain technology. In: *11ᵗʰ International Conference, SecITC 2018* (pp. 300–313). Bucharest.

14. Xu, X., Pautasso, C., Zhu, L., Gramoli, V., Ponomarev, A., & Chen, S., (2016). *The Blockchain as a Software Connector.* Volume 13 of working IEEE, IFIP Conf.

15. Beninger, P., & Ibara, M. A., (2016). Pharmacovigilance and biomedical informatics: A model for future development. *Clin. Ther., 38*(12), 2514–2525. https://doi.org/10.1016/j.clinthera.2016.11.006.

16. Makhdoom, I., Abolhasan, M., Abbas, H., & Ni, W., (2019). Blockchain's adoption in IoT: The challenges, and a way forward. *J. Netw. Comput. Appl., 125,* 251–279. https://doi.org/10.1016/j.jnca.2018.10.019.

17. Hölbl, M., Kompara, M., Kamišalić, A., & Zlatolas, L. N., (2018). A systematic review of the use of blockchain in healthcare. *Symmetry (Basel)., 10*(10). https://doi.org/10.3390/sym10100470.

18. Dinh, T. T. A., Liu, R., Zhang, M., Chen, G., Ooi, B. C., & Wang, J., (2018). Untangling Blockchain: A data processing view of blockchain systems. *IEEE Trans. Knowl. Data Eng., 30*(7), 1366–1385. https://doi.org/10.1109/TKDE.2017.2781227.

19. Schöner, M., Kourouklis, D., Sandner, P., Gonzalez, E., & Förster, J., (2017). *Blockchain Technology in the Pharmaceutical Industry.* FSBC Work. Pap.

20. Khezr, S., Moniruzzaman, M., Yassine, A., & Benlamri, R., (2019). Blockchain technology in healthcare: A comprehensive review and directions for future research. *Appl. Sci., 9*(9), 1–28. https://doi.org/10.3390/app9091736.

21. Radanović, I., & Likić, R., (2018). Opportunities for use of blockchain technology in medicine. *Appl. Health Econ. Health Policy, 16*(5), 583–590. https://doi.org/10.1007/s40258-018-0412-8.

22. Hasselgren, A., Kralevska, K., Gligoroski, D., Pedersen, S. A., & Faxvaag, A., (2020). Blockchain in healthcare and health sciences—A scoping review. *Int. J. Med. Inform., 134,* 104040. https://doi.org/10.1016/j.ijmedinf.2019.104040.

23. Drosatos, G., & Kaldoudi, E., (2019). Blockchain applications in the biomedical domain: A scoping review. *Comput. Struct. Biotechnol. J., 17,* 229–240. https://doi.org/10.1016/j.csbj.2019.01.010.

24. Chen, Y., Ding, S., Xu, Z., Zheng, H., & Yang, S., (2018). Blockchain-based medical records secure storage and medical service framework. *J. Med. Syst., 43*(1). https://doi.org/10.1007/s10916-018-1121-4.

25. Kleinaki, A. S., Mytis-Gkometh, P., Drosatos, G., Efraimidis, P. S., & Kaldoudi, E., (2018). A Blockchain-based notarization service for biomedical knowledge retrieval. *Comput. Struct. Biotechnol. J., 16,* 288–297. https://doi.org/10.1016/j.csbj.2018.08.002.

26. Gordon, W. J., & Catalini, C., (2018). Blockchain technology for healthcare: Facilitating the transition to patient-driven interoperability. *Comput. Struct. Biotechnol. J., 16,* 224–230. https://doi.org/10.1016/j.csbj.2018.06.003.

27. Clauson, K. A., Breeden, E. A., Davidson, C., & Mackey, T. K., (2018). *Leveraging Blockchain Technology to Enhance Supply Chain Management in Healthcare: Blockchain Healthc. Today,* pp. 1–12. https://doi.org/10.30953/bhty.v1.20.

28. Service, C., & Via, P., (2017). MeDShare: Trust-less medical data sharing among. *IEEE Access, 5,* 1–10. https://doi.org/10.1109/ACCESS.2017.2730843.

29. Hussein, A. F., ALZubaidi, A. K., Habash, Q. A., & Jaber, M. M., (2019). An adaptive biomedical data managing scheme based on the blockchain technique. *Appl. Sci., 9*(12). https://doi.org/10.3390/app9122494.

30. Workman, J. P., (2008). A marketplace for health: Opportunities and challenges for biomedical blockchains. *J. Bus.*, 363–369.

31. Jovic, A., Jozic, K., Kukolja, D., Friganovic, K., & Cifrek, M., (2018). Challenges in designing software architectures for web-based biomedical signal analysis. *Med. Big Data Internet Med. Things*, 81–111. https://doi.org/10.1201/9781351030380-4.

32. Mamoshina, P., Ojomoko, L., Yanovich, Y., Ostrovski, A., Botezatu, A., Prikhodko, P., Izumchenko, E., Aliper, A., Romantsov, K., Zhebrak, A., et al., (2018). Converging blockchain and next-generation artificial intelligence technologies to decentralize and accelerate biomedical research and healthcare. *Oncotarget, 9*(5), 5665–5690. https://doi.org/10.18632/oncotarget.22345.

33. Ratta, P., Kaur, A., & Sharma, S., (2020). Blockchain – Secure Decentralized Technology Blockchain-Secure Decentralized Technology. *International Journal of Advance Science and Technology, 29*(10S),

34. Seliem, M., & Elgazzar, K., (2019). BIoMT: Blockchain for the internet of medical things. In: *2019 IEEE Int. Black Sea Conf. Commun. Networking, BlackSeaCo.* https://doi.org/10.1109/BlackSeaCom.2019.8812784.

35. Albanese, G., Calbimonte, J. P., Schumacher, M., & Calvaresi, D., (2020). Dynamic consent management for clinical trials via private blockchain technology. *J. Ambient Intell. Humaniz. Comput.* https://doi.org/10.1007/s12652-020-01761-1.

36. Katuwal, G. J., Pandey, S., Hennessey, M., & Lamichhane, B., (2018). *Applications of Blockchain in Healthcare: Current Landscape & Challenges*, 1–17.

37. Abou, J. J., & George, S. R., (2019). Blockchain applications – usage in different domains. *IEEE Access, 7*, 45360–45381. https://doi.org/10.1109/ACCESS.2019.2902501.

38. McBee, M. P., & Wilcox, C., (2020). Blockchain technology: Principles and applications in medical imaging. *J. Digit. Imaging.* https://doi.org/10.1007/s10278-019-00310-3.

39. Pandey, P., & Litoriya, R., (2020). Securing and authenticating healthcare records through blockchain technology. *Cryptologia, 44*(4), 341–356. https://doi.org/10.1080/01611194.2019.1706060.

40. Modgil, S., & Sonwaney, V., (2019). Planning the application of blockchain technology in identification of counterfeit products: Sectorial prioritization. *IFAC-PapersOnLine, 52*(13), 1–5. https://doi.org/10.1016/j.ifacol.2019.11.080.

41. Deshpande, A., Stewart, K., Lepetit, L., & Gunashekar, S., (2017). Distributed ledger technologies/blockchain: Challenges, opportunities, and the prospects for standards. *Br. Stand. Inst.*, p. 82. https://doi.org/10.7249/RR2223.

CHAPTER 11

Blockchain-Enabled Secure Platforms for Management of Healthcare Data

NIMRITA KOUL and SUNILKUMAR S. MANVI

School of Computing and Information Technology, REVA University,
Bangalore, Karnataka – 560064, India,
E-mails: nimritakoul@reva.edu.in; emailnk1@gmail.com (N. Koul)

ABSTRACT

Massive adoption of technologies like wearable IoT for personal health monitoring, smart health care systems, and digitalization of health records of patients by the hospitals, known as electronic health records (HERs), has led to the generation of huge volumes of sensitive data that is liable to be misused by parties with wrong intentions. These applications are good examples of cyber-physical systems which involve an integration of functionality of physical processes, computing, and networking with built-in feedback mechanisms. Although these systems have many advantages, there have been concerns regarding the security of personal patient data that these applications handle and communicate. There do exist solutions for the security of such data, but they often prove to be too restrictive, and are even patients do not have easy access to them, they are heavily non-reciprocal in the sense that interactive capabilities have not been part of the design for security purposes. The trade-off between data security and ease of access plus interactivity is in favor of too restrictiveness in order to preserve data integrity and avoid unauthorized modifications. Blockchain technology, by design, is suitable for this kind of trustless scenario where data integrity, non-repudiation, ease of access, transactional access to data, and high availability are essential. The volume of this private health data, which contains information about a person's vital stats like name, medical history, address, clinical notes, laboratory, and imaging examination records, and screening reports, etc., is growing at an exponential rate. There are various parties involved in the generation, management, disbursement of this data, therefore there arises a risk of breach of privacy or confidentiality, unauthorized access, or manipulation of this data by any of the involved parties. Blockchain maintains data in a decentralized, transactional, replicated, un-mutable storage units called blocks. Such a platform is a very viable option to manage HERs in a secure, privacy-preserving, trusted yet interactive manner.

A blockchain-based system can ensure that only authorized parties have various access privileges to various parts of data, the data once created is un-mutable and any additions to data are preserved and accessible with history to all the stake holders like the patient, doctors, nurses, authorized caregivers, etc. In view of this growing sector of big data in need of non-repudiation and privacy preservation, many pharma companies and healthcare providers are opting for blockchain-based solutions for their services which will eliminate the need for intermediaries in the healthcare demand-supply chain. A few uses case for blockchain in healthcare involve billing and claims management, where the middle men abound, sharing of medical data among hospitals for research, securing electronic patient records, supply chain in drugs and pharmaceuticals. Every year lakhs of people lose their lives due to consumption of substandard medicines or expired drugs. With blockchain-based supply chain, we can maintain the date and location of production of drug without the possibility of unauthorized manipulation. We are already witnessing many blocks chain-based implementations for management of HERs by various big companies in the healthcare sector.

Index terms-blockchain HERs, wearable IoT devices, personal health monitoring devices, Drug supply chain, medical data analytics, anti-counterfeiting.

➢ **Objectives:**
1. In this chapter, we will discuss various use cases of blockchain in the management of healthcare data.
2. We will analyze the advantages of implementation of blockchain with a special focus on security to such use cases over conventional solutions.
3. The readers will get a broad picture of the degree of proliferation of blockchain in current space of healthcare management.

11.1 INTRODUCTION

Wearable healthcare devices are good examples of cyber-physical systems which involve an integration of functionality of physical processes, computing, and networking with built-in feedback mechanisms. The huge volume of medical data from wearable healthcare devices, electronic health records (HERs), medical imaging and genomic sequencing data is available nowadays. Since this data belongs to human patients, its security and privacy preservation is a critical requirement for any computing platform to handle this data. Although these systems have many advantages, there have been concerns regarding the security of personal patient data that these applications handle and communicate. The security requirements include standardization, regulation, privacy, integrity, confidentiality, authenticity, and availability. There is a need for secure platforms that ensures high security, high availability, non-repudiation, and privacy preservation. Blockchain technology [1] is a distributed ledger technology (DLT) [2] which by design is distributed, non-mutable, and secure. It allows the transaction to be recorded in a distributed, immutable, and encrypted form, hence provides high availability, non-repudiation, and security to data. These properties have led to blockchain being used for storage and management of transactions involving various processes and transactions involved in the healthcare system. These transactions include the steps involved in maintenance of medical records in a digital form, settling of insurance claims and billing through smart contracts (SCs). The use of blockchain enables wider access to medical records of the patients, tracking transit of drugs, tracking manufacturing dates of drugs, management of hospital assets, prescription management, management of hospital and ICU device life cycles. Blockchain can provide all these capabilities in a distributed way without a trusted third party. These factors make blockchain a good platform for the implementation of processes in the field of healthcare. Present solutions in healthcare system suffer from data fragmentation, centralized systems that act as a single point of failure (SPF), lack of security and privacy to sensitive patient data. During the last decade, we have seen an increasing use of wearable healthcare and fitness devices for remote monitoring of patients, e.g., blood pressure monitors, heart rate monitors, wearable ECG monitors, etc. Such devices are a part of the Internet of Things (IoT) [3] networks. These devices generate personal, sensitive data about the wearers, known as protected health information, therefore the transfer, storage, transactions on this data need to be handled in a secure and privacy-preserving environment. Blockchain SCs can be used for real-time management of such medical sensor networks [4] and the data generated by them. These systems can support medical interventions, store data, handle transactions, store the entire history of transactions in an immutable way [5], send notifications to stakeholders in a healthcare environment, including patients and medical practitioners. The decentralized nature of blockchain has led to its adoption in domains like finance, supply chain, and e-healthcare because of the concept of absolute truth, neutrality, and censorship resistance. It provides multiple and easy access points to distributed system with simple programming interfaces. A general architecture of a wearable cyber-physical application using embedded computers, sensors, and the internet as communication channel is given in Figure 11.1.

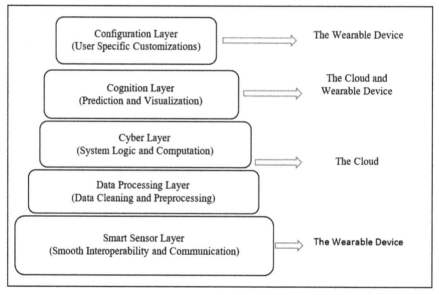

FIGURE 11.1 Layered architecture of a wearable healthcare device and placement of different layers.

The graph in Figure 11.2 shows the relative cost per record stolen due to data theft among various domains of information. While the cost of public information breach is around Rs. 1,000 per record stolen the cost of a medical data record stolen is around Rs. 9,000. The magnitude of loss itself is an indicator of the urgent need for secure platforms for maintenance, storage, communication of medical data.

The graph in Figure 11.3 shows the number of records in millions stolen and the kind of data that was revealed to unauthorized hackers. We have seen the impact in cost per record stolen in Figure 11.2.

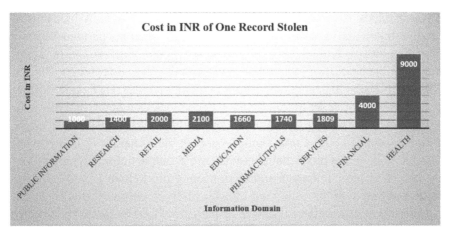

FIGURE 11.2   Comparison of cost involved in stealth of per record of various kinds of information.

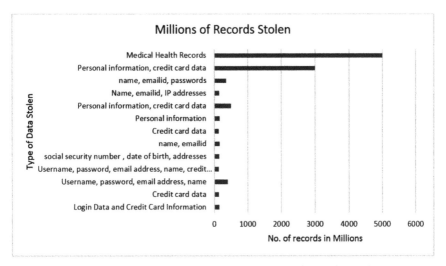

FIGURE 11.3   Number of records stolen in millions and the kind of data stolen.

11.1.1   *BLOCKCHAIN TECHNOLOGY*

The use of centralized systems to store medical data does provide control over this sensitive information, but these systems suffer from being single points of failure that are prone to data theft and other cyber-attacks [6]. Blockchain was introduced in 2008 as a platform for cryptocurrency known as Bitcoin

(BTC). A block is a package of permanently recorded data. A blockchain is a distributed and shared ledger that records transactions permanently by appending blocks. A blockchain, therefore, is a record of all transactions that happen on the block from the time it was created to the current time. Blockchain is replicated in peer-to-peer (P2P) network of nodes, each node storing a replica of the entire set of transactions. Each block except the first block in the chain is linked to the previous block by a hash of the previous block. Each block is associated with a time stamp. A user of blockchain is an owner who owns the node in the blockchain network and a unique public cryptographic key pair. User can initiate a transaction on a blockchain [22–24] These transactions include the following transfer of cryptocurrency, modification to prescription of a patient, transfer of patient's blood pressure reading to the doctor, etc. Every transaction is signed by the user initiating it, and sent to all nodes in the blockchain network. The nodes in the network validate the transaction that they receive and then record it using the public key of the original initiator of the transaction. Invalid transactions are discarded. Every candidate node has to undergo a computational review process called "mining" before it can be added to the blockchain. Mining nodes gather the valid transaction records and store these in the form of lists in time stamped blocks. Mining can be based on proof of work (PoW) concept or proof of stake concept. In PoW mining, each mining node computes a hash value of the concatenation of the last block in the blockchain and a new randomly chosen value known as a nonce. This is repeated with a different nonce till a hash value that is difficult to break is created. The first node to successfully create such a hash has the right to link its candidate block with the existing blockchain. The mining nodes having sufficient processing power only are capable of creating such hash functions. The PoW concept deters cyberattacks like distributed denial of service (DDOS) attacks [6, 7]. Mining provides two main advantages:

- Verification of legitimate transactions and avoiding duplicate transactions; and
- Creation of digital assets for miner nodes.

While in PoW, the miners break difficult cryptographic puzzles with their computational resources, the proof of stake is based on the stake of each validator in the blockchain. Validators replace miners in proof of stake-based blockchain [7]. Validators invest digital assets as their stake in the blockchain, they bet on the block next to be added to the chain. The validator is rewarded in proportion to its stake in the blockchain.

A blockchain can be public, private, or permissioned. A public chain is open source, free to access and use, e.g., the BTC chain. Once a user decides to join a blockchain network, they have to install the blockchain software and copy the blockchain into their system. User can install mining software and act as a miner node as well. The blockchain of today, known as Blockchain 2.0 offers SCs [31]. A smart contract is a program encoding the terms and conditions for enforcement of a contract between various parties in the blockchain-based transaction. E.g., the Ethereum (ETH) blockchain [7]. SCs can be executed on a server or a node and is known as a decentralized application (DApp). Since the hash between successive blocks is interdependent, the blockchain is immutable against changes hence secure and tamperproof. No block can be modified without modification to following blocks and verification by all other blocks and logging of the change. There is no single, central authority in a blockchain, hence no SPF. All blocks of the chain are replicated on all nodes of the peer network [7]. The only way the blockchain can be attacked is when an attacker gets control on a large number of nodes and mining nodes. Then corrupted data can be inserted [8]. A selfish mining node [8] can hold back transmission all newly mined blocks to blockchain. Figure 11.4 presents the general working of a blockchain based system.

Process of Secure Transactions over Blockchain

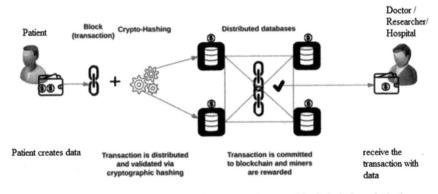

FIGURE 11.4   Process of secure transaction processing on a blockchain based platform.

Source: Created by B140970324. Reprinted from https://commons.wikimedia.org/wiki/File:Blockchain-Process.png. https://creativecommons.org/licenses/by-sa/4.0/deed.en

The next diagram, i.e., Figure 11.5 details the four important use cases of healthcare domain that can be implemented through blockchain [9, 10]. In the top left of the image, we see that healthcare companies and organizations [11] can redirect the healthcare data like blood pressure, heart rate, etc., with public ids of patients to blockchain using APIs.

In the top right of the image, we see that the blockchain platform processes incoming transactions using SCs, bottom left of the image shows the ability of query processing in blockchain which enables health organizations and institutions to query blockchain using APIs, here data analytics can be performed on the data as well. The bottom right of the image shows the scenario where the patient can securely share their data encrypted with the private key, authorized stakeholders like hospitals can access the data using public key.

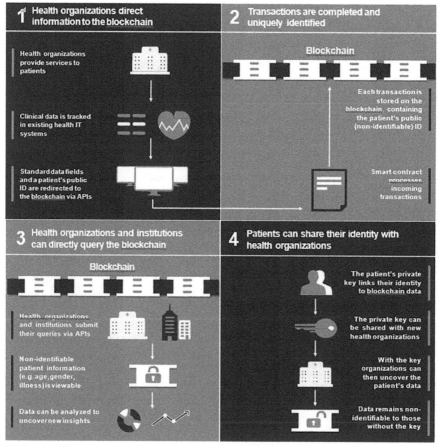

FIGURE 11.5   Applications of blockchain in healthcare domain.
Source: Reprinted from Ref. [53]. Open access.

11.1.2   MEDICAL IoT DEVICES

Before the development of IoT in healthcare, the patients had to physically visit the doctors or consult with them using telephone or text messages. Continuous monitoring always meant hospitalization for long periods. Wearable IoT devices has enabled continuous, remote monitoring of patient health. IoT in healthcare [12, 41, 42, 49] benefits all stake holders-patients, caregivers, physicians, hospitals, and insurance companies:

1. **IoT Solutions for Patients:** Wearable devices like fitness trackers, blood pressure, heart rate monitors, calorie counters, etc. These devices have helped improve the quality of life for patients by providing required alerts and reminders.
2. **IoT Solutions for Doctors:** Doctors can track the health and response of patients to drugs. It can help with administration of the best treatment approach.
3. **IoT Solutions for Hospitals:** For tracking location of medical supplies and equipment, as well as for locating medical staff. Hygiene and environment monitoring, asset management, inventory control.
4. **IoT Solutions for Insurance Companies:** The health insurance companies can use smart devices to track lifestyle of clients for dynamic premium fixing and management of claims as well as underwriting and pricing.

The data collected by wearable sensors in IoT devices can reveal important health conditions and insights about the person's health and even enable remote diagnosis and care. Some examples of medical IoT devices are:

1. A blood pressure cuff and a symptom tracking app used by cancer patients. This device sends regular updates to the physician on the response of the patient to cancer drugs.
2. A blood glucose monitoring device that continuously monitors the blood glucose levels of a diabetic patient at regular intervals over a period of days and reports them to the doctor.
3. An open-source system for closed-loop insulin delivery to bloodstream of the patient. This device acts as a continuous blood glucose monitor and delivers the required amount of insulin to the patient's blood stream.
4. Smart connected inhalers for management of asthma. This device has a sensor attached to an inhaler. The sensor can collect data that

is analyzed to detect the causes underlying the asthmatic attack or pulmonary distress. This device can track the use of medicine by the patient and forecast allergies due to allergens.

5. Ingestible sensors are embedded in medicine pills and ingested by the patient. The sensors, once inside the body, relay data to a smart phone application.

6. Google has ventured into the development of smart contact lenses. These lenses were to measure the levels of tear glucose and alert diabetes patients about the dangerous blood glucose levels.

7. Apple has created a watch that can detect depression in patients. This watch is connected to a smartphone application which allows continuous monitoring of patients with depression.

8. A device that can detect the speed of blood coagulation in patients has been developed. This device is useful to patients on blood thinners and other medicines with a risk of developing clots, strokes, or hemorrhages.

9. Apple watch with its research API can detect Parkinson's disease in wearers.

Figure 11.6 shows the scenario of a patient sharing the personal medical data on blockchain for ease of access by doctor, hospital, or analytics engines. The data is digitally signed by the patient and encrypted by a personal private encryption key before sharing. The authorized parties like the doctor, the hospital or the analytics company can use the data after decrypting it with public keys. The blockchain can incentivize the patient for sharing personal data.

11.1.3   SECURITY CONCERNS IN MEDICAL IoT

The physical and mental safety of users of medical IoT devices [42] is the primary concern closely followed by data security and privacy preservation [26]. These devices must provide reliable, secure operations and be able to resist the security threats. The data that is generated by these devices must be secured against unauthorized access and modifications. Any tampering to health measurements can prove to have non-desirable results for the patient or the doctor. Using a blockchain platform for the management of these devices will provide a fool proof environment against unauthorized manipulation of device or the data produces. The blockchain will record all the events including the data produced by the device, and securely allow

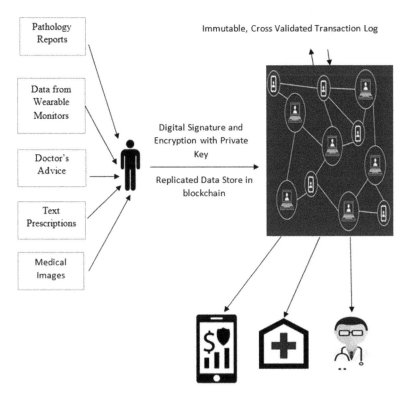

FIGURE 11.6 Scenario of the patient sharing personal healthcare data in blockchain.

the authorized stakeholders to analyze it or base decisions on it. Such blockchain can handle transactions like access to the device for configuration changes, data storing, data reading, etc. In such systems, the device will have to be assigned a hash of its unique identifier, which will be stored in the blockchain. The stakeholders in the chain will be the patient, doctor, care givers, etc. Caregivers can receive notifications of events like thresholds being crossed in vital statistics of the patient, battery life of the device, malfunctioning in the device along with the information about preventive maintenance of the device, etc. The SCs in blockchain allow the system to be secure, trustless, auditable, which are the features required in medical cyber-physical systems. The decentralization feature of blockchain eliminates the need for a third-party secure authentication system hence prevents data security compromises. It is due to this that blockchain technology is seeing a wide interest from the healthcare industry in the domains like electronic healthcare records management, patient identity management [25], insurance claims automation, drug development, medical research, patient data security, medical supply chain management (SCM), patient portals, etc. It is being seen as a technology that will enable the records of each patient to be securely logged in an open-source, trusted blockchain that can be accessed by all stakeholders with data integrity guarantees. Real time auditing on data and transactions in a blockchain minimized the possibility for unauthorized data manipulations. However, detailed implementation analysis, simulations need to be done before integration of blockchain in real applications [27], which provide tamper proof and secure data storage and commute for the sensitive data. There is a need for lightweight cryptography to suit the limited processing capabilities of wearable IoT devices. Limited processing power and memory also is a constraint when working with wearable sensors. One solution to this is the use of device dependent transaction ledgers, the use of alternative mining techniques, lightweight cryptographic algorithms.

11.2 HEALTHCARE APPLICATION OF BLOCKCHAIN

In this section, we present the applications for which blockchain [9–13, 27–29] can be applied in healthcare data management:

1. **Digital Identity Management:** The blockchain can be used to manage the identity of persons, i.e., patient, doctor, hospital, suppliers of drugs, etc., as well as identity of the devices and equipment in the hospital or medical supply chain.
2. **Insurance and Finance Management:** Blockchain can be used to manage the insurance claims, billings, purchases in the healthcare sector.
3. **Management of Clinical Research:** The clinical research data can be shared securely among the researchers [35–37] using blockchain. Monetization of personal health data can also be securely managed by the blockchain.
4. **Medical Supply Chain Management [30]:** Supply of medical devices, clinical supplies, medicines, blood cells, etc., can benefit from a blockchain-based platform that can provide an end-to-end environment for operation, compliance, financial transactions among the stakeholders like pharmacies, patients, blood banks, etc.

5. **Medical Data Storage:** Since blockchain stores data in encrypted form in a distributed fashion, the data is immutable unless all stakeholders in the network agree to the change. This enables the patient data records to remain authentic, un-tampered, and all copies of the data are synchronized [38] at each update. Table 11.1 shows the solutions provided by blockchain platform to various processes and tasks in the healthcare domain.

TABLE 11.1   Aspects of Healthcare System and the Solutions Provided by Blockchain

Aspects of Healthcare System	Blockchain Solutions
Data security and privacy preservation	Encrypted, decentralized, peer to peer, distributed ledger technology allows data security and privacy with only authorized access and permission control
Data sharing and collaboration	Stake holders-patients can share and control their data, medical professionals, researchers, hospitals can use the data and incentivize the patients for sharing over blockchain, researchers can share research data and findings, medical, and research recruiting can be done over blockchain
Incentives	Patients who share their data can receive incentives in cryptocurrency, all stake holders can be incentivized for playing their part well.
Medicine shipment tracking	Using IoT in shipments, the entire course of medical shipment from origin to end delivery point can be tracked and logged over blockchain. This will ensure the quality and authenticity of medicines and reduce drug counterfeiting.
Process integrity	Use of smart contracts enforces rules and removes the chances of fraud and manipulations.
Insurance claims and billing	Blockchain can track all procedures and medicines taken by a patient and hence can bring transparency to insurance claims and the payment system.
Analytics	Advanced analytics over the data in blockchain can make possible customized administration of drug doses at various times to patients.
Care delivery	Blockchain can track the medicine schedule of a patient and issue reminders to patient or caregivers when the medicine is due, when medicine needs to be refilled or when medicine is exceeding its expiration dates.
Real-time monitoring and advisory	In combination with IoT and wearable healthcare sensors, blockchain can acts as a storage and communication medium for this data to doctors in real-time and therefore let the physician to monitor the health of patients continuously.

11.2.1   ADVANTAGES OF BLOCKCHAIN TECHNOLOGY [34] IN HEALTHCARE DOMAIN

1. **Better Security:** The blockchain provides an immutable, secure, distributed peer-to-peer database to store patient records, healthcare data from sensors. These systems are scalable by design, secure [39, 40] against hacker attacks, the administrative costs are less, the security offered is crucial for the personal data management.

2. **Enhanced Interoperability:** Blockchain-based medical record management enhances the interoperability [35] among various stakeholders in the healthcare ecosystem-the patients, hospitals, clinics, care providers. The data sharing with blockchain in a unified database, removes the interoperability issues caused by separate record keeping systems used by the stakeholders.

3. **High Availability and Transparency:** Patients can have high availability of the personal healthcare data over blockchain. The accuracy and authenticity of data is ensured by blockchain.

4. **Medical Supply Chain Management (SCM):** Using blockchain, the medical supplies can be reliably tracked throughout the manufacturing to distribution cycle. This can reduce the drug counterfeiting problem. With the application of IoT devices [50–52] to measure conditions like temperature, humidity, blockchain can also track the storage and shipping conditions of the medicines and supplies and detect medicine quality.

5. **Prevention of False Insurance Claims:** Immutability of blockchain records can prevent false billing by hospitals and false insurance claims for the procedures that did not take place.

6. **Improving Quality of Clinical Trials:** Medical data that resides on the blockchain can be used for clinical research and for the identification of patients for clinical trials of new drugs. This will improve the clinical trial process and benefit the patients with new drugs. Blockchain can ensure data integrity for the collected information.

11.2.2   BLOCKCHAIN LIMITATIONS IN HEALTHCARE DOMAIN

There are some limitations [32] to widespread use of blockchain in the healthcare sector as of now. Some of these limitations [14] are listed here:

1. **Compliance to Regulations:** The companies willing to use the blockchain as a platform for medical services, need to comply with regulations by the government regarding data storage, sharing, and protection. This will need a higher level of privacy, access control, and customized records in blockchain network.

2. **Initial Investments:** Blockchain-based solutions need a high level of initial investment, which is a deterrent to its wide adoption.

3. **Speed:** The number of transactions that can take place per second in blockchain is much less than that can take place in a centralized database server. The patient record management system is expected to store and track data of millions of patients; this data will be in the form of text, images, and videos. Hence the speed of operation needs to be improved a lot before blockchain can be used as a total solution for medical health record management.

4. **Immutability:** This offered by blockchain is important for the prevention of fraud and building trust. However, sometimes there is a legitimate need for updates to data, e.g., when the regulations over data storage and sharing change across borders, the records need to be updated, this is not enabled in blockchain.

5. **Implementation Challenges and Integration with Current Systems:** Blockchain technology needs to be interoperable with existing healthcare solutions in order to see a wider adoption. Allowing patients to be owners of their data can pose as a security threat, so there is a need for access control at various levels.

11.3   COMPANIES USING BLOCKCHAIN IN HEALTHCARE

Ability of blockchain to reduce the costs associated with various processes in the healthcare industry like drug supply chain, management of healthcare records, inventory management, viability tracking, secure encryption of patient data, etc., has made it an interesting option to many pharma companies worldwide [10, 11, 13]. In this section, we will look at some real examples of the companies that have implemented blockchain for various tasks in the healthcare domain. Many countries [39–41] worldwide are exploring the use of blockchain in healthcare sector. Especially, Estonia has 95% of its healthcare information residing on blockchain and 99% of the medical prescriptions have been digitalized:

1. Ability of blockchain to provide decentralized, immutable, and non-repudiable logging of all transactions on data promises data security and integrity. Private Blockchain also promises data hiding and encryption along with ease of data sharing among authorized parties. A Colorado-based company called Burstiq provides blockchain solutions for storage, sale, sharing or licensing of medical data in compliance with the rules of HIPAA.

2. Texas-based Factom uses private blockchain for storing patient records that are accessible to authorized users only.

3. London-based Medicalchain [18] uses blockchain to maintain patient data while keeping the track of origin of data and maintaining a single point of truth for all stake holders in the chain. They allow patients to consult the doctors digitally and enable payments via cryptocurrency called MedTokens.

4. Massachusetts-based Simplyvital [16] Health has developed an open-source platform called Nexus Health to enable easy access to medical information for ease of coordination among medical professionals.

5. Canada-based coral health research and discovery [19] uses block-chain to automate the processes of caregiving and collaboration among medical professionals. They use smart contracts to ensure the integrity of the treatment give.

6. Russian company RoboMed uses artificial intelligence (AI) and blockchain to provide wearable medical devices and telemedicine to patients. They intend to provide a single point of care to patients with the use of chatbots. They use smart contracts to incentivize the patients who share their data.

7. Georgia-based Patientory [21] is a company using end-to-end encryption for secure storage and communication of patient data.

8. California-based chronicled [20] uses blockchain to provide a chain of custody for supply-chain management of drugs. They ensure private, efficient, safe, lawful delivery of drugs in medical supply chains. This will reduce the incidences of drug trafficking and counterfeiting.

9. French company Blockpharma [15] provides a blockchain-based supply chain to prevent drug counterfeiting by offering end-to-end drug traceability. They track all points in the supply chain shipment, and the information about quality and life cycle of the drug is provided to the patient.

10. California-based Tierion uses blockchain to maintain a history of ownership and possession of drugs along with timestamps and credentials by gathering information from various documents. They maintain a proof of ownership of drugs at all points in the supply chain.

11. Center for disease control and prevention, Georgia uses blockchain with timestamps, data processing health reporting abilities, for monitoring of diseases just like a supply chain. This platform can track outbreaks of diseases, and find patterns in the spread of the disease in real-time.

12. Massachusetts-based nebula genomics uses blockchain in genetic [43] sequencing process [44]. Generally, biotech, and pharmaceutical companies need to acquire genetic data from sequencing companies at a high cost. Nebula Genomics has built a database of genetic data that is accessible to subscribing companies without middlemen. The genetic data is safely purchased, in encrypted form, from the original producers of data.

13. Florida-based EncrypGen [17] uses blockchain for storing, sharing, and transacting genetic data using authorized and secure access.

14. California-based Doc.AI is a platform for capture and sharing of poly-omics medical data among researchers and scientists. They also provide predictive genetic modeling using the data.

11.4   CONCLUSION

In this chapter, we have seen that blockchain has application at various task levels in the healthcare domain. It can be offered as blockchain as a service in healthcare data infrastructure management, it can be used for storage and communication of electronic patient health records, advanced analytics can be performed on healthcare data over a blockchain, blockchain can be used to provide security for IoT and other medical devices and the data produced by them, blockchain can be an effective solution for digital identity management of various stakeholders in healthcare system, medical SCM can benefit from the use of blockchain platform, delivery of medicines and medical supplies can be done using blockchain as can the care delivery, blockchain can also be used for advisory and legal compliance in healthcare. We have seen a list of companies world over which have started using blockchain for various processes in healthcare.

Blockchain enables safe and immutable storage of medical data records and their sharing in a transparent medical supply chain. These systems can improve coordination among stakeholders and add incentives for desirable actions. SCs in a blockchain can automate the inventory management and ordering when stocks go low in a hospital, payments to various vendors can be automated, medical shipments can be thoroughly tracked for integrity, and transactions can be transparently reviewed by all stakeholders. Data integrity is ensured, health insurance is streamlined by elimination of false claims and false billing. Having immutable, time-stamped records of data and transactions in multiple nodes of a blockchain removes the possibility of malicious alterations to the data or records. Though blockchain technology is a promising way of performing healthcare processes yet there exist certain regulatory, logistic, and technical challenges to the implementation of blockchain-based systems. A thorough analysis of end-to-end working of healthcare uses cases needs to be carried out in order to make the implementation beneficial to the masses.

KEYWORDS

- blockchain
- decentralized application
- digitalization
- healthcare
- Internet of Things
- physical systems

REFERENCES

1. Agbo, C. C., Mahmoud, Q. H., & Eklund, J. M., (2019). Blockchain technology in healthcare: A systematic review. *Healthcare (Basel), 7*(2), 56. doi: 10.3390/healthcare7020056.
2. Bouras, M. A., Lu, Q., Zhang, F., Wan, Y., Zhang, T., & Ning, H., (2020). Distributed ledger technology for e-health identity privacy: State of the art and future perspective. *Sensors (Basel), 20*(2), 483. doi: 10.3390/s20020483.
3. Dorri, A., Kanhere, S. S., Jurdak, R., & Gauravaram, P., (2017). Blockchain for IoT security and privacy: The case study of a smart home. *Proceedings of the International Conference on Pervasive Computing and Communications Workshops (PerCom Workshops)* (pp. 618–623). IEEE.

4. Dorri, A., Kanhere, S. S., & Jurdak, R., (2017). Towards an optimized blockchain for IoT. *Proceedings of the 2017 IEEE/ACM Second International Conference on Internet-of-Things Design and Implementation IoTDI'17* (pp. 173–178). IEEE.

5. Dubovitskaya, A., Novotny, P., Xu, Z., & Wang, F., (2019). Applications of blockchain technology for data-sharing in oncology: Results from a systematic literature review [published online ahead of print. *Oncology, 2019*, 1–9. doi: 10.1159/000504325.

6. Kuo, T. T., Zavaleta, R. H., & Ohno-Machado, L., (2019). Comparison of blockchain platforms: A systematic review and healthcare examples. *J Am Med Inform Assoc., 26*(5), 462–478. doi: 10.1093/jamia/ocy185.

7. Hussien, H. M., Yasin, S. M., Udzir, S. N. I., Zaidan, A. A., & Zaidan, B. B., (2019). A systematic review for enabling of develop a blockchain technology in healthcare application: Taxonomy, substantially analysis, motivations, challenges, recommendations and future direction. *J. Med. Syst., 43*(10), 320. doi: 10.1007/s10916-019-1445-8.

8. Tracy, S. *More than 16 million Medical Records Breached in 2016 [Online].* Available: https://www.linkedin.com/pulse/more-than-16-million-medical-records-breached-2016-ilesh-dattani/ (accessed 5th August 2021).

9. Clauson, K., Breeden, E., Davidson, C., & Mackey, T., (2018). Leveraging blockchain technology to enhance supply chain management in healthcare. *Blockchain in Healthcare Today, 1.*

10. Liang, X., Zhao, J., Shetty, S., Liu, J., & Li, D., (2017). Integrating blockchain for data sharing and collaboration in mobile healthcare applications, In: *IEEE 28th Annual International Symposium on Personal, Indoor, and Mobile Radio Communications (PIMRC)* (pp. 1–5). Montreal, QC.

11. Jennifer, B., (2017). *Around 16% of Healthcare Stakeholders Plan to Use Blockchain by 2017 [Online].* Available: https://healthitanalytics.com/news/16-of-healthcare-stakeholders-plan-to-use-blockchain-by-2017 (accessed on 5th August 2021).

12. Robmenzies, (2017). *Healthcare Data Management meet Blockchain [Online].* Available: https://steemit.com/healthcare/@robmenzies/healthcare-datamanagement-meet-blockchain (accessed on 29th July 2021).

13. *Size of the Blockchain Technology Market Worldwide from 2016 to 2021 [Online].* Available: https://www.statista.com/statistics/647231/worldwideblockchain-technology-market-size/ (accessed on 29th July 2021).

14. Julia Adler-Milstein, Catherine M. DesRoches, Peter Kralovec, Gregory Foster, Chantal Worzala, Dustin Charles, Talisha Searcy, and Ashish K. Jha, Electronic Health Record Adoption In US Hospitals: Progress Continues, But Challenges Persist, *Health Affairs, Vol 34, No. 1, Affordability, Access, Models Of Care & More, 2015.* doi: 10.1377/hlthaff.2015.0992. https://www.healthaffairs.org/doi/10.1377/hlthaff.2015.0992 (accessed on 5th August 2021).

15. *Blockpharma [Online].* Available: https://www.blockpharma.com/ (accessed on 29th July 2021).

16. *Simply Vital Health [Online].* Available: https://www.f6s.com/simplyvitalhealth (accessed on 5th August 2021).

17. David, K., (2019). *The Future of Genomic Data Encryption [Online].* Available: https://encrypgen.com (accessed on 29th July 2021).

18. Medicalchain [Online]. Available: https://medicalchain.com/en/ (accessed on 29th July 2021).

19. *CoralHealth*, 2018 [Online]. Available: https://mycoralhealth.com/ (accessed on 29th July 2021).

20. *Smart Supply Chain Solutions. [Online]*. Available: https://chronicled.com/ (accessed on 29th July 2021).

21. *Patientory*. [Online]. Available: https://patientory.com/ (accessed on 29th July 2021).

22. Juneja, A., & Marefat, M., (2018). Leveraging blockchain for retraining deep learning architecture in patient-specific arrhythmia classification. In: *IEEE EMBS International Conference on Biomedical & Health Informatics (BHI)* (pp. 393–397). Las Vegas, NV, USA.

23. Liu, W., Zhu, S. S., Mundie, T., & Krieger, U., (2017). Advanced blockchain architecture for e-health systems. In: *IEEE 19ᵗʰ International Conference on e-Health Networking, Applications and Services (Healthcom)* (pp. 1–6). Dalian.

24. Gaby, G. D., Jordan, M., Matea, M., & Praneeth, B. M., (2018). Ancile: Privacy-preserving framework for access control and interoperability of electronic health records using blockchain technology. *Sustainable Cities and Society, 39*, 283–297.

25. Ramakrishnan, R., & Gaur, L., (2019). *Internet of Things: Approach and Applicability in Manufacturing*. CRC Press.

26. Dwivedi, A. D., Srivastava, G., Dhar, S., & Singh, R., (2019). A decentralized privacy-preserving healthcare blockchain for IoT. *Sensors (Basel), 19*(2), 326. doi: 10.3390/s19020326.

27. Petre, A., & Haï, N., (2018). Opportunités et enjeux de la technologie blockchain dans le secteur de la santé [Opportunities and challenges of blockchain technology in the healthcare industry]. *Med. Sci. (Paris), 34*(10), 852–856. doi: 10.1051/medsci/2018204.

28. https://openledger.info/insights/blockchain-healthcare-use-cases/ (accessed on 29th July 2021).

29. Leeming, G., Cunningham, J., & Ainsworth, J. A., (2019). Ledger of me: Personalizing healthcare using blockchain technology. *Front Med. (Lausanne), 6*, 171. doi: 10.3389/fmed.2019.00171.

30. Kamel, B. M. N., Wilson, J. T., & Clauson, K. A., (2018). Geospatial blockchain: Promises, challenges, and scenarios in health and healthcare. *Int. J. Health Geogr., 17*(1), 25. doi: 10.1186/s12942-018-0144-x.

31. Giordanengo, A., (2019). Possible usages of smart contracts (Blockchain) in healthcare and why no one is using them. *Stud Health Technol Inform., 264*, 596–600. doi: 10.3233/SHTI190292.

32. Mackey, T. K., Kuo, T. T., Gummadi, B., et al., (2019). 'Fit-for-purpose?' – challenges and opportunities for applications of blockchain technology in the future of healthcare. *BMC Med., 17*(1), 68. doi: 10.1186/s12916-019-1296-7.

33. Hylock, R. H., & Zeng, X., (2019). A blockchain framework for patient-centered health records and exchange (HealthChain): Evaluation and proof-of-concept study. *J. Med. Internet Res., 21*(8), e13592. doi: 10.2196/13592/.

34. Deloitte. (2016). *Blockchain: Opportunities for Healthcare*. [Internet] New York (NY): Deloitte; c2019. Available from: https://www2.deloitte.com/us/en/pages/public-sector/articles/blockchain-opportunities-for-health-care.html (accessed on 29th July 2021).

35. Gordon, W. J., & Catalini, C., (2018). Blockchain technology for healthcare: Facilitating the transition to patient-driven interoperability. *Comput. Struct. Biotechnol. J., 16*, 224–230.

36. Kuo, T. T., Kim, H. E., & Ohno-Machado, L., (2017). Blockchain distributed ledger technologies for biomedical and healthcare applications. *J. Am. Med. Inform. Assoc., 24*, 1211–1220.

37. Cyran, M. A., (2018). Blockchain as a foundation for sharing healthcare data. *Blockchain Healthcare Today.*

38. Kaye, J., Whitley, E. A., Lund, D., Morrison, M., Teare, H., & Melham, K., (2015). Dynamic consent: A patient interface for twenty-first century research networks. *Eur J Hum Genet., 23*(2), 141–146.

39. Lima, C., (2018). *Blockchain-GDPR Privacy by Design: How Decentralized Blockchain Internet Will Comply with GDPR Data Privacy.* Available from: https://www. semanticscholar.org/paper/Blockchain-GDPR-Privacy-by-Design-How-Decentralized-Lima/a7ac8ebc6eb4fd3e0c0d1645e685c7210ed12962 (accessed on 29th July 2021).

40. Commission Nationale de l'Informatique et des Libertes, (2018). *Blockchain and the GDPR: Solutions for a Responsible Use of the Blockchain in the Context of Personal Data.* Paris, France: Commission Nationale de l'Informatique et des Libertes. Available from: https://www.cnil.fr/en/blockchain-and-gdpr-solutions-responsible-use-blockchain-context-personal-data (accessed on 29th July 2021).

41. Innovative Medicines Initiative, (2018). *Topic: Blockchain-Enabled Healthcare, Brussels, Belgium: Innovative Medicines Initiative.* Available from: https://www. imi.europa.eu/sites/default/files/uploads/documents/apply-for-funding/future-topics/ IndicativeText_BlockchainHealthcare.pdf (accessed on 29th July 2021).

42. Singh, G., Gaur, L., & Ramakrishnan, R., (2017). *Internet of Things-Technology Adoption Model in India, 25*, 835–846.

43. Bates, M., (2018). Direct-to-consumer genetic testing: Is the public ready for simple, at-home DNA tests to detect disease risk? *IEEE Pulse, 9*(6), 11–14.

44. *San Francisco (CA): Nebular Genomics*, (2018c). Available from: http://www.nebula. org (accessed on 29th July 2021).

45. *Genomes.io*, (2018c). Available from: https://www.genomes.io (accessed on 29th July 2021).

46. *Health Wizz.* Available from: https://www.healthwizz.com (accessed on 29th July 2021).

47. *Medicalchain.* Available from: https://medicalchain.com/en/ (accessed on 29th July 2021).

48. Ketan, P., Mitchell, P., David, H., & Josip, C. (2019). *Implementation Considerations for Blockchain in Healthcare Institutions.* https://blockchainhealthcaretoday.com/index. php/journal/article/view/114/133 (accessed on 29th July 2021).

49. Pramanik, P. K. D., Solanki, A., Debnath, A., Nayyar, A., El-Sappagh, S., & Kwak, K. S., (2020). Advancing modern healthcare with nanotechnology, nanobiosensors, and internet of nano things: Taxonomies, applications, architecture, and challenges. In: *IEEE Access* (Vol. 8, pp. 65230–65266). doi: 10.1109/ACCESS.2020.2984269.

50. Rameshwar, R., Solanki, A., Nayyar, A., & Mahapatra, B., (2020). Green and smart buildings: A key to sustainable global solutions. In: *Green Building Management and Smart Automation* (pp. 146–163). IGI Global: Hershey, PA, USA.

51. Krishnamurthi, R., Nayyar, A., & Solanki, A., (2019). Innovation opportunities through the Internet of Things (IoT) for smart cities. In: *Green and Smart Technologies for Smart Cities* (pp. 261–292). CRC Press: Boca Raton, FL, USA.

52. Solanki, A., & Nayyar, A., (2019). Green Internet of Things (G-IoT): ICT technologies, principles, applications, projects, and challenges. In: *Handbook of Research on Big Data and the IoT* (pp. 379–405). IGI Global: Hershey, PA, USA.

53. Krawiec, R.J. (2016). Blockchain: Opportunities for Healthcare. http://www.rjkrawiec.com/blog/2016/9/15/blockchain-opportunities-for-healthcare

PART IV

Future Applications of Blockchain in Business and Management

CHAPTER 12

Key Drivers of Blockchain Technology for Business Transformation

SHIVANI A. TRIVEDI[1] and OKUOGUME ANTHONY[2]

[1]*S.K. Patel Institute of Management and Computer Studies-MCA,
Kadi Sarva Vishwavidyalaya, Gujrat, India, E-mail: satrivedi@gmail.com*

[2]*Lapland University of Applied Sciences, Tornio, Finland,
E-mail: Anthony.Okuogume@lapinamk.fi*

ABSTRACT

Blockchain is a disruptive technology that is igniting innovation, research, and business transformation across different business sectors from Financial, automotive, healthcare, supply chains, travels to the building of smart cities. Adopting blockchain technology in businesses and its disruptive characteristics are well understood. However, technology disruption does have two sides; it can be a source of enormous opportunity and challenges to businesses. This point is one of the key focuses of this chapter. That is, our attention on blockchain technology as key drivers of business transformation is simply said; To explore the disruptive impacts of blockchain technology on businesses through the opportunities and challenges of the disruption. The approach of the research was made through analysis of scientific articles and publications by industry practitioners. It describes the classifications of the impacts and challenges of blockchain technology from the point of its core benefits and challenges, so it is an adaptable technology, open-source vs proprietary solutions and its impact on the monetization of data.

The impacts of blockchain technology considered at three levels. At the macro level, we looked at its effects on business in general and the economics of the company. At the firm level, the analysis focused on the re-architecting of the firm or of *paradigm shift* in the conception of the firm and management science paradigms in general—lastly, a consideration of the impact of blockchain technology on firm's business model and their implementation—finally, the impact of blockchain technology in the monetization of data. Here, the analysis explored the subject at two levels. The first level is businesses, and another is consumers. The core focus is acquiring data monetization through exploiting consumers data.

In summary, how blockchain technology transforms businesses are a well-covered topic in academic research. Our analysis has identified key drivers of blockchain technology for business transformation. Our work is guiding how to change the business to achieve data monetization through blockchain technology integrated with IoT apps. The scope of this work can extend to identify implementation issues. We discovered that there is room to make contributions to the discussions and offer a more in-depth insight into the subject.

12.1   INTRODUCTION

The idea of business transformation by information technology is not new; for example, cutting the costs of transaction and interaction between companies [3, 6, 8]. Blockchain technology is the technology in question and its disruptive impact on business. The effect is so forceful because it is transforming the very basis of business architecture, i.e., it changes the foundation of value exchanges. These rules guide transactions, the currency for the trade, and it is data-driven. The big question is, why is this technology having such transformational force on the business sector? To answer that question, let us take a brief look at the technology and its characteristics. Blockchain characteristics include; security, transparency, trust, and privacy [17, 43]. With these characteristics, Blockchain has forced changes in different business sectors and transactional processes. According to Ref. [43], the effect of Blockchain disruptive-based on its capacity to reduce vagueness in transactions. The confusion of operations is uncertainty, insecurity, and ambiguity in activities by providing full transactional revelation and by generating a single truth for all network contributors. The implication of this simply stated is that business transactions will be free of what they termed as "skewed regulatory and licensing schemes," which are subjects of "hierarchical power structures" and the dubious influence of dominant interests [44] in a transactional process. Blockchain technology, apart from offering high transparency in business transactions, also provides vast opportunities through automatic business execution and disinters mediated business models and commerce

[44], provide enormous values to the participating parties in the transactional process.

Blockchain is a disruptive technology that is igniting innovation, research, and business transformation across business domains. Many businesses from tourism, healthcare, finance, airlines are adopting blockchain technology for their business opportunities along with smart city Internet of Things (IoT) business processes. Applications of blockchain technology are in banking, smart contracts (SCs), connected car, healthcare, IoTs industry, identity, marriage certificates, and many more [13–16].

The introduction of smart devices and the internet also means that the blockchain technology and its integration with IoT. The most crucial combination to look for is the merger between Blockchain and AI. Integrated technologies will transform business through Blockchain as a technology by permitting it to achieve the higher goal of being wholly self-directed and highly secured. According to Ref. [45], AI is the brain and has a more excellent capability for data crunching and learning to make better decisions, assessing datasets, and making predictions of the future.

Bitcoin (BTC) is the first application that has been implemented with blockchain technology. A block in a blockchain technology stores a value as in BTC its digital money. It is also known as crypto-currency introduced by Nakamoto [2]. This application has implemented a solution to execute a payment transaction, which is based on cryptography and trust between two parties who are involved in the payment transaction. Thus, this technology is also known as a collective bookkeeping system, which is based on hash functions. The transaction processes allow participants to meet an agreement on the approval of a transaction. Blockchain technology is storing information about executed transactions. Each block generated is verified and reviewed with consensus among parties and algorithms based on proof of concept, proof of work (PoW), etc. Then this block is added consecutively in the decentralized database in the network. It has Blockchain characteristics identified as immutability, no central authority, secure, and transparent, privacy, scalability, and availability based on the study. It quoted in the literature that "Data is the new oil," "AI/ML is new refinery," "5G is new transportation," and "Blockchain is new settlement" with its strong characteristics, immutable, transparent, and traceable. As per CilansSystem, Dubai is more than halfway towards its goal of going completely paperless and cashless by December 2021, for that they are going to implement evolving technologies like Blockchain to achieve the objective. In the banking industry, people keep their money in the bank for safety purposes. They trust the bank with their money. Thus, it is possible to make money transfers from one country to another through the banking system. In this transferring process, more than one agency comes into the picture, and charges apply to all middle agencies. Actual money being transfer between parties becomes a burden for the party, making the money transfers. Intermediate agencies eliminated and improved transaction efficiency when blockchain technology used in the money transfer process.

The process is transparent and with the consensus of parties involved in the financial transaction. Thus, Blockchain makes things transparent, safe, secure, auditable, and resistant to the outage, decentralized, and eliminates the need for a middleman. A blockchain is defined as a database where untrusted parties agree on the state of the database without using a mediator.

The blockchain-based application implemented using permissioned and permissionless mechanisms for implementing consensus. This mechanism is implemented with multiple versions of the ledger to achieve an agreement. There are many tools and technology available for implementation using various platforms and frameworks on which many tools used to deploy SCs. Such platforms are Ethereum (ETH), Truffle, Solc, Solium, Geth, Embark, Ganache, etc. And Remix is the tool used for implementing SCs using solidity language, and it is a browser-based tool.

Businesses deciding to implement blockchain technology must prepare a checklist of implementation strategies—core factors, like levels of speed, scalability, and privacy, permissioned or permissionless Blockchain. Based on the literature survey, it observed that most of the organizations decide to implement blockchain solutions in private networks, i.e., permissioned Blockchain. Data and transaction privacy is the expected outcomes in this context. This implementation strategy requires a high level of trust among the participants and uses a Byzantine fault-tolerant consensus mechanism. Businesses need not have to shift to a costly blockchain technology when trust between parties already established. Many organizations are applying Microsoft Azure Blockchain as a service, Hyperledger Fabric, IBM Hyperledger Fabric is the best example to implement business processes through blockchain technology. Many businesses have created and connected associations of Blockchain. All these consortia have made an impact on working collaboration among business competitors to set and implement blockchain technologies.

This chapter has the objective of identifying blockchain technology as a key driver in business transformation. The work is divided into five sections. The first section is the introduction to blockchain technology and its impact on business transformation. Next part, the literature review is based on the following points:

- Background studies of techniques and tools used for blockchain technology;
- Benefits and challenges of blockchain technology;
- Adoptable blockchain technologies:
- o   Open-source solutions;
- o   Proprietary solutions.
- The adoption of blockchain technology in various business sectors and its impact.

Based on the literature review findings, the next section describes blockchain technology adoption by various business sectors and in the current era. Furthermore, it describes Blockchain as a disruptive technology and presents business solutions using blockchain technology. The next section elaborates on the proposed architecture of blockchain technology in business transformation with its key drivers. The further section outlines the impact of blockchain technology in data monetization based on proposed architecture for business transformation, which includes data privacy, data tracking, and its usage, data authentication, and security. The prototype implementation based on proposed architecture is discussed with industry persons and outlines the benefits and limitations of submitted work. The last two sections are about the conclusion and future work which can be carried out based on this work.

12.2   LITERATURE REVIEW/BACKGROUND STUDIES

This section is representing a systematic literature review. We have organized literature review findings in five parts. These are blockchain technology tools and their use, which are studied as open-source solutions, proprietary solutions, and the adoption of blockchain technology in various business sectors and its impact, and Blockchain as a disruptive technology and its threats and opportunities.

12.2.1   BLOCKCHAIN TECHNOLOGY AND TOOLS

Blockchain technology is generally accepted, but this did not happen at once. The acceptance or adoption of blockchain technology in the business sector happened at different transitional stages. So, it is essential that we take a look at these transitional stages, and we do so with the Blockchain classifications presented by Erceg, Sekuloska, and Kelic [44] and added value in it as per Table 12.1.

TABLE 12.1   Blockchain Classifications

Classification	Characteristics	Application
Blockchain 1.0: Currencies	The consensus in decentralization and cost of transactions	Digital currencies-payment using digital system
Blockchain 2.0: Applications	Smart contracts with different services	The entire state of economic, market, and financial applications which are more complex than a simple cash transaction, stocks, futures, mortgages
Blockchain 3.0: DAG (directed acyclic graphs)	Decentralized application with cloud computing and storage	Application beyond organization boundary, finance, and currency, especially in government, health, and science area
Blockchain 4.0: Industry 4.00 and cross-chain	Decentralized AI and IoT	Automatic decision making: Degree of automation is high in decision making
Blockchain 5.0: World	The unique technology, high speed, innovations	Value and application: Universal smart contract, medium for financial transaction

From the blockchain categories presented in Table 12.1, we can observe that the first stage of the adoption of blockchain technology was the creation of cryptocurrency. The second phase, according to the categories, is the creation of platforms that made it possible to implement "SCs that automatically enforce their contractual conditions and terms" [44]. The third stage is where we can see a much wider application of the technology and is characterized by "decentralized applications (DApps)." According to this view, this makes it possible for people to interact with the technology in a "more systematic and accustomed ways through browsers and smartphones" [44].

Wei et al. [50] have worked on an integrity protection framework. Wei et al. [50] have implemented the virtual machine proxy model, and the unique hash value corresponding to the file generated by the Merkel Root Tree. The use of Merkel Root Tree is for monitoring the data change by implementing the smart contract on the Blockchain. In the case of data tampering, the user issues a warning message for it. For data integrity and verification scheme, a "block-and-response" mode used to construct blockchain-based cloud data.

In this research chapter, the authors are presenting a comprehensive literature review of Blockchain Technologies and its applications in various sectors. Their research represents that Blockchain is transforming and distracting organizations across all industries. Blockchain is the next significant technological invention parallel to the internet. The law and SCs are the new way of enforcing the agreement and achieving censuses without any need of the third party in blockchain technology according to Ref. [40].

Table 12.2 describes the third generation blockchain-based tools and technology depicted. It represents the business area and the impact of blockchain technology in it.

TABLE 12.2 Impact of Blockchain Technology in Various Business Areas

Technology Used	Business Area	Its Impact	Authors/ Researcher
Blockchain-based cloud data, smart contract	In general	Security and integrity of data, immutability	Wei et al. [50]
Smart contract	Finance, agriculture, social media, automotive, and many more	Enforcing agreement and censuses, without a third party.	Shrivas and Management [40]
Decentralization, smart contract	Healthcare	Digital democratization of care delivery models, personalized, and outcome-based treatment paradigm	Decentralized, Internet, and Privacy [17]
Blockchain and IoT	Healthcare	Cost reduction increased transparency, and accountability, organizations are moving beyond proof-of-concept and leveraging blockchain-based solutions	Lao et al.; Mansur and Sujatha; Khatoon; Javaid et al. [20, 35, 61, 62]
Distributed ledger	eGovernance	Governance issues in financial infrastructure, like decision rights, control mechanisms, and incentives.	Hughes et al.; Jepkemei and Kipkebut [24, 25]

TABLE 12.2   *(Continued)*

Technology Used	Business Area	Its Impact	Authors/ Researcher
PKI based blockchain technology	In general	Confidentiality of sensitive records, a group key management scheme for secure group communication	Casino, Dasaklis, and Patsakis [21]
Blockchain-based IoT, blockchain-based security, blockchain-based data management	In general	Security, data management, transparency	Lu [59]
Distributed ledger technology	Digital art management, supply chains, and healthcare	Secure and transparent, distributed accounting ledgers, cost-effective	Hughes et al. [24]
Hyperledger sawtooth, IoT	Pharmaceutical	Transparency, trust, tracking	Mansur and Sujatha [35]

Blockchain technology is making its inception with various platforms in terms of proprietary and open sources community. Irrespective of proprietary or open-source, a blockchain platform primarily built to satisfy the following development criteria:

- A first criterion is that a blockchain platform consists of a database that is transactional and working. It ensures users access in peer to peer at any point in time [9].
- A second criterion is that a blockchain platform consists of user identification labels that enable transactions between active users.
- A third criterion is the platform must verify transactions ahead of its approval. It means that the platform must have a control mechanism to stop transactions that are not validated.

Table 12.3 represents some of the open-source and proprietary platforms available for blockchain technology.

TABLE 12.3 Few Examples of Blockchain Platforms

Proprietary-Blockchain Platform	Open-Source-Blockchain Platform
Oracle blockchain platform	Ethereum
Amazon blockchain platform	Open chain
IBM blockchain platform	Hyperledger

12.2.2 ADOPTABLE BLOCKCHAIN-BASED BUSINESS SOLUTIONS AND ITS IMPACT

In this section, the authors will elaborate and present blockchain technology as an acute driver of business transformation in the modern era. We know an IoT [51, 69–70] and Blockchain has been told and researched on technology in the past decade. The business organization which has adopted blockchain technology sometimes leads to confusions and frustrations to analyze and predict business information [24]. In the beginning era of blockchain technology is more about Cryptocurrencies. Its very complex technical information meant to establish legitimacy in the technology world and cryptocurrencies as a de facto digital currency [24]. After all, Satoshi Nakamoto wrote his famous protocol for peer-to-peer (P2P) electronic cash system using Cryptocurrencies called BTC only in 2008. ETH, as a platform, on the other hand, was only created in 2015 by 19 years old Canadian Russian VitalikButerin. Unlike BTC, ETH contains powerful tools that help developers and others to develop software services that will enable the creation of distributed service-based enterprises. And we could see clearly how blockchain technology utilized by businesses of all sizes and also its transformational impact on business paradigms, processes, and economic impact [24]. Through this, it became easy to explain the outstanding benefits, application of blockchain technology, and its disruptive impact on businesses today. Blockchain to say has redeemed its promise of disrupting business paradigms and operating principles.

In this section, we consider the use of blockchain technology and its impacts from three levels. First, we will present the analysis on Blockchain from a macro perspective in the business areas and economy, the firm-level from technology impact on the re-architecting of the firm and management science, and lastly, on the business model and their successful implementation through blockchain technology. Some of the limitations are beyond the powers of an individual firm. The reason is that the adoption of blockchain technology may require a redefinition of the business architecture of a firm.

In terms of re-architecture means a firm's business strategies, organizational structures, and business process completely change. We think it will be difficult for firms to abandon their traditional business architecture for the sake of just adopting new technology no matter how promising it may be. Also, the IT-technology (enterprise resource planning (ERP) software and customer relationship management (CRM) systems) used for driving current business processes are cloud-based, and outside vendors are providing these IT-Solutions. Blockchain adoption by the software provider may require a redefinition of the base architecture of the proprietary software, and doing that will require a considerable investment in software development and for a user market that is very uncertain. So, they will most likely be reluctant to commit to such a venture. Many businesses transformed through blockchain technology and also since ETH promises blockchain technology based on a "distributed computing platform and operating system featuring smart contract" and necessary blockchain application functionality. Therefore, if a business adopts one of these Blockchains operating features, then there still lies a question of how the software would work with the current systems that support the core business process and not to mention the interoperability of such integration. We know that such huge companies like Amazon, Wal-Mart, and firms in such sectors, like banking and finance, etc., have been able to adopt blockchain technology within their business process without compromising or re-architecting their traditional business architect. Why, because they can and can also afford it. Companies can expand their capabilities by way of removing interoperability problems and drive all key business processes like customers and products, end-to-end global supply chain processes, shared services, and financial management processes by adopting blockchain technologies.

To present the analysis on the impact of Blockchain from a macro perspective relative to business and economics, in general, we will, at this point, state that the influence of blockchain technology in this context is part of the continuum of the digital transformation of business and society. From automobile, financial services, travel, healthcare care services, retail, transport, and logistics services to media and entertainment services, to mention but a few. Digital transformation has transformed society for the better. Firms today are using digital technologies to enhance critical business processes and performance and creating new business models that offer better monetization opportunities for building competitive advantages across their key industries.

The impact of blockchain technology is forceful because it is not only transforming the very basis of business architecture but also the economics itself. That is, it is changing the very foundation of value exchanges, the rules that guide transactions, and the currency of transaction [11, 12]. Simply stated, the blockchain technology is and will transform the entire structure of production, management, and governance of economic output.

So, to fully understand this, we would like to take a brief look at the essential characteristics of the technology. These characteristics include; security, transparency, trust, and privacy [43]. With these characteristics, Blockchain has forced changes in different business sectors and transactional processes. According to Ref. [43], Blockchain reduced the uncertainty, insecurity, and ambiguity in business transactions through full transactional disclosures that produce "single truth" for all in the transactional process. According to Ref. [44], the intermediaries are free of the implication of business transactions. And the related source of monetization monopolization of information and unnecessary hierarchical power structures. For example, Chowdhury [46] observed that by removing intermediaries, businesses that depended heavily on intermediaries can now enjoy better benefits from business transactions by applying blockchain technology in their business. In business, transactions can be done based on the transparency and democratization of the blockchain data structure in a distributed decentralized environment. This blockchain data structure is known to reduce the costs of searching, collaboration, and exchange of information [24]. Apart from offering high transparency in business transactions, blockchain technology also provide massive value to the parties participating in a deal through automatic business execution and disintermediated business models [44].

Let us take one characteristic of the blockchain technology that is of interest in this context. It is 'trust' or what [24] called the "TRUST PROTOCOL." Or what [53] the founder of the World Economic Forum termed the 4th Industrial revolution.' Trust is the essential currency in business and also of the digital age. In business, we go the length to build trust with stakeholders and the general public. For example, the company's brands are all there to reinforce the trust of customers and the public. Where the faith is minimum or lacking, then parties in the transaction process go through to verify information in the transaction process. And this increases the cost and inefficiency of business transactions. The verification and assurance needed to conduct the transaction further are provided by a third party in many cases. And the reason is that the question of "what is true and who to trust" will always be present. It increases transaction time and inherently, the cost of doing business.

This problem is endemic in the digital world. It led [24] to say that doing business on the internet is a "leap of faith." It is because, on the internet, it is challenging to reliably establish the fact in the information exchanged or the identity of the other party in the transaction process. According to Ref. [24], trust in a business transaction is the expectation that parties in the operations will behave with integrity in the transaction process. Integrity has four principal characteristics; honesty, consideration, accountability, and transparency. The beauty of blockchain technology is security and transparency in transactions, which are the fundamental principles and architecture of the technology. That is why it observed that Blockchain has the "capacity to reduce uncertainty, insecurity, and ambiguity in transactions by providing full transactional disclosure and by producing a single truth for all network participants" [43]. Therefore, resolving the problem of lack of trust in the transaction process according to Ref. [44] is the most important benefit of blockchain technology to businesses. According to Ref. [24], assurances achieved in a blockchain transaction context through the "implementation of consensus algorithms and cryptography" that authenticate all interactions and transactions in the network and make it easier and safer for parties to do business without concerns that the 'transaction' is dubious or corrupted [24].

On the other hand, as algorithms and cryptography manage transactions, assurances will be not be needed as transactions executed automatically. Traditionally in a highly connected B2C relationship, automatic execution is a strategy applied by the provider of services and goods, and it is where the customer surrenders all decision rights to the service provider to take care of everything on their behalf. The benefit is meeting the requirement of anticipation. It means the company can predict the needs of the customer and deliver the services before even the customer is aware of the need. The critical requirement is strong trust and rich information flowing from the customer [54] to the company providing the services as the future of connectivity is not the connection of people on the internet but the connection of smart devices interacting with each other's on the internet and executing transaction decisions automatically on behalf of the service providers and users. IBM ADEPT dishwasher functions for autonomously reordering detergent [55]. It is an example of devices independently performing various IoT transactions with blockchain technology.

The emergence of blockchain technology and networks is pioneering changes in the foundation of traditional organizations. Blockchain technology is re-architecting the firm and management science and its underlying principles of the conventional organization such as structures, authority, delegations, job description, and including how priorities in organizations set for business re-architecture. So traditional management science, as we know, is architected as an operating system for the industrial age. It is said to be incompatible in the modern era of Blockchain driven business networks [56].

Also, organizational roles are dynamics rather than on job descriptions. The principles for setting priorities based on forecast-plan-control have given way to a priority setting that is agile and dynamic. These changes are entirely democratizing management [56] and a shift to the lines of "halocracy" [3, 6]. An approach for democratic decision making in the governance of the organization has decentralized self-organizing teams [3, 6, 56].

The whole impact of Blockchain in transforming business firms is not only limited to the transformation of its underlying operating principles but also on how products and services are designed, marketed, delivered to the customers as well as how to capture value. It is referred to as the re-architecture of the firm's business model. Therefore, a business model is the architecture of value creation, delivery, and value capture mechanisms deployed by firms [57].

12.2.3 BLOCKCHAIN DISRUPTIVE EFFECTS: DIGITAL THREAT AND OPPORTUNITY

Erceg, Sekuloska, and Kelic [44] have mentioned in their work that Blockchain is influencing changes in all sectors, including tourism. Blockchain technology implemented in tourism by implementing trust with blockchain characteristics: transparency, control, influence, and recourse. The blockchain technology enables efficient communication with travelers using identity management in the entire tourism sector boundary. It will even create an impact on cost reduction in terms of currency exchange rates so that frequent travelers get an immense benefit out of it. Shrivas and Management [40] have mentioned that too many countries like Dubai, United States, China, and India are planning to implement public services like a land registry, international trades, and customs to prevent fraud and to achieve better data management. Zheng, Xie, and Dai; and Jepkemei and Kipkebut [1, 25] have mentioned that many traditional organizations are facing challenges with their current business model for sustainability. MSME, SMEs business, and startups have a new business model based on agile technology with faster decision making. The market experience says that companies have to follow the strategies for existing customers to fit in rather than looking toward the niche market. In that direction, they should come up with disruptive technology. For this new disruptive business model, Blockchain is the technology in the financial industry. Table 12.4 summarizes the critical factors of reasons to represent Blockchain as a disruptive technology.

TABLE 12.4   Reasons for Blockchain as a Disruptive Technology

Reason for Blockchain as Disrupted Technology	Explanation
Secure	It is a tamper-proof technology with transactions or ledger records that exist in Blockchain
Adopted in world	This technology is adopted across the globe and has the backing of many crowdsourcing investors from banking and nonbanking sectors
Operation automation	Software is handling all the operations automatically. Private companies are not required to manage operations.
Open	All the operations are managed through open-source technology, which is approved out through the OSS forum.
Distributed architecture	Blockchain works in a scattered mode in which records are stored in all nodes in the network if one node goes down, it does not impact any other records
Flexibility	This technology is programmable using basic programming concepts.

Table 12.4 represents the reasons to explain blockchain technology as a disruptive technology for business. It describes the critical dimensions of blockchain technology as the security adopted all over the world other than financial sectors—next key dimension transparency through automation in operations. A wide range of open-source technology platforms are available for implementation, and its distributed architecture with flexibility is making Blockchain as a disruptive technology in real sense. Bahg-Barker et al. [27] have discussed in their article about the challenges such as including throughput, latency, size, and bandwidth, security, wasted resources, usability, versioning, hard forks, and multiple chains. In Ref. [15] report represents positive and negative disrupt of blockchain

technology. Blockchain technology decentralized trust so as not to address the security of users or the applications that connected to its network, Brain wallets, dictionary attacks, traditional fishing attacks are threats in disruptive blockchain technology. In all business areas, this technology has given disruptive transformation to the current business scenario from a stable organization to the startups.

12.2.4 BLOCKCHAIN TECHNOLOGY THREATS AND OPPORTUNITIES

Bahga and Madisetti; and Drescher [4, 5] summarized the blockchain opportunities as decentralized, trustless, resilient, scalable, secure, auditable, and autonomous, and threats are as CAP and Blockchain, Smart Contract Vulnerabilities, Awareness, Regulation, Privacy, and efficiency. It has been observed that blockchain technology implemented on private, public, and consortium types. Dumas et al. [14] mentioned in the report about blockchain technology research opportunities and challenges for collaborative information systems. They have also identified some of the difficulties, like data privacy, security, legacy issues, etc.

Based on a literature review on the five sections, it observed that there are many open-source platforms and prosperity blockchain platforms are available. Based on this, we have formulated the research objective for this study to propose a blockchain application architecture and development strategies to achieve data monetization. To obtain the research objectives, we have used the blockchain system development cycle, followed by iFour Technolab Pvt. Ltd.

12.3 PROPOSED BLOCKCHAIN TECHNOLOGY ARCHITECTURE FOR DATA MONETIZATION

We are proposing seven steps for the system development strategies based on the system development cycle mentioned in the previous section. These steps identify problem and goal, define the most suitable consensus mechanism, identification of appropriate platforms, design the architecture, application development architecture, application development, and scale-up configuration and implementation. Table 12.5 depicts sub-steps for blockchain system development (Figure 12.1).

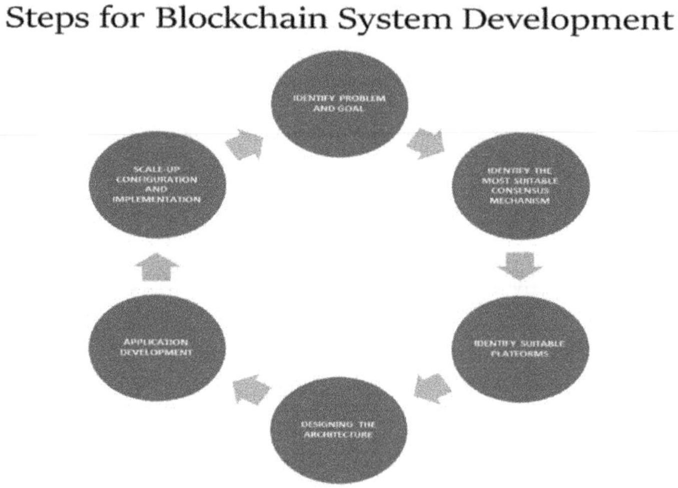

Steps for Blockchain System Development

FIGURE 12.1   Steps for blockchain system development.

TABLE 12.5   Sub-Steps for Blockchain System Development

Step 1: Identify Problem and Goal:
a.   Problem definition;
b.   Reasons for selecting Blockchain for the chosen problem;
c.   Identify the issues and risks involved.

Step 2: Identify the Most Suitable Consensus Mechanism:
a.   Selected consensus methods;
b.   Reasons for selecting consensus methods.

Step 3: Identify Suitable Platforms:
a.   About platform;
b.   Reasons for choosing a platform.

Step 4: Designing the Architecture:
a.   Architecture diagram;
b.   Hardware/software configurations.

Step 5: Application Development:
a.   Development framework used;
b.   Generating key pairs and addresses;
c.   Performing functions related to auditing;
d.   Performing data authentication using digital signatures;
e.   and hashes;
f.   Storage and retrieval of data;

TABLE 12.5   *(Continued)*

g.   Management and trigger of smart contracts relating to them;

h.   Issuance, Payment, exchange, escrow, and retirement;

i.   Coding (Python /Solidity/, etc.).

Step 6: Scale-Up Configuration and Implementation:

a.   Scale-up blockchain architecture for future needs artificial intelligence, IoT, Big Data, based on the business future need by following Step 1.

Based on system development, strategy steps are followed for the proposed architecture for blockchain technology for data monetization:

➢   **Step 1:** The problem aims to develop an application using block-chain technology to achieve data monetization for any business organization and stakeholders. Blockchain technology consisting of one of the features like identity management, and this is the feature we can use to define and achieve the goal for consumer data privacy. This problem has a risk of data privacy and how to satisfy both consumer and business owners for authentic use of consumer's data in consensus, and the consumer can trust the business party. No third party can get access to consumer's data without any authority.

The following objectives can be summarized to achieve the goal of proposed blockchain system architecture should provide:

*   Ensuring trust;
*   Preventing malicious activities;
*   Generating predictions;
*   Performing real-time data analysis;
*   Managing data sharing.

➢   **Step 2:** Mukkamala and Vatrapu [31] have discussed the consensus mechanism used in identity management is Proof-of-Work, Proof-of-Stack, and trust protocol in their proposed architecture. Proof-of-work is the consensus mechanism used in public Blockchain. It encourages participants to compete for the right to verify and settle blocks of trans-actions by solving computationally exhaustive problems. In proof-of-stake consensus mechanism sets block publishing rights based on the investment by a participant in the blockchain data privacy and mitiga-tion strategies. The consensus mechanism used to verify a participant's identity and authorization before granting the new block generation rights are known as proof-of-authority and mostly used in private Blockchain. In this step, we have examined the type of blockchain network suitable to meet the objectives of the problem definition. By comparing features of both the network type, we have concluded we will go with the hybrid/consortium blockchain type to get the advantages of both the networks, private and public. That means security, scalability, and transparency, and data sharing achieved. By way of removing centralized control and implementing Blockchain through a consortium, we would like to accomplish the reduction in counterparty risks that exist in a private chain. And it will streamline communication among each party involved in that organization.

> **Step 3:** In this step, we have examined the various platforms available to achieve the objectives achieved in step 2. Hyperledger, Corda, and multichain used to make communication between the channel and within the chain.

> **Step 4:** The platform selected as Hyperledger, Corda, and multichain the proposed architecture is also based on this to achieve the defined objectives. The basic architectural diagram is in Figure 12.1.

Figure 12.2 represents the general framework of architecture of the system with cloud; Blockchain distributed cloud, blockchain network, blockchain data communication services, device interface, and machine interface layers. Figure 12.3 represents the data workflow among the entities in the proposed architecture.

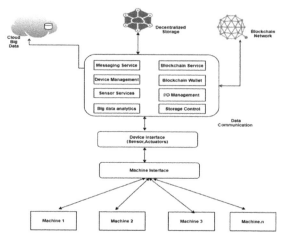

FIGURE 12.2 Blockchain architectural diagram for identity management and data monetization.

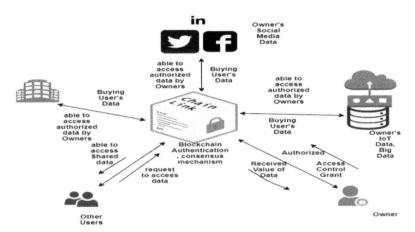

FIGURE 12.3 Dataflow in blockchain architecture of identity management and data monetization.

Figure 12.3 represents the workflow of data in the proposed architecture, which shows that using blockchain technology, owners can grant permission to access the data owned by way of implementing SCs and agreements among the parties. The owner can utilize data owned by him to monetize it.

In this architecture owner node, business organization nodes, big data, and IoT node, other interested party nodes implemented. At business organization ID, the owner's details like owner ID, Owner's Wearable device ID, etc., should be merged in the transaction details. For example, if the owner is wearing the wearable device of a particular fit band like Mi, Apple, Samsung, and fast track, etc. There should be proper data transfer among all these entities. The health records of the owner from the fitness band to the owner and the organization, i.e., form IoTD1 to OW1 to OG1, required proper transparency and value of the data. The OG1 node request for data to the OW1 node, and it is transferring from IoTD1 node with specified agreement written in the smart contract. The health data move from IoTD1 through OW1 by receiving the agreed Payment met at all participating nodes. At the very same time, the owner node receives complete information above the organization and health records obtained by the organization node.

> **Steps 5 and 6:** This section presents the various application development tools like ETH, Docker, and Docker composer application

development platform using the Hyperledger Fabric. Nodejs used for network and port information for data transfer. System requirement is older versions of Windows 10 requires Docker Toolbox and Docker version Docker 17.06.2 or larger. For Hyperledger Fabric leveraging the Hyperledger Fabric SDK for Node.js, version 8 needed. The programming languages like Python, go are necessary for implementation. Next, Hyperledger network establishment, hen implement Hyperledger Fabric smart contract (chaincode) APIs, then next implement Hyperledger fabric application SDKs, then performs certificate authority services to manage identity in the blockchain network.

12.3.1 APPLICATION DEVELOPMENT THROUGH ETHEREUM (ETH)

According to Ref. [60], ETH is a reliable tool to implement Blockchain. They have justified in their work that data points are cost-efficient than the appending them. The smart contract contains the data encoding code to reduce the cost of storing IoT sensor data (Figure 12.4).

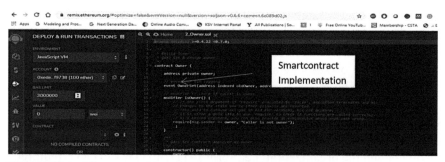

FIGURE 12.4   Ethereum tool for smart contract development.

Figure 12.5 represent Corda tool for smart contract development. The Corda platform used for

Business solutions with durability. Traditional blockchain platforms lack privacy and inevitability. It is open-source and built to address these shortcomings, enabling private transactions with immediate inevitability.

Figure 12.6 represents the docker composer tool for hyper ledger container service. Docker is used to building an application, execute, and distribute apps with containers. Application deployed using Docker containers are called containerization. Containerization used to achieve flexible, light-weight, portable, scalable, loosely coupled, and secure features in-app in this use-case based on IoT sensor data of wearable devices [68] the selection of docker for deployment.

FIGURE 12.5   Corda tool for smart contract development [63].

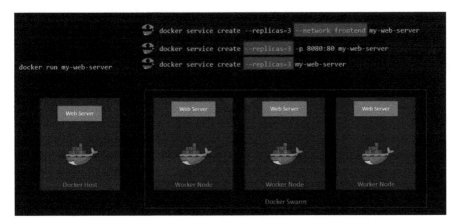

FIGURE 12.6   Docker composer hyper ledger container services [64].

Figure 12.7 represents chaincode containers of owner's and buyer's smart contract. It depicts that docker containers can contain any number of SCs according to the need.

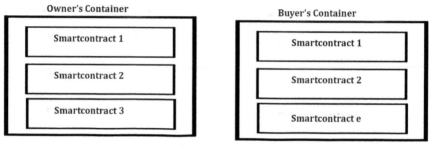

FIGURE 12.7 Hyper ledger smart contract framework for container services.

Then development strategies suggested implementing Contract names, Chaincode namespace, Transaction context, Transaction handlers, Endorsement policies, Connection Profile, Connection Options, Wallet, and Gateway to get efficient and effective solutions. After deployment of the application, further analyzes for the new requirements and any suggested changes can be scale-up, and all steps followed iteratively.

12.4 DISCUSSION AND RESULT

12.4.1 THE IMPACT OF BLOCKCHAIN TECHNOLOGY IN DATA MONETIZATION

So, what is data? Why does it deserve attention here? Well, for a more technical conception of data see the work of Tang [49], who in his book on The Data Industry: The Business and Economics of Information and Big Data, provided a very insightful and a detail description of data and its various key elements, such as, the value of data and ownership structures. In this section of the chapter, our focus is on the monetization of data in the context of information technology (Blockchain) and the business use of consumer data. So, our classification of data is made with the categories used in the European study on the nature of consumer trust and personal data published in 2014 by ORANGE is presented in Table 12.6.

TABLE 12.6 Data Classification

1. Intermediator/Monetary: Data relating to persons and contacts like their e-mail addresses, salary	2. Behavioral: Data containing facts such as place or mobile purchase history;	3. Demographic: Finally, basic demographic data, such as name, date of birth, mobile number, or marital status

So, data deserves attention because it is the next big thing of information technology. According to Ref. [47], access to data on consumer product usage and consumption data will explode [8], the data "tsunami burst." To create new customer experiences and better opportunities for data monetization, it is required to exploit these tsunamis of data to optimize key business processes. This bust or explosion is not contrary, but a chance since we have the data-crunching power of big data, AI plus Blockchain technology. For example, it has been observed that 80% of consumers would like to buy from companies that offer personalized customer experience [48]. Besides, $2.95 trillion is the value projection made by Accenture [48] that will be derived from a smart and digitally motivated personalize customers' experience. Also, it will be useful to take a glance at the value of the global market for consumer data monetization. It is beyond the scope of this section to take on such a challenge even though it is enticing. So let us begin by exploring blockchain characteristics decentralization from a spider[1] web structure to starfish structure. And it presents the impact of blockchain technology in data monetization. A vital feature of the Blockchain is that it destroys the central mechanism (the spider web) as it is no longer relevant in a blockchain system. Instead, the "participants in the blockchain network maintain the data consistency in the distributed network through a consensus algorithm" [24]. Secondly, persistence, once the transaction is entered into the Blockchain, will be impossible to delete, modified, or nullified. The third is auditability. In this view, a deal must have referenced linkages to previous unsold transactions recorded in the Blockchain then, "the status of the quoted transaction changes."

Nevertheless, with these critical characteristics blockchain technology is said to provide the 'guardrails' to customer data monetization through better security, transparency, trust, and data privacy [52], data tracking and its usage and as well as data authentication Data monetization with Blockchain, let us look at few key examples;

12.4.2 BLOCKCHAIN DEMOCRATIZE THE SHARING AND MONETIZATION OF DATA

By helping to remove intermediaries, it offers excellent opportunities for parties in the transaction to share and monetize data in a most democratic

---

[1] For a better insight on spider and starfish structures, see the book by Ori, B., & Rod, A. B. The Starfish and the Spider: The Unstoppable Power of Leaderless Organizations.

way. According to Ref. [1], Blockchain provides "common access to the same data" to all parties in the transaction. In their view, this "accelerates data acquisition, sharing, data quality, data governance and ultimately, data analytics." Regular access empowers all parties to control and own their data and how it is monetized. Companies are the big winners here in terms of better monetization. Blockchain technology improves monetization by way of reduction of data management costs [3, 6, 7, 10]. Blockchain Technology improves efficiency through faster, safer, and less expensive transaction processes.

In the use of consumer data, there are two sides to the monetization of data that are linked to the emergence of blockchain technology. Regular access to the same data in a blockchain, data tracking, and usage will give consumers insight into how their data are being used and, inherently, an estimate on the value of their data. It will amplify consumers to want to control and manage access and the use of their data. It could increase their bargaining power to negotiate compensations for their data. For example, Table 12.7 shows consumers placed a price tag on the use of their data in a study made in the UK.

TABLE 12.7 UK Specific Data: Average (Mean) Amount of Money That a Consumer Would Want to Receive to Share Their Details with Two Types of Organizations [58]

	Familiar Organization	Unfamiliar Organization
My full name or date of birth	£12.14	£15.02
My mobile number	£14.01	£16.97
My location (e.g., via mobile GPS)	£13.99	£17.66
My annual income	£15.18	£17.30
My marital status	£8.23	£12.34
My sexual orientation	£10.56	£12.82
My job	£10.70	£13.58
My children's details (e.g., sex, age)	£12.41	£16.11
Detail of my family members preferences	£16.12	£18.93
E-mail addresses of people in close personal network	£16.63	£18.80
History of purchases made on mobile phone	£14.22	£18.73
My postal address	N/A	£17.18
My main e-mail address	N/A	£15.64
Average (Mean)	£13.11	£16.24

Thus, Blockchain not only democratizes through regular access to the same data. It also democratizes the monetization of data. In this, parties in the transaction can ripe financial benefits from data. With this, blockchain technology shifts the balance of power in the appropriation of data.

Metering economy: As Blockchain enables data tracking and usage, it means data usage metered for data appropriations. Through metering, rents can easily be calculated and improve monetization of data for the parties concerned. Lastly, the Rights to creators of IPs, Blockchain technology also enables creators of intellectual properties to appropriate their IPs. Better as the assets can be "digitalized and watermarked" each copy of the IP, which in turn, transforms it into a tradeable asset that monetized.

Based on the above discussions in our big data IoT case study, we have classified data as per Table 12.8.

TABLE 12.8   General Data Classification for Social Media, IoT-based Big Data

Business Perspective	1. Intermediator/ Monetary: Data relating to persons and contacts like their e-mail addresses, salary, home automation data	2. Behavioral: The second refers to behavioral data, including information such as location or social media data sharing, likes, and comments, IoT-based health data, IoT-based business data.	3. Demographic: Finally, basic demographic data, such as name, date of birth, mobile number or marital status, ethnicity, profession, lifestyle
Blockchain Technology Perspective	Owners' node, organization node, social media account node, wearable device node, blockchain wallet, etc.	Smart contract, identity management, owner authentication, consensus mechanism, blockchain services, big data analytics services	Hyperledger fabric network node, decentralized storage node, bigdata storage node

From the proposed model, we can conclude that blockchain features like user node, consensus mechanism, authorization certificate, blockchain services, SCs, Hyperledger network nodes, decentralized storage node, machine nodes are critical drivers of blockchain technology. These are enabling business transformation with data monetization. Table 12.9 represents how data monetization increased by 2023 using blockchain technology by renowned research institutions.

TABLE 12.9   Data Monetization Survey by Research Institutions by 2023

Research Organization Report	Year	Percentage of the Increase in Blockchain Utilization	Business Benefits Data Monetization in USD
Deloitte's global blockchain survey [66]	2019	20	>$1 billion but < $10 billion from 2018 to 2019
IDC predicts blockchain spending [66]	2019	80	$2.7 forecast by 2023 it will be $15.9

Based on the above data, we can say that now more and more industry accepting blockchain technology as their business solutions and so that reducing the cost of operation, risk management, infrastructure, and utilization of data through its value, in real sense data monetization.

Table 12.10 represents the tabulation of Ref. [67] to understand the impact of IoT data in market value. According to report in Ref. [67], 8ZB data existed in 2015, and by 2025, it reaches up to 180ZB. And among it, 95% of data is unstructured data. Total users will be 440 million with 100 million connections by 2025. Figure 12.8 represents that by 2025 the total number of users in terms of smart users, smart home devices users, internet connections, IoT enabled vehicles, Drone, etc., are increasing day-by-day.

TABLE 12.10   Impact of IoT Data in Market Value by 2020

Elements in the IoT Environment	Count in Billions	Market Value in USD (in billion)
Smart devises smart home devices	60	200
Vehicle	0.2	145
Virtual reality entertainment and live	0.17	7.3
Drone	440	33.9

Figure 12.9 represents that by 2025 the market values for smart devices, smart home devices, drones, etc., are increases in a significant manner, and it expressed in USD.

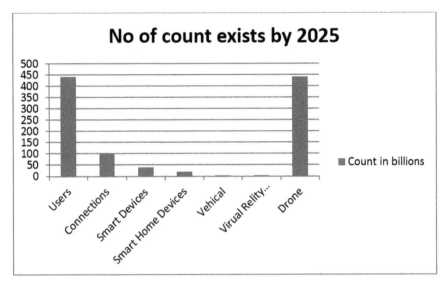

FIGURE 12.8 IoT element trends in numbers by 2025.

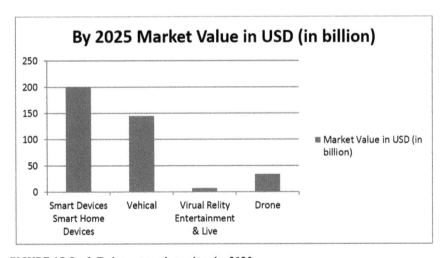

FIGURE 12.9 IoT element market values by 2025.

Figure 12.10 represents graphical representation that, in a decade, data increases in 22.5 times in the span.

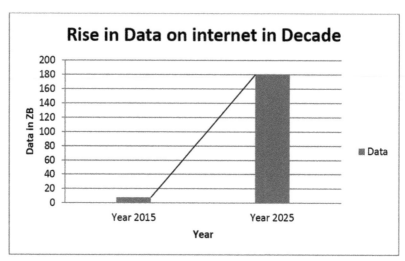

FIGURE 12.10   Growth of data in zeta bytes by 2025.

Figure 12.11 is a graphical representation of data taken from reports [66], i.e., Table 12.8 data monetization is increased by 2023 through blockchain technology.

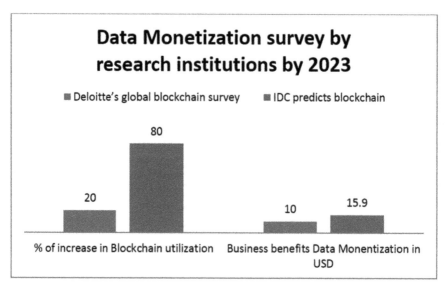

FIGURE 12.11   Data monetization survey by research institutions by 2023.

Now we can relate all this graphical representation of data to link IoT and Blockchain, and we can interpret that due to use of IoT technology increases the size of data increases. Based on that data and for its privacy, security, transaction cost and value of data become a critical business factor for transformation of business and blockchain technology satisfies the all these need to handle this business needs.

12.5 CONCLUSION

The objective of this chapter was to present the critical drivers of the academic discussions on the topic. We achieved this by first carrying out a literature review on the technologies and tools used for blockchain technology and then explored their benefits, impacts, and challenges of blockchain technology. It was made both from adoptable blockchain technologies, open-Source solutions, proprietary solutions, and the adoption of blockchain technology in various business sectors and its emergence impacts. Secondly, we explored the adoption of blockchain technology across multiple business sectors. Through that, we were able to present an analysis of business solutions using blockchain technology and their disruptive implications. Later on, we elaborated on the key characteristics of blockchain technology in the transformation of business. There we considered the use of blockchain technology and its impacts from Table 12.7 to re-architecting the business model to achieve data monetization and equally impacted management science from business areas and economics. From Tables 12.8 and 12.9; Figures 12.8 and 12.11 interpret that by 2025 IoT data also accelerating data monetization. Lastly, the effect of data monetization using blockchain technology observed for firms and the consumers derived from the exploitation of their data. Perceived limitation in the study, time constrain made it impossible to conduct empirical studies, though the limitations were solved by sourcing and exploiting secondary data for the analysis. Areas for further research are to study the challenges faced by smaller companies in the adoption of blockchain technology and the monetization of data in a blockchain-based IoT (BIoT) web context. This work extended for the fetching experimental data for IoT wearable devices and observe and collect the data for monetization.

KEYWORDS

- **blockchain technology**
- **business transformation**
- **consensus**
- **data monetization**
- **disruptive technology**
- **hyperledger fabric**
- **identity management**
- **smart contracts**

REFERENCES

1. Zheng, Z., Xie, S., & Dai, H., (2018). Blockchain challenges and opportunities: A survey. In: Zibin, Z., Shaoan, X., Hong-Ning, D., Xiangping, C., & Huaimin, W., (eds.), *Int. J. Web Grid Serv.*

2. Nakamoto, S., (2008). *Bitcoin: A Peer-to-Peer Electronic Cash System, 1–9.*

3. Tapscott, D., & Tapscott, A., (2016). The impact of the Blockchain goes beyond financial services. *Harv. Bus. Rev., 7.*

4. Bahga, A., & Madisetti, V. K., (2016). *Blockchain Platform for Industrial Internet of Things, 533–546.*

5. Drescher, D. (2017). *Blockchain Basics: A Non-Technical Introduction in 25 Steps.* United States. Apress. pp. 22, 56, 33.

6. Tapscott, A., & Tapscott, D., (2017). How Blockchain is changing finance. *Harv. Bus. Rev., 2–5.*

7. Kowalewski, D., McLaughlin, J., & Hill, A. J., (2017). Blockchain will transform customer loyalty programs. *Harvard Bus. Rev. Digit. Artic., 2–6.*

8. Iansiti, M., & Lakhani, K. R., (2017). The truth about Blockchain. *Harv. Bus. Rev., 2017.*

9. De Vega, F., Soriano, J., Jimenez, M., & Lizcano, D., (2018). *A Peer-to-Peer Architecture for Distributed Data Monetization in Fog Computing Scenarios, 2018.*

10. Jani, S., & University, P., (2018). *Applications of Blockchain Technology in Banking Finance.* https://doi.org/10.13140/rg.2.2.35237.96489, pp. 2–6.

11. Carson, B., Romanelli, G., Walsh, P., & Zhumaev, A., (2018). *Blockchain Beyond the Hype: What is the Strategic Business Value?* (pp. 1–13). McKinsey&Company. Digit. McKinsey.

12. Restuccia, F., Oro, S. D., & Kanhere, S. S., (2018). Blockchain for the Internet of Things: Present and Future. no. November.

13. Gassner, U. M., (2018). *Blockchain in EU E-Health – Blocked by the Barrier of Data Protection? 4(2), 3–20.*

14. Dumas, M., Hull, R., Mendling, J., & Weber, I., (2018). Blockchain technology for collaborative information systems. *Blockchain Technol. Collab. Inf. Syst. Dagstuhl Reports, 8(08), 67–129.*

15. McAfee, (2018). *Blockchain Threat Report*. 1–27.
16. Kshetri, N., & Voas, J., (2018). Blockchain-enabled. *IEEE Softw.,* 1–5.
17. Decentralized, H., Internet, B., & Privacy, G. D., (2018). *Blockchain-GDPR Privacy by Design* (Vol. 1, pp. 1–5). Blockchain-GDPR Priv. by Des. (C. Lima).
18. Gatteschi, V., Lamberti, F., Demartini, C., Pranteda, C., & Santamaria, V., (2018). *To Blockchain or Not to Blockchain: That Is the Question* (Vol. 20, No. 2). IT Prof.
19. Lee, J. Y., (2019). A decentralized token economy: How Blockchain and cryptocurrency can revolutionize business. *Bus. Horiz., 62*(6), 773–784.
20. Lao, L., Li, Z., Hou, S., Xiao, B. I. N., Guo, S., & Yang, Y., (2019). *A Survey of IoT Applications in Blockchain Systems: Architecture, Consensus and Traffic Modeling, 1*(1).
21. Casino, F., Dasaklis, T. K., & Patsakis, C., (2019). A systematic literature review of blockchain-based applications: Current status, classification, and open issues. *Telemat. Informatics, 36*, 55–81.
22. Tandon, A., (2019). *An Empirical Analysis of using Blockchain Technology with the Internet of Things and its Application. 9*, 1469–1475.
23. Biswas, B., & Gupta, R., (2019). Analysis of barriers to implementing Blockchain in industry and service sectors. *Comput. Ind. Eng., 136*, 225–241.
24. Hughes, A., Park, A., Kietzmann, J., & Archer-Brown, C., (2019). Beyond bitcoin: What Blockchain and distributed ledger technologies mean for firms. *Bus. Horiz., 62*(3), 273–281.
25. Jepkemei, B., & Kipkebut, A., (2019). *Blockchain – A Disruptive Technology in Financial Assets, 2*(9), 38–47.
26. Drosatos, G., & Kaldoudi, E., (2019). Blockchain applications in the biomedical domain: A scoping review. *Comput. Struct. Biotechnol. J., 17*, 229–240.
27. Bahg-Barker, J., Frizzo-Barker, J., Chow-White, P. A., Adams, P. R., Mentanko, J., & Ha, D., (2019). Blockchain as a disruptive technology for business: A systematic review. *Int. J. Inf. Ianage.* pp. 1–8.
28. Chalmers, D., Matthews, R., & Hyslop, A., (2019). Blockchain as an external enabler of new venture ideas: Digital entrepreneurs and the disintermediation of the global music industry. *J. Bus. Res.,* 1–15.
29. Deng, R. H. (2017). Handbook of Blockchain, Digital Finance, and Inclusion, Volume 1: Cryptocurrency, FinTech, InsurTech, and Regulation. United Kingdom: Elsevier Science. pp. 1–450.
30. Blossey, G., (2019). *Blockchain Technology in Supply Chain Management: An Application Perspective, 6*, 6885–6893.
31. Faber, B., Michelet, G., Weidmann, N., Mukkamala, R. R., &Vatrapu, R. (2019). BPDIMS: A Blockchain-based Personal Data and Identity Management System. In *Proceedings of the 52nd Hawaii International Conference on System Sciences*, https://doi.org/10125/60121, pp. 6855–6864).
32. Emmanouil, & Santos, (2019). EIB Working Paper 2019/01-Blockchain, FinTechs, and Their Relevance for International Financial Institutions, pp. 1–40.
33. Wang, Y., (2019). Designing a blockchain-enabled supply chain. *IFAC-PapersOnLine, 52*(13), 6–11.
34. Davradakis, R. W., Emmanouil, & Santos, (2019). *EIB Working Paper 2019/01-Blockchain, FinTechs, and Their Relevance for International Financial Institutions.*

35. Mansur, V., & Sujatha, R., (2019). *Hyperledger Sawtooth Blockchain-IoT E-Provenance Platform for Pharmaceuticals,* (1), 268–273.
36. Hassan, (2019) N. U. L., Member, S., Yuen, C., & Member, S. *Blockchain Technologies for Smart Energy Systems: Fundamentals, Challenges and Solutions.* 61750110529.
37. PWC, (2020). Financial services technology 2020 and beyond: Embracing disruption. *PwC Finance. Serv. Technol. 2020 Beyond, 19.* www.pwc.co (accessed on 29th July 2021).
38. Makhdoom, I., Zhou, I., Abolhasan, M., Lipman, J., & Ni, W., (2020). PrivySharing: A blockchain-based framework for privacy-preserving and secure data sharing in smart cities. *Comput. Secur., 88,* 101653.
39. Frizzo-Barker, J., Chow-White, P. A., Adams, P. R., Mentanko, J., Ha, D., & Green, S., (2020). Blockchain as a disruptive technology for business: A systematic review. *Int. J. Inf. Manage., 51,* 102029.
40. Shrivas, Mahendra & Yeboah, Dr. (2018). *The Disruptive Blockchain: Types, Platforms and Applications.* 10.21522/TIJAR.2014.SE.19.02.Art003.
41. Interviewed, G., Technology, W., & Are, D., (2020). Blockchain Gaining Ground in the Enterprise (pp. 1–7). https://itchronicles.com/technology/blockchain-gaining-ground-enterprise/.
42. Pan, X., Pan, X., Song, M., Ai, B., & Ming, Y., (2020). Blockchain technology and enterprise operational capabilities: An empirical test. *Int. J. Inf. Manage., 52,* 101946.
43. Ertz, M., & Boily, É., (2019). The rise of the digital economy thoughts on blockchain technology and cryptocurrencies for the collaborative economy. *Int. J. Innov. Stud., 3*(4), 84–93.
44. Erceg, A., Sekuloska, J. D., & Kelic, I., (2020). Blockchain in the tourism industry: A review of the situation in Croatia and Macedonia. *Informatics, 7*(1),1–16.
45. Hassan, F., Ali, A., Latif, S., Qadir, J., Kanhere, S., & Singh, J.(2019). *Blockchain and the Future of the Internet: A Comprehensive Review,* 1–21.
46. Chowdhury, E. K., (2019). Transformation of business model through blockchain technology transformation of business model through blockchain technology. *Cost Manag., 47,* 1–9.
47. William, S., & Hitachi, V., (2017). *Is Blockchain the Ultimate Enabler of Data Monetization?* https://www.kdnuggets.com/2017/04/blockchain-ultimate-enabler-data-monetization.html, pp. 1–4.
48. Singh, G., Gaur, L., & Agarwal, M., (2017). Factors influencing the digital business strategy. *Pertanika J. Soc. Sci. Humanit,* 1489–1500.
49. Tang, C., (2016). *The Data Industry and the Business and Economics of Information and Big Data.* John Wiley & Sons.
50. Wei, P., Wang, D., Zhao, Y., et al., (2020). Blockchain data-based cloud data integrity protection mechanism. *Future Generation Computer Systems, 102,* 902–911.
51. Solanki, A., & Nayyar, A., (2019). Green Internet of Things (G-IoT): ICT technologies, principles, applications, projects, and challenges. In: *Handbook of Research on Big Data and the IoT* (pp. 379–405). IGI Global: Hershey, PA, USA.
52. Singh, G., Gaur, L., & Ramakrishnan, R., (2017). *Internet of Things-Technology Adoption Model in India, 25,* 835–846.
53. Klaus, S., & Nicholas, D., (2015). *Shaping the Future of the Fourth Industrial Revolution a Guide to Building a Better World.* Klaus Schwab the fourth industrial revolution: What it means and how to respond. SNAPSHOT.

54. Nicolaj, S., & Christian, T., (2019). *The Age of Continuous Connection*. HBR.
55. Veena, P., Sanjay, P., & Sumabala, N., (2015). *Empowering the Edge: Use Case Abstract for the ADEPT Proof-of-Concept*. IBM global business services.
56. Vyas, G., Gaur, L., & Singh, G., (2016). Evolution of payments bank and impact from M-PESA: A case of mobile banking services in India. *Proceedings of the Second International Conference on Information and Communication Technology for Competitive Strategies*, 1–4.
57. Vida, J. M., Jeannette, P., & Edward, B., (2018). How blockchain technologies impact your business model. *Business Horizons*. *62*(3), 295–306, ISSN 0007-6813, https://doi.org/10.1016/j.bushor.2019.01.009.
58. Orange, (2014). *The Future of Digital Trust*. A European study on the nature of consumer trust and personal data.
59. Lu, Y., (2019). The Blockchain: State-of-the-art and research challenges. *Journal of Industrial Information Integration, 15*. https://doi.org/10.1016/j.jii.2019.04.002.
60. Peker, Y. K., Rodriguez, X., Ericsson, J., Lee, S. J., & Perez, A. J., (2020). A cost analysis of Internet of Things sensor data storage on the Blockchain via smart contracts. *Electron., 9*(2).
61. Khatoon, A., (2020). A blockchain-based smart contract system for healthcare management. *Electron., 9*(1).
62. Javaid, A., Zahid, M., Ali, I., Khan, R. J. U. H., Noshad, Z., & Javaid, N., (2020). *Reputation System for IoT Data Monetization Using Blockchain, 97*. Springer International Publishing,
63. https://www.corda.net/blog/improving-the-corda-developer-experience-4-tools-for-your-coding-pleasure/ (accessed on 29th July 2021).
64. https://hyperledger-fabric.readthedocs.io/en/release-2.0/developapps/contractname.html (accessed on 29th July 2021).
65. https://docs.docker.com/ (accessed on 29th July 2021).
66. https://www.hyperledger.org/learn/research (accessed on 29th July 2021).
67. https://www.huawei.com/minisite/giv/Files/whitepaper_en_2018.pdf (accessed on 29th July 2021).
68. Pramanik, P. K. D., Solanki, A., Debnath, A., Nayyar, A., El-Sappagh, S., & Kwak, K. S., (2020). Advancing modern healthcare with nanotechnology, nanobiosensors, and internet of nano things: Taxonomies, applications, architecture, and challenges. In: *IEEE Access* (Vol. 8, pp. 65230–65266). doi: 10.1109/ACCESS.2020.2984269.
69. Rameshwar, R., Solanki, A., Nayyar, A., & Mahapatra, B., (2020). Green and smart buildings: A key to sustainable global solutions. In: *Green Building Management and Smart Automation* (pp. 146–163). IGI Global: Hershey, PA, USA.
70. Krishnamurthi, R., Nayyar, A., & Solanki, A., (2019). Innovation opportunities through the Internet of Things (IoT) for smart cities. In: *Green and Smart Technologies for Smart Cities* (pp. 261–292). CRC Press: Boca Raton, FL, USA.

CHAPTER 13

Development of Blockchain-Based Cryptocurrency

DEEPAK KUMAR SHARMA, ANUJ GUPTA, and TEJAS GUPTA

Department of Information Technology, Netaji Subhas University of Technology (Formerly known as Netaji Subhas Institute of Technology), New Delhi, India, E-mails: dk.sharma1982@yahoo.com (D. K. Sharma); ganuj32@gmail.com (A. Gupta)

ABSTRACT

The first-ever cryptocurrency invented was bitcoin (BTC) in 2008. With this, the first-ever blockchain network was launched all over the globe. The greatest advantage that Blockchain backed cryptocurrencies offer is the security and non-dependency on a third-party organization for trust issues. It allows independent verifiability of transactions while keeping the transactions anonymous but recognized by a virtual ID. Blockchains can also process transactions at a much lower cost than banks and other financial companies. In May 2016, when the Blockchain Revolution went to print, the entire crypto asset market had a value of $9 billion. Fewer than two years later, the crypto asset market was $420 billion in size. The development of cryptocurrency faces legal challenges as well. The currency developed has to develop trust among the parties involved in transactions. The blockchain technology is continuously evolving as new consensus algorithms are being developed, which may prove to be better than the currently used algorithms. Many different ways have been introduced to modify the modern-day currency into digital currency such as BTC. However, the trust remains an important factor to overcome. As the government of any country cannot monitor the use of cryptocurrencies, they are hesitant to legalizing them as a regular mode of exchange. Cryptocurrencies are heavily affected by mobilization amongst its users. In case the demand of cryptocurrency comes down to level zero, then the value of it is also zero.

13.1 INTRODUCTION

13.1.1 INTRODUCTION TO CRYPTOCURRENCY

Currency is one thing you are aware of because, without the use of currency, you might not be able to buy this book. Currency in the most general sense is a system of money. Banknotes and coins are the most common type we have seen in them. We may also describe currency as the unit of value of an entity or a product or a service. A cryptocurrency is a computerized capital designed to serve as a medium of trade that uses powerful cryptography to encrypt financial transactions, monitor additional unit development, and validate asset transfer. Cryptocurrency got attention with the introduction of bitcoin (BTC) by fictitious Satoshi Nakamoto in 2009 in a paper marked as "Bitcoin: A peer-to-peer (P2P) electronic cash system." Cryptocurrencies typically use a decentralized control to allow encrypted transactions, through technologies such as Blockchain, and strong cryptography. There are many cryptocurrencies available along with BTC, such as Ethereum (ETH), Dash, Monero, etc., which are managed by collective consensus and a mechanism is used to hold a description of and ownership of the virtual currency units. Cryptocurrency has the power to revolutionize the traditional trading system where people depend on an institution that governs their trading. It will also empower normal citizens as the exchange of cryptocurrency requires no centralized authority at all [21].

13.1.2 UNFOLDING OF CRYPTOCURRENCY

Cryptocurrency is an online trading mechanism which uses cryptographic functions to carry out financial transactions. Cryptography was earlier used mostly for military and intelligence communication to prevent leakage of classified information. David Chaum, a cryptographer, invented an electronic cryptographic currency called "Ecash" in 1983. Later, in 1989, he reimplemented it by "Digicash," an early method of digitalized cryptographic currency exchange that enabled consumer software to retrieve notes from a bank and assign unique encryption keys until they could be transmitted to a receiver. It permitted the electronic money to be undetectable by the providing bank, government or other third parties. The NSA in 1996 issued a paper marked "How to Make a Mint: The Cryptography of Anonymous Electronic Cash," exp a Cryptocurrency system. Wei Dai released an overview of "b-money" in 1998, which was described as a private, distributed online cash network. Nick Szabo described bit gold shortly thereafter. Bitgold was characterized as a digital currency program that encouraged users to complete a proof of work (PoW) challenge with solutions being cryptographically installed and distributed. In 2009, probably pseudonymous researcher Satoshi Nakamoto invented the first decentralized cryptocurrency, BTC. For its proof-of-work, it used the "SHA-256" cryptographic hashing algorithm. Namecoin was founded in April 2011 as an effort to develop a decentralized DNS (domain name system) which would exacerbate internet censorship. Litecoin, launched in October 2011, was the first successful cryptocurrency to be using scrypt (A key derivation function based on passwords) as a replacement to SHA-256, as its hash function. Peercoin, another significant cryptocurrency, was the first to use a hybrid of both, the PoW (proof-of-work) and the PoS (proof-of-stake) [12]. The market value of some popular cryptocurrencies as of May 2020 are shown in Table 13.1 [14].

TABLE 13.1 Market Capitalization of Some Cryptocurrencies as of May 2020

Name	Price	Market Capital	Circulating Supply
Bitcoin	$9,503.43	$174,666,022,341	18,379,268 BTC
Ethereum	$201.78	$22,389,683,959	110,962,274 ETH
Bitcoin cash	$238.23	$4,385,650,306	18,409,350 BCH
Litecoin	$43.95	$2,845,449,846	64,748,368 LTC

13.1.3 BLOCKCHAIN IN CRYPTOCURRENCY

The Blockchain is an imperishable encrypted archive of financial transactions that can be prepared for tracking not just effective transactions but virtually everything that carries a price. There is no central authority for the technology to look at or any regulatory mechanism to provide valid rules. Blockchain-based cryptocurrencies empower normal citizens as the exchange of these requires no centralized authority at all. Decentralization, transparency, and immutability are the pivotal characteristics of any Blockchain-based cryptocurrency. These concepts were introduced by the aforementioned Satoshi Nakamoto with BTC. BTC rapidly gained popularity all over the world, and soon ETH was introduced with the application of smart contracts

(SCs), yet another disruptive innovation in the history of technological innovations. Depending on whether a cryptocurrency is developed on an existing blockchain or is independent of any other blockchain; it is of two types: a Token and a Coin. A cryptocurrency which is developed on an existing blockchain is called Token. It does not have a blockchain of its own. For example, BAT (basic attention token) and BNT (Bancor) happen on the ETH blockchain. Conversely, a cryptocurrency which is completely independent of any other existing blockchain is called Coin. It functions as a native currency of a given financial system. For an example of coin there is BTC, ETH and XRP (Ripple). They all function on frameworks which were specifically built for their use. Cryptocurrencies can be directly transmitted between two parties using the public keys assigned to them. Ironically, this transaction requires very low transaction fees and acknowledges the issue of high fees paid to traditional financial institutions. Blockchain takes all the entities involved to a reasonable degree of accountability without the risk of making the transactions destroyed, no consumer or network error, or not even an unrecognized transaction swap. On top of everything, the most important field where Blockchain is useful is to maintain record consistency by recording it not just on a central registry, but also a similar distributed ledger network, both of which are linked by a safe validation mechanism. Blockchain is deemed necessary for shaping the infrastructure to ensure improved protection and privacy in other realms, such as the IoT ecosphere [1]. The key advantages of cryptocurrencies centered on Blockchain are the independent authenticity of transactions and the anonymity they enable.

13.1.4 BENEFITS OF BLOCKCHAIN FOR CRYPTOCURRENCY

The key advantages that Blockchain offers are Immutability, Decentralization, Transparency, and Security. These advantages turned out to be an enormous revolution in data handling or big data and led to the foundation of the first cryptocurrency, i.e., BTC. In BTC, or generally referring, in any Cryptocurrency, the major aim is to take away power from the hands of a single governing agency and distribute it in the hands of users having them trust the technology. Also, the cryptographic nature of storing transactions almost makes it impossible to intercept any transaction without the authorization and authentication of both the parties involved in the transaction. Below are some listed advantages of using blockchain in cryptocurrency:

- The blockchain mechanism enables authentication without the need to rely on a third party.

- The data structure empowering the blockchain technology is "append-only." So, you cannot modify or erase any transaction that occurred in the past.
- The data ledgers are maintained cryptographically. Also, the current block is dependent upon the completed block adjacent to the current block to complete the process.
- After the authentication phase, all transactions and data are added to the frame. Both parties in the ledger decide on what is to be appended in the series.
- The transactions are registered in time order. Therefore, all blocks in the Blockchain ledger are marked with time, that is they are 'Time Stamped.'
- The ledger is distributed to all the participants of the Blockchain, thus making it decentralized.
- The decentralized nature of the blockchain storage mechanism makes sure the data is recoverable easily if the storage at a few nodes gets corrupted.
- The transactions which occur are transparent. The transaction is viewable by the individuals who are granted authority.
- The root of any ledger may be traced to its place of origin in the chain.
- Since different consensus protocols are used to validate the entry, the possibility of duplicate entry or fraud is minimized, and thus, the issue of double-spending is solved.
- The cryptocurrency companies will pre-set requirements on the Blockchain for the SCs. The automated transfers are only activated when requirements are met.

13.2 CONCEPTS OF BLOCKCHAIN

13.2.1 ARCHITECTURE

In this section, we are going to discuss various requisitions needed to formulate or generate or even apply the Blockchain. "We can define Blockchain as a mechanism that enables us to trade with any person without knowing the actual identity." In other words, imagine an e-commerce website that enables us to put any product and allows us to sell it to any other person. We need to trust this e-commerce website which assures us that the second person and the product are authentic. Blockchain is exactly the same, but instead, there is no 'middleman.' All we have to do is follow some set of rules, and we can trade securely with any person without the problem of authorization, authentication, and reliability.

More formally, Blockchain can be defined as a distributed ledger of all transactions taking place in a network of computers. The name derives from two words, one is 'block,' which means that there are blocks of individual data containing all the transaction information and the other is a chain, which means that all such data are grouped in a single sequence, here referred to as a chain [1]. A blockchain is a "chain of blocks" that is append-only. Blockchain is a distributed, decentralized public ledger. This simply means digital information ("the block") stored in a public database ("the chain") [2]. The network of Blockchain is P2P, i.e., every two nodes in the network of Blockchain are directly connected. The "Distributed Ledger" (Database) of transactions gets updated every time any new transaction takes place. This transaction (or block) gets appended in the chain and is updated at every node, hence, getting verified by everyone in the network. So basically, A blockchain is a software code that runs independently on every node in the network and governs all the working, removing the requirement of any institution to validate this exchange of information.

13.2.1.1 BLOCK

Block is the starting part of any blockchain. The functionalities of any blockchain depend greatly on the structure of a block. Block consists of information related to transactions such as the timestamp, the amount exchanged, and other relevant details. The block which has no parent block is known as the genesis block.

The structure of the block is divided into two parts. First is the block header and the second is the block body. Block header is that part which gives some information about itself and is majorly important for connecting to the other blocks in the chain. Figure 13.1 shows the basic block structure.

Block header consists of:

1. **Block Version:** There might be different sets of rules to be followed for block validation. This part indicates the set of rules to follow for block validation for the particular block.
2. **Merkle Tree Root Hash:** Each transaction has its own unique code which sets it apart from other transactions. It essentially separates all transactions from others, so that it would be easy to identify that

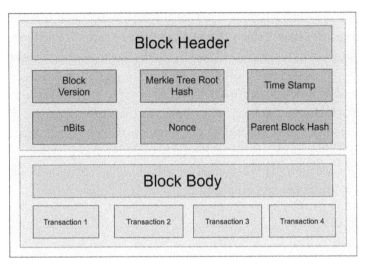

FIGURE 13.1   Block structure.

particular transaction at the time of the inquiry. The unique identifier is called 'Hash.'

3. **Timestamp:** This is one of the most important parts of a transaction that needs to be stored with the details of the transaction. It stores a single value which refers to time as seconds since January 1, 1970. This reflects the chronological order in which the transaction has occurred.

4. **nBits:** The goal is the point at which a block header hash needs to be to validate the block, and nBits is the goal threshold encoded type as it occurs in the object header.

5. **Nonce:** A nonce is an acronym for "number used only once," and is a number added to a block in a blockchain which meets the difficulty level constraints when rehashed. The nonce is the proportion the miners of Blockchain solve for. This is a 4-byte space, normally beginning at 0 and increasing with any miners hash calculation. If the answer is found, cryptocurrency is presented in return [4].

6. **Parent Block Hash:** It is a 256-bit hash value that corresponds to the previous block in the chain, but there is one block in the chain which does not contain this hash value which is the first block of the chain. This block is not having any parent block, and this block is named as the Genesis Block. A hash value is a key, which is generated using a hashing algorithm that operates on the entire data and represents

it into a single line of fixed length depending upon the algorithm used. This hash value will be unique for every block, and the original values cannot be reversed back from the hash value, i.e., hashing is a one-way encryption technique. Some hashing techniques used in the field of Blockchain are SHA-256, SHA-512, MD-5, MD-6, etc.

7. **Block Body:** The body of the block consists of the count of transactions and the transactions. The amount of transactions that can be held in a block are fixed by the size of the block and the size of a transaction [3]. Formation of a basic blockchain and linking of blocks is shown in Figure 13.2.

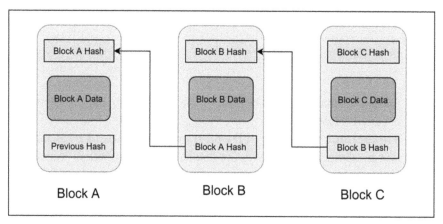

FIGURE 13.2   Basic blockchain.

13.2.1.2   DIGITAL SIGNATURE

It is the unique identification key that is allotted to the user at every single node that is connected to the network. This acts as the means to identify users who participated in a transaction.

13.2.1.3   KEY CHARACTERISTICS OF BLOCKCHAIN

Talking about the Blockchain, there are few essential characteristics (Figure 13.3) that have already been introduced to you in the introduction. Let us describe these in detail now.

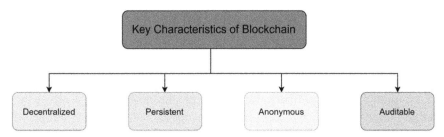

FIGURE 13.3   Key characteristics of blockchain.

13.2.1.3.1   *Decentralization*

Blockchain has no central location to store any of its information. Instead, the ledger is replicated over a network and decentralized. Each node updates the Blockchain if a new block is added to the Blockchain to accommodate the transition. By distributing the knowledge around a network, Blockchain is more difficult to control, rather than holding it in a single ledger. If a copy of the Blockchain fell into a hacker's hands, then only one copy of the information would be compromised, rather than the entire network.

All transactions have to be verified by a trusted agency in conventional centralized transaction systems, resulting inevitably in the cost and performance bottlenecks on the servers. In contrast to the centralized model, Blockchain no longer requires third parties. We can understand this with the help of Figure 13.4. Consensus algorithms (to be discussed later) are used in Blockchain to maintain consistency of the data in the distributed network [24].

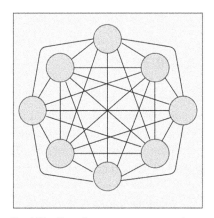

FIGURE 13.4   Decentralized/distributed peer-to-peer network.

13.2.1.3.2 Persistence

Transactions can be validated swiftly, and honest miners would not admit invalid transactions. Once they are included in the Blockchain, it is almost impossible to delete or rollback transactions. Blocks with invalid transactions can be immediately identified. Tampering at one node will not affect the database due to the decentralized nature. It implies that blocks, if tampered at someplace, the primordial copy is available on other nodes of the network, hence making it almost impossible to tamper overall Blockchain.

13.2.1.3.3 Anonymity

Every user can communicate with a created address within the Blockchain, which does not expose the user's true identity. Note that because of the intrinsic constraint, Blockchain cannot guarantee the perfect preservation of privacy. To explain this in simple words, I can say that blockchain information continues to chain the address for a given wallet within its network. This discourages the ability to trace payments made to a given wallet.

13.2.1.3.4 Auditability

Since each of the blockchain transactions is verified and stored with a time-stamp, nodes can conveniently check and track the before held records by visiting any node throughout the decentralized network. Each transaction may be tracked iteratively to previous transactions. This facilitates trace-ability and data store accountability within the Blockchain.

13.2.2 CONSENSUS ALGORITHMS

The consensus algorithm, which is underlying in a blockchain system is highly significant for the performance and security of the system. Consensus algorithms allow updation to a distributed shared state securely. There are predefined state alteration rules which are defined by the state machine that is running on all the replicas. The updates that take place in all the replicas are governed by these rules. This technique is known as state machine replication (SMR). Even if one or more nodes of a system crash, the state of the system is not lost as the replica of the state is available at all the nodes at all the times. The fundamental concept behind SMR is that it can represent a computing machine as a deterministic state machine. The computer accepts an input request, carries out its predefined computation, and may generate an output/response. Essentially, these acts alter the state. SMR conceptualizes that such a state machine can be mirrored between various nodes, with an initial condition. The alteration in the state of a state machine of every participating node must take place in the same fashion. To ensure this, each node must receive the same set of input messages and in the exact same order. Every node will produce the same result if they get the same set of input messages in the same order. This ensures the stability and accuracy of the network state across all the nodes even though node failures occur.

There are three main properties of a consensus protocol, depending upon which the applicability and the efficacy of the protocol can be determined [5]:

1. **Safety:** A safe consensus protocol is one in which every node produces the same output, and the outputs are valid as per the protocol rules. Another reference used for this is the consistency of the shared state.
2. **Liveness:** This means that all the participating, non-faulty nodes produce a value eventually.
3. **Fault Tolerance:** This can be described as the ability to recover from the failure of a participating node in consensus.

Consensus mechanisms used for the different blockchain networks can be categorized on the basis of the mechanism of payout that participating nodes may receive. The two categories into which they are classified:

1. **Incentivized Consensus:** These algorithms reward the participants for the creation and addition of new blocks into the Blockchain. They are exclusive for public blockchains.
2. **Non-Incentivized Consensus:** These algorithms do not depend on any incentive mechanism for participants for the creation and addition of new blocks into the Blockchain. They are deployed in private blockchain systems. With the absence of rewards, the participants are considered to be trustworthy as only authorized participants are allowed the creation and addition of blocks to the Blockchain.

Now let us understand the two most prevalent consensus algorithms, The Proof-of-Work and The Proof-of-Stake. These algorithms are of utmost importance in terms of cryptocurrency applications. There are numerous other consensus algorithms that are developed and are being developed to be used in blockchain applications but are not relevant as most cryptocurrencies are based on these two consensus algorithms.

13.2.2.1 PROOF OF WORK (POW)

Somebody has to be chosen in a decentralized network to record the transactions. The simplest way is through random selection. Random selection makes the system vulnerable to attack. Before publishing a block of transactions, the willing node has to prove that it is not likely to attack the system. For proof, a lot of work needs to be done. The work here means computation. Every PoW mechanism is tied to a threshold, known in many blockchain systems as the parameter of difficulty. The prover would perform the computational task in several rounds until a PoW is generated that matches the required threshold, and each round is known as a single attempt at proof. In the attempts, a hash value of the block header is computed by each node of the network. The block header includes a nonce; the miners periodically change the nonce to obtain various hash values. The majority needs the measured value to be equivalent to, or less than, a specified value. When one node hits the target value, the block would be transmitted to other nodes, and all other nodes would validate the accuracy of the hash value to one another. All miners will add this latest block to their own blockchains if the block is authenticated. Nodes computing the hash values are called miners, and in BTC, the PoW process is called mining. The computation consumes really high power which leads to the wastage of resources. Here lies an opportunity. Some PoW algorithms have been modeled to mitigate the loss in which the output could have certain side-applications. Primecoin, for example, searches for a chain of prime numbers with special properties that may be useful for research in the field of mathematics [18].

PoW is by far the most commonly used method for achieving a distributed consensus among the members regarding the order of the block and the chain state. PoW helps avoid a Sybil Attack. Sybil Attack is a form of attack seen in P2P networks where a network node deliberately runs several identities concurrently and compromises the authority of credibility schemes. A PoW mechanism in a blockchain especially serves two critical purposes:

1. **Prevent Sybil Attack:** To launch a Sybil attack, an attacker's monetary cost would be directly corresponding to the count of Sybil identities, which may overpower any value achieved from deploying a Sybil attack.

2. **Achieve Distributed Consensus:** The PoW algorithm is seen as a requirement to a feature that is eventually used to obtain the desired distributed consensus whenever a trigger occurs inside a blockchain.

Limitations of PoW:

1. **Energy Consumption:** As discussed earlier, PoW requires the nodes to do some work that proves the, to be honest. This makes the process of adding a block computationally very expensive.
2. **Mining Centralization:** As mining requires a lot of work to be done, miners need a large number of CPUs and GPUs to carry out mining activity. Hence, it is not possible for every node to mine the cryptocurrency. This leads to the centralization of mining ability to a very few numbers of users who can afford the expensive mining equipment.
3. **Tragedy of Commons:** It refers to a potential business disruption that may arise in the far future that may lead to the drop in the value of the capital earned through mining cryptocurrencies to zero.
4. **Absence of Penalty:** The proof-of-work algorithm favors the behaving miners by giving them a fee for mining. However, the problem is that PoW does not penalize the misbehaving miners, miners trying to intrude the cryptocurrency.

13.2.2.2   PROOF OF STAKE (POS)

The principle of PoS algorithms is to overcome the biggest disadvantage of PoW algorithms which is the huge amount of electricity consumption taking place in mining operations. Miners in PoS must prove their ownership of the amount of currency. People with more currencies are thought less likely to attack the network. Therefore, to engage in the block formation process, they must deposit a certain amount of their currencies, considered as a stake, into an escrow account. The stake serves as an assurance that the node will operate in compliance with the laws of the protocol. In PoS terminology, these participants can be called stakeholders, leaders, forgers, or minters. If a minter misbehaves, then it can lose its stake.

Comparing Pow and PoS, if any person wishes to earn rewards for block creation in a Blockchain-based cryptocurrency network, then there are two options. One can invest in mining equipment and start using it for mining and win a reward. The other option is to use that same amount in buying cryptocurrencies and using it as a stake to buy proportionate block creation chances in the system by becoming a validator. Since the technological requirements of mining equipment might change in the future but the stake will definitely increase with an increase in investment with time. Investing in buying stake is a more suitable option than buying mining equipment.

The selection of the next creator just based on the account balance can become unfair as the richest single person will only benefit from it. To determine the next generator, Blockchain uses randomization. It uses a formula which searches for the lowest hash value in combination with the stake size [19]. In the event that the stakeholder gets an opportunity to establish a new block, the stakeholder will be awarded in one of two forms. Either it may obtain the transaction fees inside the block, or a certain amount of currencies is given against their stake as a form of interest. Advantages of PoS:

- Energy efficiency;
- Mitigation of centralization;
- Explicit economic security.

Bootstrap problem—a major obstacle in a PoS mechanism is the generation of initial coins and fairly dividing them among the stakeholders so that they can be used as stakes. This hurdle is termed as the bootstrap problem. Two strategies can be useful to address this problem:

1. **Pre-Mining:** A few coins are pre-mined at the time of the integration of the system. These pre-mined coins are then sold in an initial public offering (IPO) or initial coin offering (ICO) before the launch of the Crypto-network.
2. **PoW-PoS:** Initially, the system is started with PoW to divide the cryptocurrency fairly amongst the stakeholders and then the system slowly transits towards the PoS. For instance, ETH is planning to move from Ethash (PoW) to Casper (PoS).

Limitations of PoS:

1. **Collusion:** If the number of users/nodes is small, it might be easier for a person to accommodate resources which may account for more than 50% computing power, making attacks easier. This is known as collusion.
2. **Wealth Effect:** Users with a higher number of coins have a larger influence of the PoS algorithm, and hence, they are favored.

13.2.3   CASE STUDY: BITCOIN (BTC)

13.2.3.1   INTRODUCTION

Although the mechanism employed for handling currency exchange, including banks, governments, master authorities, etc., is operating well enough for most transactions, the intrinsic shortcomings of the centralized model are still present. Completely irreversible contracts are impossible to execute because financial firms are powerless to prevent dispute resolution. There is a need for a mechanism in which the user, or the owner of the amount, is the master and handles all the transactions himself, without any centralized organization handling it or monitoring it. As stated earlier, in 2009, Satoshi Nakamoto came with such a mechanism [22].

According to Satoshi Nakamoto, a digital payment system is needed which relies on cryptic evidence rather than confidence/trust, which allows any two users to make direct transactions with one another without trying to get a reliable third party. Computationally expensive transactions which are hard to undo. This would prevent dealers from fraud which should be applied by regular escrow procedures to cover buyers. Figure 13.5 shows the growth of the market capitalization of BTC since its introduction.

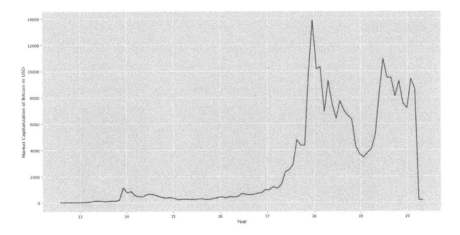

FIGURE 13.5   Market capitalization growth of bitcoin [20].

In this case study, we are going to discuss how Satoshi Nakamoto solved the double-spending problem employing a P2P decentralized time stamp system to produce sequential order statistical proof of transactions. The network is stable as long as the honest nodes, all together manage more computing power than any community of intruder nodes that cooperate.

13.2.3.2 Transactions

Satoshi Nakamoto describes BTC as a digital signature chain. By digitally signing the previous transaction's hash, then using the cryptographic hash of the next owner, each owner sends the coin to the next user by appending it to the coin. A receiver may inspect signatures to check ownership chain. The question, of course, is that the payee cannot guarantee whether the coin was not double-spend by either of the owners. What was needed was a mechanism for the payee to realize that no prior transactions were signed by the former owners [23].

The earliest transaction, for our purposes, is one that counts, and we do not worry for later double-spending attempts. The best way to prove a transaction's absence is to remain aware of all transactions. Transactions will be officially reported to do that without a third party, so we require a mechanism for participants to rely on a common ledger in which the transactions were registered in chronological order. The recipient requires evidence that the majority of nodes agreed it was the first obtained at the time of each transaction [11]. Figure 13.6 depicts the mechanism of storing transactions in a blockchain.

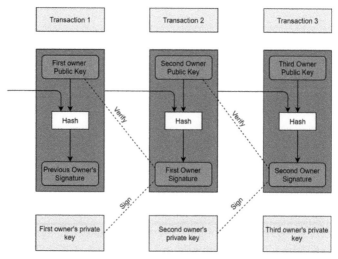

FIGURE 13.6   Transactions in a blockchain.

13.2.3.3   TIMESTAMP SERVER

The method that was suggested to accomplish the above task starts with a server consisting of a timestamp. A timestamp server operates by accepting a hash value of a set of objects to get timestamped and broadcasting the hash value. The timestamp implies the transaction would have occurred at the moment to get through the hash value. Each block incorporates the previously held timestamp in its hash, creating a chain as shown in Figure 13.7 [25].

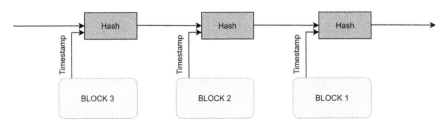

FIGURE 13.7   Timestamping.

13.2.3.4   PROOF-OF-WORK

The PoW algorithm for BTC involves looking for a value whose hash begins with a number of zero bits. The average work required is exponential in the amount of required zero bits, which can be verified by the execution of a single hash. The PoW is achieved by incrementing a nonce in the block until a specific value is reached that gives the necessary zero bits to the block hash. After the expenditure of the CPU effort to fulfill the proof of operation, the block cannot be changed without redoing the job. The work to modify the block will require redoing all the blocks after it, as subsequent blocks are clustered after it.

The PoW also addresses the question of evaluating participation in decision-making by the majority. The majority judgment is reflected by the longest chain, which has given the greatest effort in PoW. If honest nodes hold a majority of the computing resources, the honest chain should expand the quickest and outrun any competing chains [29].

An attacker would have to redo the block's proof-of-work and all the blocks after it to alter the network, and then catch up and conquer the work performed by truthful nodes. When additional blocks are inserted, the likelihood of a slower intruder catching up reduces exponentially.

13.2.3.5   NETWORKING

The network execution phases are listed below:

- Latest transactions are broadcasted to every node of the network.
- Every node collects the latest transactions and registers them into a block.
- Every node will then have to compute a computationally extensive PoW for the block.
- As soon as a node computes a PoW, it publishes the block to every other node.
- Every other node validates the block only when all transactions in it are authentic and not spent previously.
- Every node expresses their acceptance of the latest block by keeping the hash value of the validated block hash value of the previous block to construct the next block in the Blockchain.

Nodes always consider the longest chain to be the right one and will continue to function to expand it. When two nodes concurrently transmit separate copies of the next row, then certain nodes could first obtain one or the other. In that scenario, they operate on the first one they got, but in case it gets longer, they save the other section. If the next PoW is located and one branch gets longer, the tie will be broken; the nodes that were operating on the alternate branch will then turn to the longer chain. Block broadcasts are tolerant of lost blocks. If a block is missing from a node when the next block is received, it can order the block, and it recognizes it has missed one [26].

13.2.3.6   INCENTIVE

The initial transaction in a block is by default a unique transaction, which initializes a new coin that belongs to the builder of the block. It provides an incentive for nodes to join the BTC network, which offers a means to immediately disperse coins into flow since no single body remains to release them. The gradual introduction of a continuous volume of fresh coins is similar to the spending of money by gold mines to bring gold to circulation. For our situation, it is the cost of CPU energy and time. A transaction wage is also added to the incentive if the displayed value of a transaction is less than the intake value. If a set amount of coins has reached circulation, the incentive will fully shift to transaction fees and be safe from inflation [30].

Incentive force plays an important role in urging nodes to remain fair. If a gatecrasher could accumulate more computing assets than every single genuine node, he would need to pick between utilizing it to swindle residents by reclaiming his coins or utilizing it to create new coins. He would think of it as progressively beneficial to work by the guidelines, guidelines that benefit him with unmistakably more new coins than all others consolidated, then devastating the framework and the credibility of his riches.

13.2.3.7   Reclaiming Disk Space

If a coin's new transaction is hidden below a number of blocks, the transactions expended before it can be dumped to conserve storage. To simplify this without breaching the hash of the block, the Merkle tree is used to hash the transactions, with only the Merkle root used in the hash of the block. Old blocks will then be compressed by dislocating tree branches, shown in Figure 13.8. There is no need to store internal hashes.

A block header is about 80 bytes without transactions. If we presume that, every 10 minutes, blocks are produced, that implies that 4.2 MB storage is required annually. This implies that even though block headers are placed in memory, storage is not a concern.

13.2.3.8   PAYMENT VERIFICATION

Payments may be checked without running a complete node to the network. A user just wants to hold a copy of the block headers of the largest PoW series, which can be accessed by network nodes before he is sure that he has the largest series, and get the Merkle branch that connects the transaction to the block in which it is timestamped. He cannot verify the transaction on his own, but by connecting it to a point in the Blockchain, he will see that this has been authorized by a network node, and blocks attached after the network validates it. The validation is effective as long as the network is managed by honest nodes, but becomes more fragile when an intruder overpowers the network. The simplistic approach may be tricked by transactions created by an attacker for as long as the attacker will manage to dominate the network.

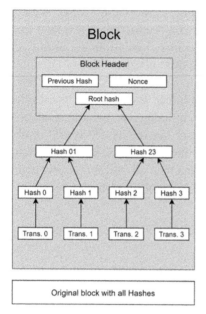 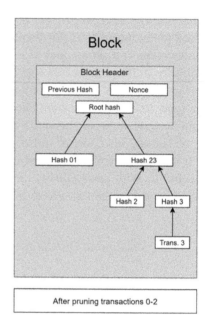

FIGURE 13.8 Storage optimization.

13.2.3.9 COMBINING AND SPLITTING VALUE

While coins may be treated separately, having a different transaction with any cent in a trade would be unmanageable. The transactions involve several inputs and outputs to enable the value to be separated and integrated. Typically, there will be a sole input from a greater previous transaction or several inputs incorporating smaller numbers, and at most two outputs would be available: First will be for payment, and other, if any, returns the difference to the sender.

13.2.3.10 PRIVACY

Privacy may be protected by controlling the flow of information somewhere else: by holding public keys anonymously. The public can view that somebody transferred an amount to someone else but without the details that connect the transaction to anyone [28].

For each transaction, a new pair of keys will be used as an external firewall to avoid them to get linked to a similar user. Any linking with transactions having multiple inputs is still inevitable and ultimately shows that the same owner-operated their inputs. A possibility is that linking might expose certain transactions that belong to the same owner if the owner of a key is exposed. From Figure 13.9, we can see the difference between the new and old privacy models.

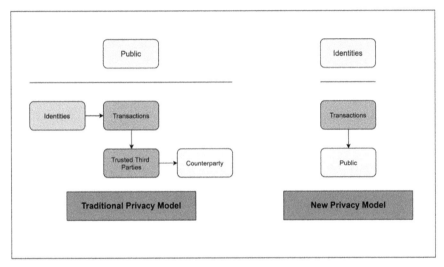

FIGURE 13.9   Traditional vs. new privacy model.

13.2.4   ADVANCES IN ETHEREUM (ETH) OVER BITCOIN (BTC)

ETH is a decentralized software framework built on a public blockchain based open-source platform, introducing SCs. It provides transaction-based state transformations to an updated variant of Nakamoto consensus, i.e., BTC PoW Consensus. Although BTC attempts to challenge electronic banking, ETH seeks to substitute internet third parties with a blockchain — those that store records, pass mortgages, and maintain track of complicated financial instruments. However, as ETH and BTC have certain parallels, the two systems have separate goals. Where BTC is simply a digital currency, intended to act as a medium of payment, ETH is taking a more ambitious approach. ETH serves as a network from which people can build and operate applications using Ether tokens, and more specifically, SCs. In brief, ETH aims to be a tool that will decentralize the current client-server architecture and democratize it [7].

While all blockchains have the potential to process data, most of them are heavily limited. ETH is something unique. ETH helps developers to build any operations they choose, rather than providing a collection of specific operations. This ensures developers will create thousands of diverse apps that go far beyond anything we have foreseen before. Figure 13.10 shows the growth of the market capitalization [15] of BTC since its introduction.

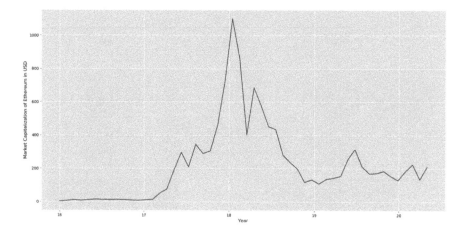

FIGURE 13.10   Market capitalization growth of Ethereum [20].

The ETH replaces servers and databases with thousands of so-called "nodes" operated by volunteers from across the globe, just as in BTC. ETH, like BTC, is a decentralized network of public blockchains. While there are some major technical differences between the two, the most critical contrast to remember is that the purpose and functionality of BTC and ETH are radically different. BTC provides one specific use of blockchain technology, an automated P2P cash network that facilitates BTC transfers electronically. Even though the BTC blockchain is used to control digital currency possession, the ETH blockchain focuses on running every decentralized application's (DApp) programming code [6].

13.2.4.1   *SMART CONTRACTS (SCS)*

A smart contract is indeed a term used to define programming code that will make capital, information, properties, shares or something of interest easier to trade. When operating on the Blockchain, a smart contract is like a software code which runs when certain conditions are satisfied. Since SCs are operating on the Blockchain, they operate as designed without the risk of surveillance, interruption, theft, or intervention by third parties. SCs are code-written contracts which are submitted to the Blockchain by the developers. Each time one of those contracts is performed, every node on the network executes it, added to the Blockchain; hence, it is deposited in the potentially tamper-proof public ledger as discussed earlier in this chapter. By fact, SCs are formulated as declarations 'if-then.' If those requirements are fulfilled, the contract clauses are enforced by the project [16].

One of the widely quoted benefits of SCs is that "middlemen" such as lawyers or notaries are not required. In principle, this ensures you can conduct deals without the processing periods found in paper documents, often without charging commissions to someone who will normally manage such a deal. This is important for people living in countries with a legal system that is corrupt or inefficient—such as Venezuela, Cuba, Zimbabwe, and China, etc.

13.3 CRYPTOCURRENCY: ADVANTAGES AND DISADVANTAGES

Till now, we have understood the majority of concepts related to Blockchain in Cryptocurrency. There are some other significant aspects related to the study of cryptocurrency. This section discusses the advantages and disadvantages of using cryptocurrency. An important question that stays is are Cryptocurrencies actually advantageous or disadvantageous. There are distinct and divergent perspectives on the future prospects of cryptocurrencies in general. While those with libertarian views of life are positive and endorse the cryptocurrency scheme, other researchers, economists, and academics from this area are not enthusiastic about using cryptocurrency throughout the payment and financial transaction framework [8].

Like every other debatable topic, the emergence of Blockchain-based cryptocurrency has two different viewpoints. The positive view of the use of cryptocurrencies is highlighted by the fact that they allow easy exchange of funds between two parties in a transaction. Earlier, we have seen the public and private keys, which are fundamental in making transfer easy and effective. Also, the fact that these transactions are processed at very low processing fees is a good approach to avoid the steep fees charged by banks for processing transactions. Several nations have already started adopting BTC as a legitimate currency. Countries particularly that want to get rid of cash have a very friendly approach to cryptocurrencies. A point that proponents of BTC use is BTC's, ETH's and other cryptocurrencies' market capitalization, arguing that the cryptocurrency market has grown quite large and strong, so prohibiting it would be very costly for any nation. There may be a case that a country's government does not allow a cryptocurrency to be legitimate because it may be used for money laundering and also financing illegal activities since there is no governing authority. This is the reason what the opponents of cryptocurrencies have with them. Another argument in this regard is the high value of price volatility of cryptocurrencies. Additionally, it lacks a central issuing agency and also it does not have any economic and financial basis on which it is created.

Let us list out some advantages and disadvantages of using cryptocurrencies as a medium of trade amongst the general population.

> **Advantages:**

1. **Anonymous Identity:** In a blockchain cryptocurrency network, the information about a transaction which is available to everyone does not contain any private data of the users. At the maximum, a miner can get to know the public keys of the two parties involved, but this can in no way lead the miner to the personal details of the involved parties.

2. **Decentralized and User-Controlled:** The P2P network in a cryptocurrency system abandons the requirement of a master server which is in charge of all operations. An exchange that takes place in this system is between two or more software clients. There is no authority that can put restrictions on occurring transactions. All transactions can be executed with full freedom.

3. **Immutable:** There are no boundaries. Canceling a payment is impossible in this system and even faking, copying, and double-spending are impossible. The transactions, once registered successfully, are immutable. This guarantees the integrity of the system.

4. **Economical:** Extremely low operation costs. The need of paying commission and fees to banks and other organizations is shunned.

5. **User Empowered:** Power in the hands of a participant rather than a central authority otherwise. The only rules that a participant needs to follow are those that are set by the developers.

6. **Ease of Access:** Easy to use is the most productive advantage of cryptocurrencies. An organization may get ready for trade making use of cryptocurrency in as little as 5 minutes. Compare this to a weeklong procedure of opening a bank account with limitless rules and regulations.

7. **Transparency:** Since the miners have access to all the transactions that are stored in the ledger. They get to know the UIDs of the involved parties. This makes it transparent while they do not know the real person behind the UID, thus keeping it anonymous.

8. **Swift Transactions:** The speed of transactions is really fast. It is just a matter of a few minutes that are required for processing a transaction and successfully storing it in the decentralized network.

➢ **Disadvantages:**

1. **Lack of Education:** Just from the rumor that people are earning from BTCs (or any other cryptocurrency) they start investing in it and lose their valuable wealth. It is very important to educate the people about this new framework of currency before they start investing in it. Also, it is the responsibility of the investors of their own to educate themselves before investing in cryptocurrencies.

2. **Less Acceptance:** The usage of cryptocurrencies is limited even though they are highly used in the market. The reason being the lack of an official approval from the government to legalize it for general public use. This confines its usage to a restricted set of domains.

3. **Possibility of Deflation:** The value of cryptocurrency is highly volatile and is totally dependent on supply and demand, i.e., if the cryptocurrency is not regulated or circulated, chances of deflation increase. Hence, Mobility of cryptocurrency plays a vital role in governing its value.

4. **The Government May Veto:** As we said, the government cannot regulate cryptocurrencies, but they have the ability to prohibit

transactions or illegalize them. This casts a shadow, of course, on these bold, unfettered movements. Yet the government cannot actually ban the notion of cryptocurrency entirely. The latest news suggests that governments may come to an agreement with cryptocurrencies for more straightforward fashion in managing their economic system.

5. **Irreversible Payment:** When you pay someone unintentionally using cryptocurrencies, then there really is no way to reverse the fee. This is statutory that a cryptocurrency exchange has taken place. Regardless of its permanence, you cannot rollback the contract at any cost. The only way to revert the payment is to ask for your cryptocurrency back from the person you paid.

6. **Non-Recoverable Key:** Due to the reason that all Blockchain-based cryptocurrencies lack a central regulatory authority, it is, therefore, the duty of each person to keep their account secure and confidential. No one will help them to recover the wallet key in case they forget it.

7. **Funding for Illegal Activities:** The cryptocurrency's secrecy makes it appealing to both the drug market and money launderers. Since the name is not disclosed, there are several occasions where violations are recorded. Popular two are the "silk road" website selling illicit drugs and other illegal things payable via BTC and having to scream cyber-attack more recently [27].

13.4   RESEARCH OBSTACLES AND FUTURE PROSPECTS

Blockchain will automate the middle and back-office functions increasing the efficiency and also creating new business opportunities. The basic principle behind every Blockchain is fairly straightforward-every transaction is cryptographically signed, and they consume inputs (transfer of cryptocurrency from the initiator) and produce new output (transfer of cryptocurrency to the receiver). To check that a transaction is valid, every transaction of the networks is broadcasted over the P2P network. Simply broadcasting is, however, an incomplete solution. With this, it is possible that the copy of the history of transactions that is stored at each node is not identical due to the propagation delays in the network and the nodes going on and off. There are two properties that a distributed system design must satisfy to prevent inconsistency in the nodes. First, the transaction activity frequency must be much lower than the propagation delay in the network. Second, each part of the history of the transaction must reference the previous part of the transaction history so that a chain of transaction history is created. An unanswered question still resides-How will the distributed network decide on which block is the right one? Mining solves this problem. Miners have to compete with each other to find the next authentic block. A related to block, hard to solve, easy to check computing problem is required to be solved. This problem of expensive computation is what restricts the number of legitimate blocks occurring on the Blockchain network, and hence a global consensus is established on the current transaction history [9].

The consensus process used in public blockchain networks remains a progressing research problem. A significant volume of power is used by proof-of-work systems to run machines to compute difficult to solve but easy to validate technical challenges that have little practical use other than protecting the Blockchain. In order to mitigate this issue, the study has been undertaken in two ways:

- Discarding the need of PoW algorithms; and
- Find problems involving high computation for the PoW whose outcome may be beneficial for the society.

In addition to the vast quantities of energy needed for mining PoW crypto-currencies, the incredibly high computing power is necessary for successful cryptocurrency mining. This produces a quite high entry barrier to becoming a miner, and hence, the cryptocurrency framework built to be decentralized tends toward centralization. Think of how BTC mining progresses. First, mining by the CPU. Then, GPU mining took place. This was immediately followed by FPGA mining. Following that came ASIC mining. As of today, BTC will now profitably be mined on massive ASIC farms only. It can also be concluded that BTC is heading towards greater centralization because more and more of the hashing capacity of the network is being focused on a limited number of parties [13].

One existing research goal for proof-of-work is to consider a collection of hard to simplify but easy to validate scientific problems. Many problems in scientific computation can only be tested by reproducing the actions taken to get to the answer. Consequently, the present research goal for these problems is to make sure a valid answer can be determined without needing all nodes to measure the computationally expensive solution [17].

The system scalability of Blockchain is also of concern since the Blockchain follows an append-only approach; its size is ever-growing. For instance, the size of the BTC blockchain at the end of March 2020 was 270 GigaBytes [10]. Blockchain pruning is one solution to reducing the database space needs for maintaining a whole blockchain node. To enforce blockchain pruning properly, the hash of the new block must indicate the current state of the wallets of all the users. Try to understand blockchain pruning with a small example. We have 3 nodes N1, N2, and N3 each having a total of 10 coins. The transaction in the first block denotes a transition from N1 to N2 of 1 coin. The network's present wallet condition is 9 coins in N1, 11 coins in N2 and 10 coins in N3. Another block on the Blockchain now emerges, indicating a transfer from N2 to N3 of 1 coin. The new wallet state is 9 coins in N1, 10 coins in N2 and 11 coins in N3. So, the transactions from N1 to N2 and N2 to N3 can be substituted with a single transaction from N1 to N3.

Hard drive space needed by an ever-growing network is not the only question regarding network scalability. The blockchain transaction transfer rate is often constrained by the computing capacity and network's capability of the P2P nodes. The BTC network is reportedly able to handle transactions at an average rate of 7 transactions in one second, a number that is around 1,000 times lesser than the VISA transaction processing network's peak capacity.

To avoid wasting too much computing power through SCs, ETH introduces a limit, thereby restricting the number of operations on instruction that can be done by the smart contract. For more computationally expensive smart contracts, both nodes will need to reach consensus on the result of the execution of the smart contract without any node needing to operate the smart contract. This is a research problem that is still being solved.

Since all operations that exist on a blockchain, for example, the BTC P2P network are openly viewable, numerous questions have been raised in respect to consumer privacy. Except for criminal behavior, many consumers object to the exposure of their transaction background for fear of consumer monitoring or targeted advertising. Due to these questions regarding privacy, research has been and is to be carried out into introducing confidentiality security to blockchain networks.

13.5   CONCLUSION

Blockchain will automate the middle and back-office functions increasing the efficiency and also creating new business opportunities. The basic principle behind every Blockchain-based cryptocurrency is fairly straightforward-every transaction is cryptographically signed, and they consume inputs (transfer of cryptocurrency from the initiator) and produce new output (transfer of cryptocurrency to the receiver). The process for validating a transaction is to broadcast the transaction over the P2P network. It should be ensured that at every node the previous transaction is synchronized with all the peers, i.e., all the nodes have the exact same copy of the chain of transactions. An unanswered question still resides-How will the distributed network decide on which block is the correct one? Mining solves this issue. Miners have to compete to find the next block that is authentic. A related block that is hard to solve, simple to check computing issues are required to be solved. This problem of uneconomical computation is what restricts the number of legitimate blocks occurring on the Blockchain network, and therefore, on current transaction history, a global consensus can be established.

KEYWORDS

- **bitcoin**
- **blockchain**
- **cryptocurrency**
- **cryptography**
- **decentralization**
- **ethereum**
- **hash**
- **immutable**
- **incentive**
- **ledger**
- **mining**
- **node**
- **proof of work**
- **timestamp**
- **transaction block**

REFERENCES

1. Hameed, B. I., (2019). Blockchain and cryptocurrencies technology: A survey. *JOIV: International Journal on Informatics Visualization, 3*(4). https://doi.org/10.30630/joiv.3.4.293.

2. Emmanuel, C., (2020). *Basics of Blockchain and Cryptocurrencies.* https://www. researchgate.net/publication/340174600_Basics_of_Blockchain_and_Cryptocurrencie s?channel=doi&linkId=5e7c6a6d92851caef49da0cc&showFulltext=true (accessed on 29th July 2021).

3. Zheng, Z., Shaoan, X., Hongning, D., Xiangping, C., & Huaimin, W., (2017). An overview of blockchain technology: Architecture, consensus, and future trends. In: *2017 IEEE International Congress on Big Data (BigData Congress).* https://doi.org/10.1109/ bigdatacongress.2017.85.

4. Frankenfield, J., (2020). *Nonce Definition.* Investopedia. https://www.investopedia. com/terms/n/nonce.asp (accessed on 29th July 2021).

5. Baliga, A., (2017). *Understanding Blockchain Consensus Models.* Persistent Systems Ltd. https://www.persistent.com/wp-content/uploads/2017/04/WP-Understanding-Blockchain-Consensus-Models.pdf (accessed on 29th July 2021).

6. Buterin, V., Daniel, R., Stefanos, L., & Georgios, P., (2019). Incentives in Ethereum's hybrid Casper protocol. In: *2019 IEEE International Conference on Blockchain and Cryptocurrency (ICBC).* https://doi.org/10.1109/bloc.2019.8751241.

7. Epan (2018). *Ethereum Case Study – Understanding What, Who, Why, When, and How – Part 1* (What). Steemit. https://steemit.com/ethereum/@epan35/ethereum-case-study-understanding-what-who-why-when-and-how-part-1-what (accessed on 29th July 2021).

8. Bunjaku, F., Gorgieva-Trajkovska, O., & Miteva-Kacarski, E., (2017). Cryptocurrencies – advantages and disadvantages. *Journal of Economics, 2*(1). ISSN 1857-9973 http:// js.ugd.edu.mk/index.php/JE/article/view/1933 (accessed on 29th July 2021).

9. Mahmoud, Q. H., Michael, L., & May, A., (2019). Research challenges and opportunities in blockchain and cryptocurrencies. *Internet Technology Letters, 2*(2). https://doi. org/10.1002/itl2.93.

10. *Bitcoin Blockchain Size: YCharts.* https://ycharts.com/indicators/bitcoin_blockchain_ size (accessed on 29th July 2021).

11. Wright, C. S., (2008). Bitcoin: A peer-to-peer electronic cash system. *SSRN Electronic Journal.* https://doi.org/10.2139/ssrn.3440802.

12. Deepak, K. S., Shrid, P., Mehul, S., & Shikha, B., (2020). Cryptocurrency mechanisms for blockchains: Models, characteristics, challenges, and applications. *Handbook of Research on Blockchain Technology* (pp. 323–348). Academic Press, Elsevier.

13. Deepak, K. S., Ajay, K. K., Aarti, G., & Saakshi, B., (2020). Internet of things and blockchain: Integration, need, challenges, applications, and future scope. *Handbook of Research on Blockchain Technology* (pp. 271–294). Academic Press, Elsevier.

14. Chan, S., Jeffrey, C., Saralees, N., & Joerg, O., (2017). A statistical analysis of cryptocurrencies. *Journal of Risk and Financial Management, 10*(2), 12. https://doi. org/10.3390/jrfm10020012.

15. Jurić, V., Vanja, Š., & Domagoj, K., (2019). Statistical analysis of the most influential cryptocurrencies. *SSRN Electronic Journal.* https://doi.org/10.2139/ssrn.3490485.

16. Brownsword, R., (2019). Smart Contracts. Regulating Blockchain, pp. 311–326. https:// doi.org/10.1093/oso/9780198842187.003.0018.

17. Meva, D., (2018). Issues and challenges with blockchain a survey. *International Journal of Computer Sciences and Engineering, 6*(12), 488–491. https://doi.org/10.26438/ijcse/ v6i12.488491.

18. Katarya, R., & Aamir, M., (2020). Blockchain and consensus algorithms. *SSRN Electronic Journal.* https://doi.org/10.2139/ssrn.3562974.

19. Sharma, K., & Deepakshi, J., (2019). Consensus algorithms in blockchain technology: A survey. In: *2019 10th International Conference on Computing, Communication and Networking Technologies (ICCCNT).* https://doi.org/10.1109/icccnt45670.2019.8944509.

20. *Cryptocurrency Market Capitalizations.* CoinMarketCap, https://www.coinmarketcap.com/ (accessed on 29th July 2021).

21. Jhanwar, N., (2017). The growth of crypto-currencies. *International Journal of Recent Trends in Engineering and Research, 3*(6), 368–379. https://doi.org/10.23883/ijrter.2017.3319.evnx3.

22. Vujičić, D., Dijana, J., & Siniša, R., (2018). Blockchain technology, bitcoin, and Ethereum: A brief overview. In: *2018 17th International Symposium Infoteh-Jahorina (INFOTEH).* https://doi.org/10.1109/infoteh.2018.8345547.

23. Shrivas, M. K., (2017). A critical review of cryptocurrency systems. *Texila International Journal of Academic Research, 4*(2), 116–131. https://doi.org/10.21522/tijar.2014.04.02.art012.

24. Bhatia, R., Praveen, K., Shilpi, B., & Seema, R., (2018). Blockchain -the technology of cryptocurrencies. In: *2018 International Conference on Advances in Computing and Communication Engineering (ICACCE).* https://doi.org/10.1109/icacce.2018.8441738.

25. Szalachowski, P., (2018). (Short Paper) towards more reliable bitcoin timestamps. In: *2018 Crypto Valley Conference on Blockchain Technology (CVCBT).* https://doi.org/10.1109/cvcbt.2018.00018.

26. Donet, J. A. D., Pérez-Solà, C., & Herrera-Joancomartí, J., (2014). The bitcoin P2P network. *Financial Cryptography and Data Security Lecture Notes in Computer Science,* 87–102. https://doi.org/10.1007/978-3-662-44774-1_7.

27. Greeshma, K. V., (2015). Cryptocurrencies and cybercrime. *Recent Trends for Privacy Preservation Techniques in Data Mining, 3*(30). https://www.ijert.org/research/crypto-currencies-and-cybercrime-IJERTCONV3IS30022.pdf (accessed on 29th July 2021).

28. Conti, M., Sandeep, K. E., Chhagan, L., & Sushmita, R., (2018). A survey on security and privacy issues of bitcoin. *IEEE Communications Surveys & Tutorials, 20*(4), 3416–3452. https://doi.org/10.1109/comst.2018.2842460.

29. Chowdhury, N., (2019). Consensus mechanisms of blockchain. *Inside Blockchain, Bitcoin, and Cryptocurrencies,* 49–60. https://doi.org/10.1201/9780429325533-3.

30. Ghimire, S., & Henry, S., (2018). A survey on bitcoin cryptocurrency and its mining. In: *2018 26th International Conference on Systems Engineering (ICSEng).* https://doi.org/10.1109/icseng.2018.8638208.

CHAPTER 14

Blockchain as a Facilitator Technology in the Digital Era: Analysis of Enablers

RAVINDER KUMAR

*Department of Mechanical Engineering, Amity University, Noida,
Uttar Pradesh, India, E-mail: rkumar19@amity.edu*

ABSTRACT

In today's era of customization and personalization, the digitalization of services has become highly important. The kinetics of the society and organizations are changing due to technologies like artificial intelligence (AI), IoT, augmented/virtual reality, blockchain, etc. These technologies have brought new features with different challenges. While dealing with significant issues and challenges of new technologies, blockchain give an edge of balance in the current scenario. Its properties of fixity, temper evident and circumvent fraud make this technology very important for digitalization in the modern system.

In this chapter, the author has analyzed the enablers of blockchain technology by using DEMATEL techniques. Finding of the study states high security of passwords, multi-node storage of data, mutable by hashing power, smart clauses of the contract are main causing enablers of blockchain technology. Similarly enhance cybersecurity, temper evident, efficient, and prompt reply, improve the transparency of data and circumvent fraud/double-spending are effect group enablers determined by cause group enablers. Blockchain finds application majorly in the financial sector; still, there is a need to explore its scope in other sectors like healthcare, education domain, supply chain management (SCM) and data analytics, etc.

14.1 INTRODUCTION

Organizations of the modern era are under intense pressure due to mass customization, personalization, and short product life-cycle of products [13].

Technological advancement is the ray of hope in this difficult time. Different technologies like IoT, cloud computing, big data analytics, augmented, and virtual reality with AI, and cyber-physical systems are very trending these days [6, 27]. Technologies like IoT and IIoT generates data in abundance. By adopting these technologies, the issue of data security arises. Blockchain gives secure and personalized data exchange in all fields like finance, manufacturing, and health. Blockchain is a digital technology working on a cryptographically linked chain of data blocks. It has the qualities of being coherent, incremental, effectual, and digital [4]. There are many applications other than the bitcoin (BTC), like online tickets booking, where blockchain makes the full process of financial transactions secure. Application of blockchain technology in supply chain management (SCM) has minimized the interference of the third party between the retailers and consumers. Blockchain improves the transparency and hardiness of transactions in the supply chain. Blockchain allows decentralized operations, with enhanced functionality and reliability [23].

Security and efficiency of all transaction will improve by the use of blockchain in the supply chain [2]. Incorporation of blockchain with the supply chain enables in accomplishing sustainability [31]. Blockchain improves the randomness of a system improves with security and robustness [15]. Cryptocurrencies and blockchain technology help in enhancing unanimity and social connection in financial dealings [19]. Blockchain has the scope of innovation and application in many fields. It helps in reducing the interference of middle parties in financial and other vital areas. But blockchain needs technology application with continuous up-gradation [1]. In this chapter, the author has identified the eminent enablers of blockchain technology and analyzed them using DEMATEL technique. The chapter is organized as follows. Section 14.2 discusses colligated literature. Research methodology applied has been discussed in Section 14.3. Section 14.4 discusses the finding of the study. Finally, Section 14.5 concludes the chapter with implications, limitations, and future ways of research.

14.2 LITERATURE REVIEW

In this section, the author has discussed the related literature and identification of enablers of blockchain. Blockchain digitally stores information in the form of blocks. It consists of a series of blocks which are cryptographically attached to the preceding blocks. This technology has recolonized the digital transaction process without third-party intervention. Blockchain allow decentralized operations, with improved functionality and reliability [2]. Security and efficiency of all transactions have been enhanced by the use of blockchain in the system. Blockchain helps in maintaining diaphanous and trustable documentation between different information sharing organizations [23]. Changing company financial records by altering electronic files, documents, and transaction information is widespread fraud these days [5]. Blockchain can give prevention from all types of scams and cyber-attacks. Healthcare systems could be made simple and affordable by improving the management of patient records, involving insurance companies and technology like blockchain [24]. Feng et al. [7] developed the blockchain-enabled traceability system for food supply chains. IoT-enabled traceability system improves the monitoring, security, and transparency of the food supply chain.

The resource sharing mechanism in a blockchain-based cyber-physical system strengthens cybersecurity and privacy protection (initially used in BTC). These days' industries are using IoT devices for improved control and connect in the cyber-physical system [32]. Excessive data generated by different IoT devices require cloud space for storage. Blockchain technology facilitates in privacy and data security on cloud spaces. Blockchain could secure a lightweight wireless communication of IoT devices in industries. Blockchain enhances the facilities of IoT devices with the decentralized network, resulting in low susceptibility to manipulation of data and illegal copying by venomous participants as compared to ordinary IIoT devices [22].

There is a connection between blockchain and IoT. Authors also learned the resources utilization, sharing, automation, privacy in a blockchain-enabled IoT system [3]. Blockchain technology is tamper-evident. It means that if there is any attempt to temper the information, the system rejects the process [4]. The authors studied the influence of blockchain on data security in public sector units and its process. Authors also studied the security, governance, and regulatory implications of blockchain technology [30]. Blockchain had reduced the interference of go-betweens in the tourism sector. The culture of online tourism has given the power in the hand of consumers [18].

In blockchain, information is hived on many computers constituting the blocks [29]. Blocks are further joined in chains, which cannot be altered or deleted by any single member. Polvora et al. [16] studied the challenges and opportunity of blockchain considering it on knottiness of policy, technical, legal, and sustainability factors. Khan et al. [10] studied the application of blockchain technologies in different non-financial sectors like health, automation of industries, energy, security, and smart grids. In healthcare, authors studied the four-layer model of precision medicine, clinical trials, and automation of patient records with security [10]. Blockchain improves the transparency of the supply chain. Authors observed that knowledge exchange and supply chain partners pressure help in the adoption of blockchain [28].

Blockchain provides high security in the form of multi-signature to operate, and multi-keys are required to finalize the process or transaction. If the hacker tries to enter into the system or tries to steal information, then there are lots of other back-ups (the interconnected computer), by which it could be easily retrieved [25]. In the blockchain, hackers have to hack more than 50% of the computers in the network simultaneously to get success in operation, which seems impossible most of the time. The process of multi-signature matching in blockchain, improve its security from threats. The compulsory process of ledgers and digital signature certification for each transaction in blockchain makes it fool proof technology [17]. A blockchain-enabled system can replace costly systems of banks, insurance, and many government departments with a lot of cost-saving. The stochasticity of the system could be improved with security and robustness by using blockchain [15]. Authors studied the connect between the blockchain and sustainable supply chain management (SSCM). Also explored different factors enabling connect of blockchain and sustainable supply chain like data decentralization with safety, handiness, secure rules, and policy with minimum corroboration [31]. Blockchain is a disseminated database having invariant legers which rely on cryptography for security. Authors studied its classification and application areas [8]. Blockchain allows a de-concentrated and disseminated public ledger for all the members involved. The author developed a framework of blockchain including innovation, organizational, environmental, and user espousal properties [26]. Application of blockchain makes SCM efficient and transparent [9]. The middlemen problem can also be eradicated. The application areas of blockchain technology identified by the author from literature are graphically shown in Figure 14.1. The key enablers of blockchain technology identified from literature are also shown graphically in Figure 14.2. The key enabling technologies of blockchain has been summarized in Table 14.1.

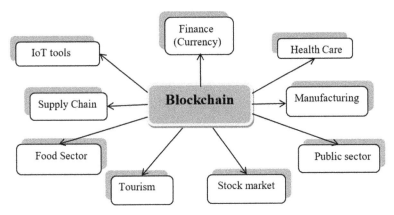

FIGURE 14.1   Key application areas of blockchain technology.

TABLE 14.1   Enablers of Blockchain

SL. No.	Enablers	Descriptions	References
E1	Multi-node storage of data	As blockchain is a decentralized network and every data has to be stored in the form of nodes	[2, 8, 15]
E2	Enhance cybersecurity	Blockchain gives security from various cyber-attacks which leads to data theft and network malfunction.	[23, 29, 31]
E3	Efficient and prompt processing	Blockchain allows fast and secure encryption and decryption	[4, 22, 29]
E4	Enhance transparency of data	In the blockchain, every transaction can be examined, inspect, and scan publicly.	[17, 25, 31]
E5	Tamper evident	Any attempt to temper the information/security can be tracked in the blockchain.	[4, 5, 23]
E6	High security of passwords	Specific keys are provided to operate with coins, which serves as high secure passwords	[2, 23, 29]
E7	Smart clauses of the contract	Blockchain allows the storage of smart clauses on codes, which in the future reduces the need for intermediaries in transactions between parties.	[4, 17, 22]
E8	Mutable-by-hashing-power	Transaction in a blockchain can only be modified by a sufficient amount of computing power and collaboration	[4, 15, 17]
E9	Circumvent fraud/double spending	There are specified procedures to find out the frauds like double-spending, which is why the blockchain leads to faithless unanimity.	[4, 22, 29, 31]

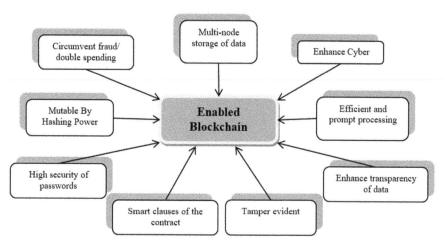

FIGURE 14.2   Key enablers of blockchain.

14.3   RESEARCH METHODOLOGY

DEMATEL approach has been habituated in the chapter to analyze the enablers of blockchain technology. DEMATEL is the preferred technique over many multi-criteria decision making (MCDM) methods as it differentiates factors/enablers into cause-and-effect group and indicates the inclemency of their effects. It has more range to respond as (0, 1, 2, 3, and 4) to explore the cause-effect relationship among the enablers. Categorization of factors further helps decision takers in devising effective strategies to manage them. The DEMATEL is the multiple criteria decisions making (MCDM) technique, which helps in developing interrelationship among the enablers, challenges, or barriers. This tool works as a strong tool for decision making. Rajput and Singh [20] grouped the enablers and barriers of circular economy (CE) and Industry 4.0 using DEMATEL techniques. Yadav and Singh [31] habituated fuzzy-DEMATEL for grouping the blockchain factors in cause-and-effect groups. Kumar and Dixit [11] applied DEMATEL techniques in designing a model to analyze the challenges of e-waste in an effective way. Singh et al. [21] habituated the DEMATEL technique for information and communication technology (ICT) use in the Indian food sector SMEs. Kumar [14] have applied the DEMATEL technique to study enablers of Industry 4.0 in Indian manufacturing organizations. In the present chapter author has used this technique to categorize the enablers in cause-and-effect group. Followed research methodology has been shown graphically in Figure 14.3.

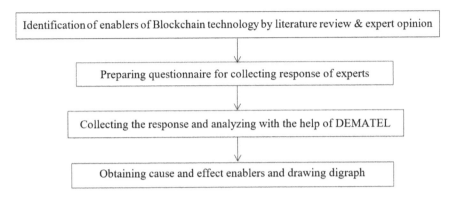

FIGURE 14.3   Research methodology.

The strides of DEMATEL methodology are like this:

➢ **Stride 1: Fill the Relationship Matrix by the Experts of That Field:** In this method, the matrix is being filled by the experts according to their knowledge and experience. The matrix is filled to evaluate the influence of one factor on the rest of the factors. Rating of the factors is measured on a scale of 0 to 4. Where 0 shows the least significance and four significance max influence. The scale used is called the DEMATEL scale, as shown in Table 14.2, the matrix which we obtained is called relationship matrix.

➢ **Stride 2: Estimating the Mean Matrix:** After obtaining a relationship matrix, the sum is calculated and divided by each element to get the average matrix table. The average matrix is shown in Table 14.3.

$$N_{avg} = \begin{pmatrix} N_{1i} & N_{1n} \\ N_1 & N_{nn} \end{pmatrix}$$

➢ **Stride 3: The Matrix is Being Normalized:** After obtaining the average matrix, normalizing has done by ranking each element of the matrix between 0 and 1. In this type of matrix, each diagonal element of the normalized matrix is made zero, also known as 'Fuzzy cognitive method.' The symbol is being used to denote the matrix, which is marked by $Y = [Y_{ij}]_{n \times n}$. The Y shows the influence of the matrix in-between the elements. It is shown in Table 14.4.

➢ **Stride 4: Estimating the Full Direct/Indirect Influence Matrix:** After obtaining the initial influence matrix in the above step now, we calculate the full direct/indirect influence matrix with the help of Eqns. (1) and (2). Symbol N_{ij} is used in this equation where i, j represents elements of the matrix.

$$X = p*N \tag{1}$$

where; $p = \min \{1/\max 1 \leq i \leq n \Sigma^n_{j=1} N_{ij}; 1/\max_{1 \leq i \leq n} \Sigma^n_{i=1} N_{ij}\}$

$$\mathrm{Lim}_{h \to 0} = X^h [0]n*n, 0 \leq x_{ij} \leq 1 \tag{2}$$

➢ **Stride 5: Calculating Total Influence Matrix:** It is calculated by Eqn. (3), where I represent the identity matrix.

$$T_O = Y(I - Y)^{-1} \tag{3}$$

It is shown in Table 14.5.

➢ **Stride 6: Obtaining Prominence Factor and Cause and Effect Values:** In the Total relation matrix, the sum of the all 'i'th row elements represented as *Di* and Sum of the entire 'j'th row element represented as *Rj*. Now *Di+Rj* and *Di-Rj* values are obtained, in which *Di+Rj* indicates the relation of one enabler to the other enablers and *Di-Rj* suggests the kind of relation in the form cause and effect with other challenges where positive values are considered as cause group. Negative values are considered as effect group of challenges (shown in Table 14.6). The negative value of D-R shows effect group indicating receiver, i.e., it gets affected by other criteria. Figure 14.4 shows the causal diagram for blockchain enablers.

14.4   FINDING OF STUDY

The authors applied the research methodology as per steps (Tables 14.2–14.7) mentioned in Section 14.3 and obtained the results as discussed in this section. The D-R value provides us with the knowledge of classifying enablers in cause-and-effect group. As a result, obtained from the matrix above, we can conclude that the negative value of D-R represents effect enablers and a positive value represents the cause enablers. From the finding of the study, the author observed that enablers E6 (High security of pass-words), E1 (Multi-node storage of data), E8 (Mutable by hashing power), E7 (Smart clauses of contract) are cause enablers. These finding of the study implies that blockchain is very secure technology and mentioned enablers' help in achieving secure, efficient, and prompt reply and transparency of data. In blockchain, information is decentralized and hived on multi-computer in the blocks too [29]. Blockchain makes supply chain efficient and transparent too [9]. Characteristics of blockchain-like data decentralization with safety, handiness, secure rules, and policy with minimum corroboration make it enabling technology for SSCM [31]. Blockchain characteristics of interoperability of data related to patient and hospital make it more versatile in asset utilization. It improves routine dealings of the healthcare system like patient records, health insurance billing, with security and administered record of all information [24].

During the further study, the author observed that enablers E2 (Enhance cybersecurity), E5 (Temper evident), E3 (Efficient and prompt reply), E4 (Enhance transparency of data) and E9 (Circumvent fraud/double-spending) are effect group enablers. These finding of the study implies that mentioned enablers are affected by cause group enablers and helps in making blockchain a secure and fraud-proof technology. Blockchain improves the transparency and hardiness of transactions in the supply chain [23]. Blockchain allow decentralized operations and security; the efficiency of all transaction has improved by the use of blockchain in the system [2]. Blockchain could give prevention from all types of fraud and cyber-attack on the organization [5]. Conte et al. [4] also observed that blockchain technology is tamper-evident.

TABLE 14.2   Initial Influence Matrix for the Blockchain Enablers

Enablers		E1	E2	E3	E4	E5	E6	E7	E8	E9
Multi-node storage of data	E1	0	1	2	3	0	4	0	0	0
Enhance cybersecurity	E2	0	0	1	2	1	0	0	1	3
Efficient and prompt processing	E3	0	0	0	1	3	2	1	1	0
Enhance transparency of data	E4	0	4	1	0	2	0	0	2	1
Tamper evident	E5	3	0	4	1	0	0	1	0	0
High security of passwords	E6	3	4	3	3	3	0	2	0	3
Smart clauses of the contract	E7	0	3	1	0	2	1	0	0	3
Mutable-by-hashing-power	E8	0	3	0	0	3	0	3	0	3
Circumvent fraud/double spending	E9	0	2	2	1	2	0	2	3	0

TABLE 14.3 Average of All Influence Matrices for Blockchain Enablers

	E1	E2	E3	E4	E5	E6	E7	E8	E9
E1	0	0.04762	0.09524	0.14286	0	0.19048	0	0	0
E2	0	0	0.04762	0.09524	0.04762	0	0	0.04762	0.14286
E3	0	0	0	0.04762	0.14286	0.09524	0.04762	0.04762	0
E4	0	0.19048	0.04762	0	0.09524	0	0	0.09524	0.04762
E5	0.14286	0	0.19048	0.04762	0	0	0.04762	0	0
E6	0.14286	0.19048	0.14286	0.14286	0.14286	0	0.09524	0	0.14286
E7	0	0.14286	0.04762	0	0.09524	0.04762	0	0	0.14286
E8	0	0.14286	0	0	0.14286	0	0.14286	0	0.14286
E9	0	0.09524	0.09524	0.04762	0.09524	0	0.09524	0.14286	0

TABLE 14.4   Total Influence Matrix for Blockchain Enablers

	0.05	0.15	0.18	0.21	0.1	0.22	0.05	0.05	0.08	1.09
	0.02	0.06	0.1	0.12	0.11	0.02	0.04	0.09	0.18	0.74
	0.05	0.07	0.08	0.09	0.21	0.12	0.09	0.07	0.05	0.83
	0.03	0.24	0.11	0.05	0.16	0.02	0.04	0.13	0.11	0.89
	0.16	0.06	0.24	0.1	0.07	0.06	0.08	0.03	0.04	0.84
	0.19	0.32	0.28	0.25	0.28	0.07	0.17	0.09	0.25	1.9
	0.03	0.2	0.13	0.06	0.17	0.07	0.05	0.05	0.2	0.96
	0.03	0.21	0.09	0.06	0.22	0.02	0.19	0.05	0.21	1.08
	0.03	0.17	0.17	0.09	0.19	0.03	0.15	0.18	0.08	1.09
R	0.59	1.48	1.38	1.03	1.51	0.63	0.86	0.74	1.2	–
D + R	1.68	2.22	2.21	1.92	2.35	2.53	1.82	1.82	2.29	–
D – R	0.5	–0.74	–0.55	–0.14	–0.67	1.27	0.1	0.34	–0.11	–

TABLE 14.5   Summation of Influences Given and Received Among the Enablers of Blockchain

Enablers	D	R	D + R	D – R
E1	1.09	0.59	1.68	0.5
E2	0.74	1.48	2.22	–0.74
E3	0.83	1.38	2.21	–0.55
E4	0.89	1.03	1.92	–0.14
E5	0.84	1.51	2.35	–0.67
E6	1.9	0.63	2.53	1.27
E7	0.96	0.86	1.82	0.1
E8	1.08	0.74	1.82	0.34
E9	1.09	1.2	2.29	–0.11

TABLE 14.6   Enablers of Blockchain in Cause-and-Effect Groups

Rank	Enablers of Cause Group	D – R	Enablers of Effect Group	D – R
01	E6	1.27	E2	–0.74
02	E1	0.5	E5	–0.67
03	E8	0.34	E3	–0.55
04	E7	0.1	E4	–0.14
			E9	–0.11

TABLE 14.7   Ordering of Enablers of Blockchain According to Prominence Vector

Rank	Enablers	D + R
1	E6	2.53
2	E5	2.35
3	E9	2.29
4	E2	2.22
5	E3	2.21
6	E4	1.92
7	E7	1.82
8	E8	1.82
9	E1	1.68

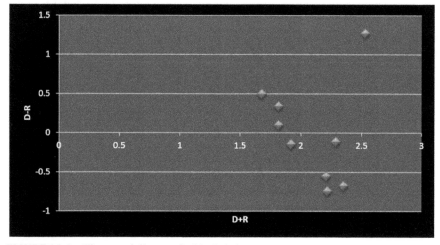

FIGURE 14.4   The causal diagram for blockchain.

14.5   CONCLUSION, IMPLICATIONS, LIMITATION, AND FUTURE RESEARCH DIRECTIONS

In this chapter, the author has first identified the enablers of blockchain technology by literature review. Further by DEMATEL technique, the enablers have been divided into two groups. The cause group enablers such as E6 (High security of passwords), E1 (Multi-node storage of data), E8 (Mutable by hashing power), E7 (Smart clauses of contract) helps in making blockchain a secure, tamper-proof, and transparent technology. Cause group enablers determine effect group enablers such as E2 (Enhance cybersecurity), E5 (Temper evident), E3 (efficient and prompt reply), E4 (Enhance transparency of data) and E9 (Circumvent fraud/double-spending). Finding adds a valuable contribution to making blockchain technology of the modern era. Finding of study like multimode storage of data, smart clauses of the contract, high security of password make blockchain a ready and upgraded technology for enhancing cybersecurity, transparent data, temper-evident, and capable to an efficient and prompt reply. Blockchain as modern technology finds application in many fields like healthcare [10, 24], manufacturing with IoT and quality improvement [12, 16, 32], public sector process [30], finance [8], tourism [18], food supply chain [7] and improving knowledge exchange and transparency of SC [28].

Research and implementation of blockchain technology are in its refinement stage, and this study will motivate researchers and industrial experts to apply this technology in many other fields and sectors. The finding of the chapter is constructive for both academia and industry. The author has applied the DEMATEL technique for analyzing the enablers of blockchain technology. DEMATEL methodology too has its limitations, and results are highly dependent on experts' opinions. Finally, the results of this study may give valuable guidance to all stakeholders in planning effective strategies for blockchain-based technologies. The findings may also motivate organizations from many sectors to shift on safe and secure technology to excel in global competition. Blockchain as new and safe technology has proved its applications in the financial sector. From here onward, there is a need to do more research and advancement in this technology to make this applicable to other areas such as in healthcare, education domain, data analytics, etc.

KEYWORDS

- **blockchain**
- **digital era**
- **enablers**
- **information communication technology**
- **multi-criteria decision making**
- **sustainable supply chain management**

REFERENCES

1. Adams, R., Parry, G., Godsiff, P., & Ward, P., (2017). The future of money and further applications of the blockchain. *Strategic Changes: Briefing in Entrepreneurial Finance, 26*(5), 417–422.

2. Casado-Vara, R., Javier, P., Fernando De, L. P., & Juan, M. C., (2018). How blockchain improves the supply chain: Case study alimentary supply chain. *Procedia Computer Science, 134*, 393–398.

3. Christidis, K., & Devetsikiotis, M., (2016). Blockchains and smart contracts for the Internet of Things. *IEEE Access, 4*, 2292–2303. doi: 10.1109/access.2016.2566339.

4. Conte De, L. D., Stalick, A. Q., Jillepalli, A. A., Haney, M. A., & Sheldon, F. T., (2017). Blockchain: Properties and misconceptions. *Asia Pacific Journal of Innovation and Entrepreneurship, 11*(3), 286–300.

5. Dai, J., Wang, Y., & Vasarhelyi, M. A., (2017). Blockchain: An emerging solution for fraud prevention. *The CPA Journal, 87*(6), 12–14.

6. Dutta, G., Kumar, R., Sindhwani, R., & Singh, R., (2020). Digital transformation priorities of India's discrete manufacturing SMEs – a conceptual study in perspective of industry 4.0. *Competitiveness Review: An International Business Journal.* https://doi.org/10.1108/CR-03-2019-0031.

7. Feng, H., Wang, X., Duan, Y., Zhang, J., & Zhang, X., (2020). Applying blockchain technology to improve agri-food traceability: A review of development methods benefits and challenges. *Journal of Cleaner Production, 260*, 121031.

8. Ghosh, A., Gupta, S., Dua, A., & Kumar, N., (2020). Security of cryptocurrencies in blockchain technology: State-of-art, challenges and future prospects. *Journal of Network and Computer Applications, 163*, 102635.

9. Gurtu, A., & Johny, J., (2019). Potential of blockchain technology in supply chain management: A literature review. *International Journal of Physical Distribution & Logistics Management, 49*(9), 881–900.

10. Khan, F. A., Asif, M., Ahmad, A., Alharbi, M., & Aljuaid, H., (2020). Blockchain technology, improvement suggestions, security challenges on smart grid and its application in healthcare for sustainable development. *Sustainable Cities and Society, 55*, 102018.

11. Kumar, A., & Dixit, G., (2018). An analysis of barriers affecting the implementation of e-waste management practices in India: A novel ISM-DEMATEL approach. *Sustainable Production and Consumption, 14*, 36–52.

12. Kumar, R., (2019). Kaizen is a tool for continuous quality improvement in Indian manufacturing organization. *International Journal of Mathematical, Engineering and Management Sciences, 4*(2), 452–459.

13. Kumar, R., (2020a). Sustainable supply chain management in the era of digitalization: Issues and challenges. *Handbook of Research on Social and Organizational Dynamics in the Digital Era.* IGI Global: 446–460. doi: 10.4018/978-1-5225-8933-4.ch021.

14. Kumar, R., (2020b). Espousal of Industry 4.0 in Indian manufacturing organizations: Analysis of enablers. *Handbook of Research on Engineering Innovations and Technology Management in Organizations.* IGI Global. doi: 10.4018/978-1-7998-2772-6.ch013.

15. Parjapati, P., & Chaudhari, K., (2020). KBC: Multiple key generation using key block chaining. *Procedia Computer Science, 167*, 1960–1969.

16. Polvora, A., Nascimento, S., Lourenco, J. S., & Scapolo, F., (2020). Blockchain for industrial transformations: A forward-looking approach with multi-stakeholder engagement for policy advice. *Technological Forecasting and Social Change, 157,* 120091.

17. Ramkumar, M., (2018). Executing large-scale processes in a blockchain. *Journal of Capital Markets Studies, 2*(2), 106–120.

18. Rashideh, W., (2020). Blockchain technology framework: Current and future perspectives for the tourism industry. *Tourism Management, 80,* 104125.

19. Scott, B., Loonam, J., & Kumar, V., (2017). Exploring the rise of blockchain technology: Towards distributed collaborative organizations. *Strategic changes: Briefing in Entrepreneurial Finance, 26*(5), 423–428.

20. Rajput, S., & Singh, S. P., (2019). Connecting circular economy and industry 4.0. *International Journal of Information Management, 49,* 98–113.

21. Singh, R. K., Luthra, S., Mangla, S. K., & Uniyal, S., (2019). Applications of information and communication technology for sustainable growth of SMEs in India food industry. *Resources, Conservation & Recycling, 147,* 10–18.

22. Skwarek, V., (2017). Blockchains as security-enabler for industrial IoT-applications. *Asia Pacific Journal of Innovation and Entrepreneurship, 11*(3), 301–311.

23. Sund, T., Loof, C., Nadjm-Tehrani, S., & Asplund, M., (2020). Blockchain-based event processing in supply chains—A case study at IKEA. *Robotics and Computer Integrated Manufacturing, 65,* 101971.

24. Tanwar, S., Parekh, K., & Evans, R., (2020). Blockchain-based electronic healthcare record system for healthcare 4.0 applications. *Journal of Information Security and Applications, 50,* 102407.

25. Treiblmaier, H., (2018). The impact of the blockchain on the supply chain: A theory-based research framework and a call for action. *Supply Chain Management: An International Journal, 23*(6), 545–559.

26. Upadhyay, N., (2020). Demystifying Blockchain: A critical analysis of challenges, applications and opportunities. *International Journal of Information Management, 54,* 102120.

27. Vaidya, S., Ambad, P., & Bhosle, S., (2018). Industry 4.0-a glimpse. *Procedia Manufacturing, 20,* 233–238.

28. Wamba, S. F., Queiroz, M. M., & Trinchera, L., (2020). Dynamics between blockchain adoption determinants and supply chain performance: An empirical investigation. *International Journal of Production Economics, 229,* 107791.

29. Wang, Y., (2019). Designing a blockchain-enabled supply chain. *IFAC Papers On-Line, 52*(13), 6–11.

30. Warkentin, M., & Orgeron, C., (2020). Using the security triad to assess blockchain technology in public sector applications. *International Journal of Information Management, 52,* 102090.

31. Yadav, S., & Singh, S. P., (2020). Blockchain critical success factors for sustainable supply chain. *Resources, Conservation and Recycling, 152,* 10450.

32. Yu, C., Jiang, X., Yu, S., & Yang, C., (2020). Blockchain-based shared manufacturing in support of cyber-physical systems: Concept, framework, and operation. *Robotics and Computer-Integrated Manufacturing, 64,* 101931.

CHAPTER 15

Identifying Applications of Blockchain Technology in the Construction Industry and Project Management

PRIYANKA SINGH

Department of Civil Engineering, Amity School of Engineering and Technology, Amity University Uttar Pradesh, Noida, Uttar Pradesh, India, E-mail: priyanka24978@gmail.com

ABSTRACT

The forerunners of the technology of Blockchain are the concept of crypto-currency, i.e., bitcoin that has a widespread application and usage in present-day life with its base platform of digital currency. Blockchain may be said as the decentralized ledger registering the possible sources of information in an encrypted form of a significant data source. The credibility of the Blockchain concepts covers under it, the arenas of cybersecurity, enforcing account-ability and, if used to its maximum potential, can also be a driving force in the maintaining of the cooperative chain of management in the construction industries. This book chapter presents an overview and motive of enabling blockchain technology in the construction industry and project manage-ment for the smooth functioning without much duplicity in the system. The collaboration of Blockchain technologies and IoT (Internet of Things) improves and opens up more applicability in real-time. The Blockchain concept neglects the interference of the third party, which makes the transac-tion smooth and provides strengthened security. Thus, this chapter shows the potential applications of Blockchain with IoT in the construction industry and the overall management. This book chapter mainly focuses on the use of blockchain technology and IoT for transparent and smooth conduction of the construction project. Introduction of Blockchain Technology and, descrip-tion of its elements, is presented in the first section. Whereas, in the second section, the Structure of the Blockchain is explained. In the last section, a detailed literature review of over two decades about the implementation of both Blockchain and BIM in the construction industry and project management is explored. Finally, the shortcomings of Blockchain technology and its application and recommendations to overcome these shortcomings are elucidated.

15.1 INTRODUCTION

The forerunners of the technology of Blockchain are the concept of cryptocurrency, i.e., bitcoin (BTC) that has a widespread application and usage in present-day life with its base platform of digital currency. Blockchain may be said as the decentralized ledger registering the possible sources of information in an encrypted form of a significant data source. The credibility of the Blockchain concepts covers under it, the arenas of cybersecurity, enforcing accountability, and if used to its maximum potential, can also be a driving force in the maintaining of the cooperative chain of management in the construction industries. This book chapter presents an overview and motive of enabling blockchain technology in the construction industry and project management for the smooth functioning without much duplicity in the system. The collaboration of Blockchain technologies and IoT (Internet of Things) improves and opens up more applicability in real-time. The Blockchain concept neglects the interference of the third party, which makes the transaction smooth and provides strengthened security. Thus, this chapter shows the potential applications of Blockchain with IoT in the construction industry and the overall management. This book chapter mainly focuses on the use of blockchain technology and IoT for transparent and smooth conduction of the construction project. Introduction of blockchain technology and description of its elements, is presented in the first section. Whereas, in the second section, the Structure of the Blockchain is explained. In the last section, a detailed literature review of over two decades about the implementation of both Blockchain and BIM in the construction industry and project management is explored. Finally, the shortcomings of Blockchain technology and its application and recommendations to overcome these shortcomings are elucidated.

The construction industry is the only sector that grows regularly and contributes to the central part of GDP. The construction industry contributes 7.43% of GDP in India. The countries' gross domestic product (GDP) plays a substantial role in the implementation of smart ledger technologies such as the Blockchain concept. The decentralized ledger of data collected from sources involves the transactions and accounts over a linked network [1]. It depends on the agreement between the node points, which in turn enhances the transparency, traceability, and collaborative nature of Blockchain. The evolution of smart contracts (SCs) has also gained an impetus in the world of the digital finance sector, enabling transparency and trustable security [2].

The main problem faced by the industry is that they are not able to maintain the accounting records and squeezed the margin profits frequently. Potential applications of the Blockchain in the construction industry are: (i) payment and project management; (ii) building information and asset management; and (iii) supply of chain management. However, the problem faced by the construction industry is overdue payments. Many firms cannot tolerate late payment [3]. According to the 2012 survey, data shows that 97% of 250 small firms experienced unfair trade and overdue payment; this practice directly affects their business. After 2015, the percentage of late payments increased to 27%. The minimum period for late payment is 82 days, which rises to 120 days [4]. This practice can be mitigated with the help of the blockchain concept. SCs help to resolve this dispute, which is very useful and provides a tender process. Generally, for making contracts first notary is mandatory for which some lawyer fees are required. But it can be quickly be processed with the help of SCs. The decentralized ledger of data collected from sources involves the transactions and accounts from over a linked network [5, 6]. The online management of data achieves significant goals of saving labor, cost, redundant payments, labor, easy sharing of data aligning the hierarchy of data flow in an organization, thus linking various teams of a project, hence reducing the load of storing and hoarding of the information. Blockchain reduced the problem faced by the construction industry like overdue payments, overflow cash, project management, proper record of employees, transparent fees [7].

The most significant element of the Blockchain is transparency, which is very useful for the construction industry. It makes it comfortable to control everything very easily (SCs) and boost the trust between the parties. The Blockchain provides an excellent opportunity to maintain a more trusted business with proper flow, and beneficially for both the parties [8]. In this industry, the mega construction projects require extensive work for supervision and scheduling [9]. Lack of accountability causes delay and affect the profit margins of the project, thus posing the loss in forms and others all over. The induction of blockchain-enabled payment and project managing techniques can easily make shift significantly [10]. It provides options to shape a more trustable business environment, ensuring transparency inflow of value, as established in a construction project [11].

Therefore, it can be stated that the idea of Blockchain served as a helping tool in the construction industry. It provides transparent payments and traces the work progress with the collaboration of all team members [12]. The collaboration of Blockchain technologies, IoTs, and building information modeling (BIM) improves and opens up more applicability in real-time [13, 35, 68, 69].

15.2   BLOCKCHAIN TECHNOLOGY

The Blockchain technology is a set of various block transactions between two communities, and all the transactions are recorded in the form of ledgers. The Blockchain was discovered approximately 10 years ago by Satoshi Nakamoto [14]. Conceptually, this theory of Blockchain is based on BTC. BTC is a transparent currency. Trying to eliminate the interference of a third party, the concept of Blockchain was introduced in the market. Every single transaction is mentioned in the decentralized ledger. Highly advanced computers maintain the whole set up of the Blockchain. Up to 8 decimal places, a single BTC can be separated [15]. All the transaction is highly augmented by cryptography. One of the most important motives is to maintain a trust relationship between parties, and an accounting ledger is shared, which reduces the level of conflicts and maintain a secure network [16]. Figure 15.1 illustrates the possible applications of Blockchain Technology in different fields.

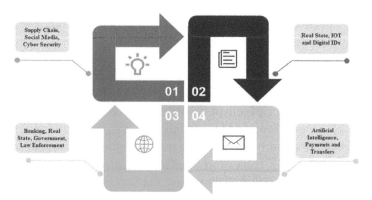

FIGURE 15.1   Applications of blockchain technology.

Blockchain technology is different from banks, where no third party helps to commute the transactions between two or more than two communities. It is a decentralized ledger that issued every single purchase in the encrypted form. Example: If 'Hello' is the passcode and it is converted into encrypted form (puo2810jhy), like this way every time a new encrypted passcode is formed. Adopting of blockchain technology decreases the various problems and save the transaction amount, time, human error [17]. A sequential block of encrypted data or discrete information that has a header and block for the entirety of the chain. The preceding block to that of the present block is called the "parent block."

An example of a simple transaction shows the working of the blockchain:

- The transaction is made between two parties;
- This transaction creates a separate data and circulates to the network for the validation;
- Throughout the process of validation, it is investigated that the sender and receiver are correct or not. It is also checked whether the same transaction is not being used for another one. After all the confirmation, the transaction is considered as a block;
- Blockchain is updated, and all the transactions between the same parties are recoded as a separate block. The whole Blockchain is made up of different blocks. In this way, every party has its network;
- The transaction is wind up with the details of origin to the reached point. All the details are highly secured by cryptography.

If we conclude this process, no interference of third party (bank, government), the process of sharing data provides an appreciable trust between parties. Suppose, a transaction is carried out in a company of five stakeholders; then, everyone has their own, but a similar record of incoming and outgoing money [18, 19].

15.2.1 CLASSIFICATION OF BLOCKCHAIN NETWORK

Blockchain networks are classified into Private and public blockchain network. In general, in a Public blockchain network, anyone can link to the system and interact, which is not safe, and all the transactions are shown to everybody [20]. Anyone may access a blockchain network, which ensures that they can read, compose, or engage in a shared blockchain. Public blockchains are decentralized, no one has control over the network, and they are secure, and data cannot be changed once it has been validated on the Blockchain. It makes it very tough to make changes in the system, and it cannot be easily hacked.

The public network also charged some transaction fees. The public and the private Blockchain vary on the point of giving access rights to the system. The main idea of the public Blockchain is an open type, whereas the close type is of the private Blockchain. The Ethereum (ETH) is of "open type" [21].

In private, only shareholders/group members can interact, which is safe, and the transactions are shown to the members only. In this, system settings can be easily carried out. The transaction fee can be eliminated. Example: In the construction industry, a private network is required because it is confidential data, and it is not shared with the competitors; hence, the transaction data is only visible to clients or group members [22]. The sequential series of code blocks that has the complete list of the transactions of a public ledger system is the present-day Blockchain. The Blockchain serves as a tool for the upcoming times about its vast applicability in business [23].

Both public and private blockchain are P2P networks and one can participate in the transaction as the ledger is distributed among themselves. However, being public or private Blockchain can be both permissionless and permissioned. Example of permissionless Blockchain is BTC or ETH and of permissioned is Hyperledger blockchain framework.

Thus, there are two kinds of blockchain network—public and private with permissions or permissionless. They are categorized as public and permissionless, public and permissioned, private and permissionless, and private and permissioned, which is shown in Figure 15.2 with the accessibility features.

In general, the Blockchain consists of cryptography, peer-to-peer networking, consensus theory, ledger, and the rules and regulations attached to it [24]. It is concluded that the maintaining of confidentiality requires the system to use cryptography. The cryptography uses keys for the encryption and decryption of the blocks [25]. Thus, in turn, making the blockchain technology a unique implementation to the industry.

'Blockchain is the technology that underlies the concept of BTC Technology. The studies have led to conclusions that the main motive of the BTCs is facilitating digital payments and transactions. The network of BTC automatically generates a fair and accurate register of owners and stakeholders in specified timeframes, thus removing the bloat of third parties [26].

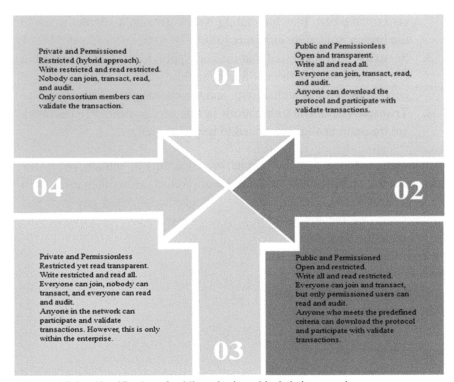

FIGURE 15.2   Classification of public and private blockchain network.

Platforms such as the ETH also provide a user-friendly network that helps use the blockchain network. Mentioning, a case where Microsoft has announced the project using Blockchain as a Service that is flexible with open access [27].

15.2.2   SALIENT FEATURES OF BLOCKCHAIN

1. **Decentralization:** The process of decentralization is the distribution and dispersion of power from a central authority. Its core infrastructure, the Blockchain, facilitates this decentralization as it gives any single customer an ability to become one of the multiple payment processors in the network. Blockchain invokes transparency and robustness in the system.

2. **Anonymity:** The private and the public keys help to verify the identity of the users in the chains, enabling the anonymous third party

also to take part, thus interacting with a generated address to avoid the breach of identity exposure [28].

3. **Security:** It is suggested that the encryption technology involved in Blockchain reduces risks of fraud and false identity. The provision of encryption ensures authenticity and maintaining data integrity [29].

4. **Transparency:** The transactions in this peer-to-peer (P2P) network are transparent and announced to be public [30].

It is demonstrated that the purchasing capability is influenced by the blockchain technology, the cycle analysis of products rather than estimation of the product value. The induction of Blockchain-enabled payment and project managing techniques can easily make shift significantly. It could also monitor the long-standing problems in case of payment management and cash flow issues serving handy to follow up overdue payments [31]. It provides options to shape a more trustable business environment, ensuring transparency in the flow of value and thus beneficial in the whole running of the construction projects [32]. The layered architecture of Blockchain is depicted in Figure 15.3.

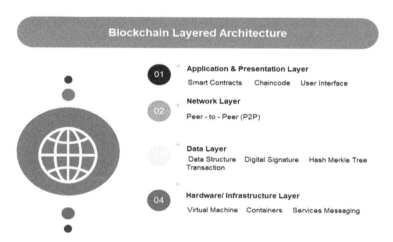

FIGURE 15.3   Pictorial representation of layered architecture of blockchain.

15.3   BLOCKCHAIN TECHNOLOGY

The emergence from the pit-houses construction and grass shelters; the construction industry has come across a long way. The advancements in the

20th century brought about a significant turning point to the industry—the introduction of AutoCAD and other automated technology brought in the digitization in the field of detailing, required for construction industry [33].

The advent of Blockchain served as a financial innovation forming a ledger system that records transactions, store documents, and documentation. The big giant companies are taking into account the management due to the publicity of Blockchain and also its investments, interests, and the predicted outcomes [34]. Figure 15.4 demonstrates the possible application of blockchain technology.

The construction industry is the only sector that grows regularly and contributes to the central part of GDP. The main problem faced by the industry is that they are not able to maintain the accounting records and squeezed the margin profits frequently [9]. Figure 15.4 shows the potential applications of blockchain technology in the construction industry. Broadly, it is applied in these fields:

- Payment and project management;
- Building information and asset management;
- Supply of chain management.

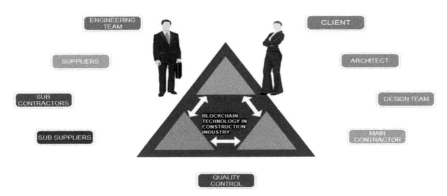

FIGURE 15.4   Application of blockchain technology in the construction industry.

15.3.1   PAYMENT AND PROJECT MANAGEMENT

The most significant element of the Blockchain is transparency, which is very useful for the construction industry. The major problem faced by the construction industry is overdue payments. Many firms cannot tolerate late payment [36]. According to a 2012 survey, data shows that 97% of 250 small firms experienced unfair trade and overdue payment; this practice directly affects their business. After 2015, the percentage of late payments increased to 27%. The minimum time for late payment is 82 days, which rises up to 120 days. This practice can be deleted with the help of a blockchain concept. SCs help to resolve this dispute, which is very useful and provides a tender process. Generally, for making contracts first notary is mandatory for which some lawyer fees are applicable. But it can be quickly be processed with the help of SCs [37].

15.3.2 ACTIVATE CONTRACT COLLABORATION

Employee management is one of the problems in this industry, as it hires most of the labor, and it is complicated to manage. So SCs resolve this problem, input all the details of the agreement, employment details, working hours of employees, and their wages. Automatically at the end of the day (working hour), payments are directly transferred to the employees, and separate entry (separate block) is made for each day, and the Blockchain continues. In this way, all the employees get their payments on time, and records are easily maintained. However, all the blockchain details can be shared only between the directors, contractors, and consultants [38].

The Blockchain embedded systems help to get business reports needed by the regulators and owners through keeping the record of the contracts that require a check and monitoring the contract terms put up by different parties [39].

Since BTC serves the platform to blockchain technology, these cryptocurrencies have allowed beneficiaries to complete transactions without any intermediate party or even a bank, thus finding its use in the digital asset management, remittance, and online payment transactions of employees and labors in construction field [40].

15.3.3 BLOCKCHAIN GENERATIONS

The impact of Blockchain and its field of application in society is enormous. Blockchain as a mode of the cryptocurrency serves to validate and to store information of transactions. The versions of Blockchain involved in the economic, market, and commercial applicability is the blockchain 2.0. Blockchain 3.0 applies to accountability in the construction supply chain [41, 42].

15.3.4   DISTRIBUTED LEDGER: BLOCKCHAIN

The accounting involves the transactions using the concept of nodes that are serving as the entry points to new data fed in systems [43]. The nodes involved are authorized for the allowing of the requests to make secure transactions, thus making the selected genuine transactions legitimate. When the consensus is achieved, from both ends involved in the operations, then it is valid. This work is called mining [44]. The working of the P2P network that included the achieving of the consensus in the medium thus provided the transaction is valid, thereby constituted the main parts of the block ledger format [45].

15.3.5   DESIGNING OF STRUCTURE WITH THE HELP OF BLOCKCHAIN

Firstly, the designing and detailing of the structure are completed, and thereafter, it is submitted to be checked by the necessary parties through SCs. The ID of the SCs is also registered with a digital signature [46]. The transactions made to each person is recorded, and the timely payments are updated. In this process, the public network is used. The main motive for this collaboration is to maintain the management with all transaction proofs. It minimizes the mistakes while making the record and facilitate with accurate data which is traceable and transparent. Implementation of this process, enhance productivity by 9% and save up to 7% cost [47].

15.3.6   CRANE MANAGEMENT WITH BLOCKCHAIN

At the construction site, for high leveled projects, crane plays an important role. But sometimes, due to overloading, some accidents can lead to serious health issues. This problem can also be reduced by using a blockchain. Sensors automatically register the value of load in the SCs. Safety alarm rings when the value of the shipment exceeds. Values are recorded in the form of blocks and form a blockchain [48, 49].

15.4   APPLICATIONS OF BLOCKCHAIN IN PROJECT MANAGEMENT

In project management, the supply chain has to be set up, which provides all the transactions with full transparency, traceability, and supportive management. It can enhance the transparency from currency transactions to each detail of materials between shareholders and provide traceability function. The main reason for the spreading of the Blockchain is that everyone can see the documentation, and there are no hiding possibilities. Traceability is one of the advanced features provided by the Blockchain through which we can trace the route of materials supplying. But this function has shown two problems [50]. The first problem is that we can check the location where it is originated. But it is impossible to check, whether the material is loaded is suitable or not. The only way is to visit the site from where the material is loaded. The second problem is that the full shipment of material can be traced, but the need for real-time tracing cannot be fulfilled. Due to this reason, blockchain technology has to be improved for locating the real-time tracking of shipment [51]. If blockchain technology is applied in the construction industry, then much cooperation and collaboration are required between the parties. Potential Application of Blockchain technology in Construction Project and Management is shown in Figure 15.5.

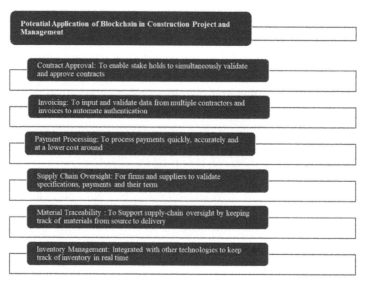

FIGURE 15.5   Potential application of blockchain in construction project and management [52].

Although it is a new concept for the industry and to understand it, it takes a reasonable amount of time, so cooperation is a must. It provides several potential executions:

- Eliminate third-party organization;
- Provides traceability and transparency function;
- The whole sector depends on who owns the blockchain management;
- Enhance the company reputation;
- Verification of all data can be done by using Blockchain;
- Have chances to combine many supply chains in a single supply chain.

15.4.1 SMART CONTRACTS (SCS) ON BLOCKCHAIN TO ENHANCE EFFICIENCY

The digital contract or the smart contract are competent to carry out the underlying or terms as mentioned above and conditions to meet the target. The very first concept ran on the ETH platform and thus had a series of codes that were immutable on the blockchain data sources [53]. The process involved in the execution can be improved, automatized, and be more industrially oriented.

It can easily include the registration of data and also the interpretation of data initiating contract management and also maintaining the payment schedule involved as well that, in turn, makes the feature of transparency more pronounced [54].

But the critical challenge is to make the whole network of data flow tamperproof. Because there is the entire possibility that there can be a hack, files can get corrupt, and wrong decisions may occur. Therefore, as the advent of the Blockchain exits, it tries to secure all the possible risks, thus removing human errors [55].

15.5 BIM AND BLOCKCHAIN

The emergence of BIM has led to a new platform. It binds together, the industry and gathering the different data, be it finance budgets, project plan, geometrical dimension, and planning them all in one through various plug-ins and compliance-check [56]. The various field in which BIM and Blockchain can be applied is shown in Figure 15.6.

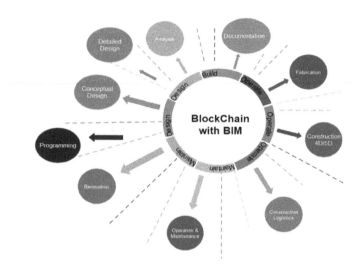

FIGURE 15.6   BIM and blockchain in construction industry.

BIM provides advanced engineering frameworks. It has one source of flow for data while putting in series the audits, design approval, data verification, and total project management decisions in every single aspect that involves from day one to its completion day. BIM acts efficiently and is able to work as a source and also a dashboard of all the information [57]. It may combine the Blockchain with supply chain, material provenance and also payment details during the ongoing project of construction. It is also capable of assigning information to the Blockchain like a model decision, and modification of orders passed [58].

15.5.1   BIM AND SMART ASSET MANAGEMENT

BIM is a smart 3D model-based operation, which provides expertise and instruments to design, engineering, and construction (DEC) experts in planning, creating, constructing, and managing buildings and infrastructures more efficiently. The BIM deals in digital and real-life assets. The various sectors involved in the construction industries have a wide array of information in a model. Usually, BIM gives collected information as a "dimension" that are the dimensions of 3D geometry, project management, scheduling of time, labor management, the costs, and the overall maintenance of the projects [59].

As BIM plays an essential role in the digitalization of data workflow involved in the construction industry for better management during the operation phase; the exchange of information is done by importing data from one source to another [60, 61]. Process of BIM and blockchain in SCs and supply chain with transparent payment information is shown in Figure 15.7. Therefore, it is significant to bridge the implementation of BIM for design and the construction stages in the Construction Supply Chain.

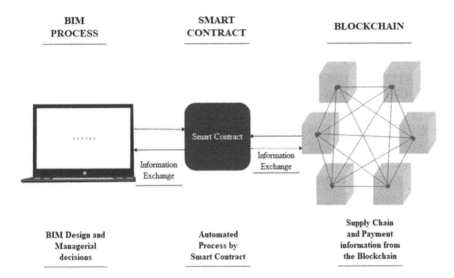

FIGURE 15.7   BIM and blockchain technology in smart contract [62].

15.6 CHALLENGES OF BLOCKCHAIN TECHNOLOGY

Though there are positive potentials of the Blockchain yet leaving some loopholes, the privacy leakage is possible, thus hindering the transactions and payments only with the use of the public and the private key, even the I.P. address is trackable. The supply chain, as well as the logistics, are involved in the production and the delivery of goods. But the main obstacle in the path of implementation of Blockchain is maintaining the utmost transparency and traceability, quality management, and growing the trust between the parties involved in the supply chain layer due to such challenges, the customers do not find faith in the transactions [63–65]. Further, if Blockchain is made to be immutable, it provides coordination, integration, and logistics, thus improving the procurement of supply chain management (SCM).

The companies implementing the blockchain ideas supporting the BIM concept has identified the possible obstacles faced by the company at the initiation level and also towards the level of completion. The new challenges are regulatory uncertainty and also the lack of trust among its users. The next difficulties that surface is the cost of maintenance and the lack of governance [66].

Furthermore, it is found that the primary need is the trust at the different levels involved in the Blockchain, it must show on the tool of the decentralized ledger. The next important issue is the implementation of the standard of the different companies that, in turn, facilitates the use of private blockchains along with its rules [67].

15.7 CONCLUSION AND FUTURE WORK

The implementation of SCs in real-time, in the business processes and administrative tasks, can be automatized to increase efficiency and the agreement on the terms in the contracts. Also, by the project management paradigm, if applied, there is the potential reduction of late payments, remediation, and disputes, thus making the company a more trusted entity of business with a regular flow of cash and trusted verified data. Application of Blockchain is necessary for the procurement of the project management and also to the transactions involved in the SCM Together with BIM, and the Blockchain can reshape the aspects of truth in the construction industry. Such a model can become a digital twin of an asset supporting the ideas, maintenance, and overall operation. Thus, this concept is a must for the construction industry. It facilitates the digital transformation of the construction industry, thus influencing the growth and productivity of the industry.

KEYWORDS

- **blockchain**
- **building information modeling**
- **chain management**
- **construction industry**
- **Internet of Things**
- **project management**

REFERENCES

1. Aghimien, D., Aigbavboa, C., Oke, A., & Koloko, N., (2018). Digitalization in construction industry: Construction professionals perspective. *Proceedings of International Structural Engineering and Construction, 5.*

2. Berger, M., (1999). *Digitization for Preservation and Access: A Case Study* (Vol. 17, pp. 146–151). Library Hi-Tech.

3. Peck, M., (2017). Blockchain world – do you need a blockchain? This chart will tell you if the technology can solve your problem. *IEEE Spectrum, 54,* 38–60.

4. Franke, B., Gao, Q., & Stenzel, A., (2019). Can you trust the blockchain? The (Limited) power of peer-to-peer networks for information provision. *SSRN Electronic Journal.*

5. Shackelford, S., & Myers, S., (2016). Block-by-block: Leveraging the power of blockchain technology to build trust and promote cyber peace. *SSRN Electronic Journal.*

6. Wood, D., (2018). A future history of international blockchain standards. *The Journal of the British Blockchain Association, 1,* 1–10.

7. Demirkan, S., Demirkan, I., & McKee, A. (2020). Blockchain technology in the future of business cyber security and accounting. *Journal of Management Analytics, 7*(2), 189–208.

8. Sutton, H., (2018). Blockchain 101 for the registrar's office: Recordkeeping for the future. *The Successful Registrar, 18,* 1–7.

9. Zwitter, A., & Hazenberg, J., (2020). Decentralized network governance: Blockchain technology and the future of regulation. *Frontiers in Blockchain, 3.*

10. Schedlbauer, M., & Wagner, K., (2018). Blockchain beyond digital currencies – a structured literature review on blockchain applications. *SSRN Electronic Journal.*

11. Jirgensons, M., & Kapenieks, J., (2018). Blockchain and the future of digital learning credential assessment and management. *Journal of Teacher Education for Sustainability, 20,* 145–156.

12. Rella, L., (2019). Blockchain technologies and remittances: From financial inclusion to correspondent banking. *Frontiers in Blockchain, 2.*

13. Cocco, L., Pinna, A., & Marchesi, M., (2017). Banking on blockchain: Costs savings thanks to the blockchain technology. *Future Internet, 9,* 25.

14. Buchwalter, B., (2018). Decrypting crypto assets: An introduction to blockchain. *SSRN Electronic Journal.*

15. Ma, Z., Jiang, M., Gao, H., & Wang, Z., (2018). Blockchain for digital rights management. *Future Generation Computer Systems, 89,* 746–764.

16. Banerjee, M., Lee, J., & Choo, K., (2018). A blockchain future for Internet of Things security: A position paper. *Digital Communications and Networks, 4,* 149–160.

17. Chang, V., Baudier, P., Zhang, H., Xu, Q., Zhang, J., & Arami, M. (2020). How Blockchain can impact financial services – The overview, challenges and recommendations from expert interviewees. *Technological Forecasting and Social Change, 158,* 120166. https://doi.org/10.1016/j.techfore.2020.120166.

18. Vermeulen, E., Fenwick, M., & Kaal, W., (2018). Why blockchain will disrupt corporate organizations: What can be learned from the digital transformation. *The Journal of the British Blockchain Association, 1,* 1–10.

19. Clack, C., (2018). A blockchain grand challenge: Smart financial derivatives. *Frontiers in Blockchain, 1.*

20. Hill, T., (2018). *Blockchain for research: Review* (Vol. 31, pp. 421, 422). Learned Publishing.
21. Wolfskehl, M., (2018). Why and how blockchain? *The Journal of the British Blockchain Association, 1,* 1–5.
22. Wang, J., Wu, P., Wang, X., & Shou, W., (2017). The outlook of blockchain technology for construction engineering management. *Frontiers of Engineering Management, 4,* 67.
23. Tezel, A., Papadonikolaki, E., Yitmen, I., & Hilletofth, P., (2020). Preparing construction supply chains for blockchain technology: An investigation of its potential and future directions. *Frontiers of Engineering Management.*
24. Turk, Ž., & Klinc, R., (2017). Potentials of blockchain technology for construction management. *Procedia Engineering, 196,* 638–645.
25. Hewavitharana, F., & Perera, A., (2019). Gap analysis between ERP procedures and construction procedures. *MATEC Web of Conferences, 266,* 03011.
26. Ramakrishnan, R., & Gaur, L., (2016). Smart electricity distribution in residential areas. *Internet of Things based Advanced Metering.*
27. Batwa, A., & Norrman, A., (2020). A framework for exploring blockchain technology in supply chain management. *Operations and Supply Chain Management: An International Journal,* 294–306.
28. Sheel, A., & Nath, V., (2019). Effect of blockchain technology adoption on supply chain adaptability, agility, alignment and performance. *Management Research Review, 42,* 1353–1374.
29. Vyas, G., Gaur, L., & Singh, G., (2016). Evolution of payments bank and impact from M-PESA: A case of mobile banking services in India. *Proceedings of the Second International Conference on Information and Communication Technology for Competitive Strategies.*
30. Nakamoto, N., (2017). Centralized bitcoin: A secure and high performance electronic cash system. *SSRN Electronic Journal.*
31. Taylor, P. J., Dargahi, T., Dehghantanha, A., Parizi, R. M., & Choo, K. K. R., (2019). A systematic literature review of Blockchain cybersecurity. *Digital Commun. Networks.* doi: 10.1016/j.dcan.2019.01.005.
32. Mohsin, A. H., Zaidan, A. A., Zaidan, B. B., Albahri, O. S., Albahri, A. S., Alsalem, M. A., & Mohammed, K. I., (2018). Blockchain authentication of network applications: Taxonomy, classification, capabilities, open challenges, motivations, recommendations and future directions. *Comput. Stand. Interfaces.* doi: 10.1016/j.csi.2018.12.002.
33. Qian, X., & Papadonikolaki, E., (2020). Shifting trust in construction supply chains through blockchain technology. *Engineering, Construction and Architectural Management.*
34. Summer, S., Herold, D., Dobrovnik, M., Mikl, J., & Schäfer, N., (2020). A systematic review of blockchain literature in logistics and supply chain management: Identifying research questions and future directions. *Future Internet, 12,* 60.
35. Solanki, A., & Nayyar, A., (2019). Green Internet of Things (G-IoT): ICT technologies, principles, applications, projects, and challenges. In: *Handbook of Research on Big Data and the IoT* (pp. 379–405). IGI Global: Hershey, PA, USA.
36. Wierzbowski, P., (2018). Application of blockchain technology in information management in supply chains. *Transport Economics and Logistics, 78,* 179–191.

37. Kottler, F., (2018). Potential and barriers to the implementation of blockchain technology in supply chain management. *SSRN Electronic Journal*.

38. Wang, X., (2012). *BIM Handbook: A Guide to Building Information Modeling for Owners, Managers, Designers, Engineers and Contractors* (Vol. 12, pp. 101, 102). Construction economics and building.

39. Azhar, S. (2011). Building information modeling (BIM): Trends, benefits, risks, and challenges for the AEC industry. *Leadership and Management in Engineering, 11*(3), 241–252.

40. Zhang, S., Teizer, J., Lee, J., Eastman, C., & Venugopal, M., (2013). Building information modeling (BIM) and safety: Automatic safety checking of construction models and schedules. *Automation in Construction, 29*, 183–195.

41. Mohammad, W. N. S. W., Abdullah, M. R., Ismail, S., &Takim, R. (2020). An exploratory factor analysis (EFA) of building information modelling uses towards BIM adoption for BIM-based projects: contractors perspective. *Journal of Critical Reviews, 7*(5), 101–108.

42. Bhatija, V. P., Thomas, N., & Dawood, N. (2017). A preliminary approach towards integrating knowledge management with building information modeling (K BIM) for the construction industry. *International Journal of Innovation, Management and Technology, 8*(1), 64–70.

43. Choi, J., & Ryu, H., (2015). Application and effects analysis of BIM (Building Information Modeling) for construction management of a construction field. *Journal of the Korea Institute of Building Construction, 15*, 115–121.

44. Salinas, J., & Prado, G., (2019). Building information modeling (BIM) to manage design and construction phases of Peruvian public projects = building information modeling (BIM) para la gestión del diseño y construcción de proyectos públicos peruanos. *Building & Management, 3*, 48.

45. Perera, S., Nanayakkara, S., Rodrigo, M., Senaratne, S., & Weinand, R., (2020). Blockchain technology: Is it hype or real in the construction industry? *Journal of Industrial Information Integration, 17*, 100125.

46. San, K., Choy, C., & Fung, W., (2019). The potentials and impacts of blockchain technology in construction industry: A literature review. *IOP Conference Series: Materials Science and Engineering, 495*, 012005.

47. Sreckovic, M., Sibenik, G., Breitfuß, D., & Preindl, T., (2020). Analysis of design phase processes with BIM for blockchain implementation. *SSRN Electronic Journal*.

48. Halaburda, H., (2017). Blockchain revolution without the blockchain. *SSRN Electronic Journal*.

49. Oduyemi, O., Okoroh, M., & Fajana, O., (2017). The application and barriers of BIM in sustainable building design. *Journal of Facilities Management, 15*, 15–34.

50. Singh, B., Sharma, K., & Sharma, N., (2020). Blockchain applications, opportunities, challenges and risks: A survey. *SSRN Electronic Journal*.

51. Zeadally, S., & Abdo, J., (2019). Blockchain: Trends and future opportunities. *Internet Technology Letters, 2*, 130.

52. Chandrasekaran, C., Somanah, D., Rughoo, D., Dreepaul, R., Cunden, T., & Demkah, M., (2019). Digital transformation from leveraging blockchain technology, artificial intelligence, machine learning and deep learning. *Proceedings of Fifth International Conference India 2018* (Vol. 2). 10.1007/978-981-13-3338-5_25.

53. Mahmoud, Q., Lescisin, M., & AlTaei, M., (2019). Research challenges and opportunities in blockchain and cryptocurrencies. *Internet Technology Letters, 2*, 93.
54. Shojaei, A., (2019). Exploring applications of blockchain technology in the construction industry. *Proceedings of International Structural Engineering and Construction, 6*.
55. Li, J., Greenwood, D., & Kassem, M., (2019). Blockchain in the built environment and construction industry: A systematic review, conceptual models and practical use cases. *Automation in Construction, 102*, 288–307.
56. Nawari, N., & Ravindran, S., (2019). Blockchain and building information modeling (BIM): Review and applications in post-disaster recovery. *Buildings, 9*, 149.
57. Matvienko, V., (2019). Integration of BIM technologies with blockchain. *Scientific Development Trends and Education*.
58. Wilfer, G., (2019). Regulation of sustainable housing industry through blockchain technology. *SSRN Electronic Journal*.
59. Aste, T., Tasca, P., & Di Matteo, T., (2017). Blockchain technologies: The foreseeable impact on society and industry. *Computer, 50*, 18–28.
60. Cole, R., Stevenson, M., & Aitken, J., (2019). Blockchain technology: Implications for operations and supply chain management. *Supply Chain Management: An International Journal, 24*, 469–483.
61. Miraz, M., (2020). Blockchain in automotive supply chain. *International Supply Chain Technology Journal, 6*.
62. Di Giuda, G., Pattini, G., Seghezzi, E., Schievano, M., & Paleari, F., (2019). The construction contract execution through the integration of blockchain technology. *Digital Transformation of the Design, Construction and Management Processes of the Built Environment*, 27–36.
63. Wu, H., Li, Z., King, B., Ben, M. Z., Wassick, J., & Tazelaar, J., (2017). A distributed ledger for supply chain physical distribution visibility. *Integrating Process-Oriented, Event-Based and Data-Driven Systems, 8*(4).
64. Boyes, H., (2014). *Building Information Modeling (BIM): Addressing the Cyber Security Issues*. Institution of cybersecurity (IET) Consortium Report: London, UK.
65. Eastman, C., Eastman, C. M., Teicholz, P., Sacks, R., & Liston, K., (2011). *BIM Handbook: Building Information Modeling for Owners, Managers, Designers, Engineers and Contractors*. John Wiley & Sons.
66. Hughes, D. (2017). The impact of blockchain technology on the construction industry. Medium. URL https://medium. com/the-basics-of-blockchain/the-impact-of-blockchain-technology-on-theconstruction-industry-85ab78c4aba6 (accessed 6 April 2019).
67. Cardeira, H., (2015). *Smart Contracts and their Applications in the Construction Industry*. New perspectives in construction law conference, Bucharest.
68. Rameshwar, R., Solanki, A., Nayyar, A., & Mahapatra, B., (2020). Green and smart buildings: A key to sustainable global solutions. In: *Green Building Management and Smart Automation* (pp. 146–163). IGI Global: Hershey, PA, USA.
69. Krishnamurthi, R., Nayyar, A., & Solanki, A., (2019). Innovation opportunities through the Internet of Things (IoT) for smart cities. In: *Green and Smart Technologies for Smart Cities* (pp. 261–292). CRC Press: Boca Raton, FL, USA.

CHAPTER 16

Mitigating the Supply Chain Wastages Using Blockchain Technology

DHRITIMAN CHANDA,[1] SHANTASHREE DAS,[2]
NILANJAN MAZUMDAR,[3] and D. GHOSE[4]

[1]*Assistant Professor, Faculty of Commerce and Management, Vishwakarma University, Pune, India, Email: operationsdchanda@gmail.com*

[2]*Software Research Analyst, SelectHub, Denver, United States, Email: dshantashree26@gmail.com*

[3]*Assistant Professor, Department of Business Administration, University of Science and Technology Management, Guwahati, India, Email: nilanjanmazumdar@ustm.ac.in*

[4]*Associate Professor, Department of Business Administration, Assam University, Silchar, India, E-mail: operationsdghosh@gmail.com*

ABSTRACT

Supply chains are crucial for the overall well-being of any business. The present systems of supply chain management are antiquated and need to be revamped. Different kinds of unnecessary wastages are generated across the supply chain, which reduces the performance as well as the responsiveness of the entire supply chain. For a supply chain management system to be efficient, it must maintain a network which is transparent, cost-effective, efficient, responsive, robust, and there is proper end-to-end communication. The present study aims to apprehend the effect of blockchain technology in reducing the different wastages produced in the supply chain. This study into the mitigation of the various supply chain generated wastages first highlights the various features of blockchain. It then tries to understand how these characteristics can aid in reducing the different supply chain wastages generated across the diverse domains of the supply chain thereby resulting in improved product tracking and visibility, reduced overhead costs, removal of intermediaries and better forecasting data for inventory demand.

16.1 INTRODUCTION

Supply chains are the foundation of the macroeconomy and the global markets. The traditional supply chain process comprises of accumulating raw materials in an orderly fashion and transforming them into physical products, which then gets transferred to the customer. However, managing supply chains nowadays has become exceptionally complicated and uncertain as the various stakeholders maintain paper-based trials. Uncertainty and complexity have become a natural part of the supply chain ecosystem, which can quickly turn into issues and risks that snowball through the entire network, causing significant problems. Based on the kind of the produce, the delivery chain consists of several stages, manifold stakeholders, and entities, varied geographical locations, a myriad of expenses and invoice.

Supply chains networks are not defined by the traditional OEM networks or suppliers anymore [3]. Today's supply chain operations are quite dynamic, having much shorter product life cycles. Though the concept of supply chains has transformed over the years, most of the companies have failed to update themselves and adjust to the changes technologically. In a scenario where blockchain integrated supply chain management (SCM) may exist, companies can revamp their approach to producing and delivering their products, benefit valuable insights, and have a more integrated and exhaustive view of their complex environments.

Blockchain, occasionally known as distributed ledger technology (DLT), makes the transaction history of any digital benefit irreversible and clear by using transference and cryptographic hashing. Due to the lack of lucidity across the supply chain, the blockchain brings up potential opportunities to reduce the different wastages generated in the supply chain. The use of blockchain can be a vital factor in reducing stock waste, improving responsiveness and efficiency, and allowing companies to have better control over their entire supply chain network.

Regardless of the fact to the blockchain technology's prospective potential meant for the creation of the future internet systems, it is in front of several technical challenges. Firstly, scalability is a huge concern [49]. Secondly, it has been established that miners can gain more substantial returns through self-interested mining strategy [13]. A particular phenomenon where the miners bury their mined blocks intended for other returns in the expectation of future prospects may result in branches creation taking place recurrently, and consequently, the development process of the blockchain gets hindered [50]. Finally, it has been revealed that even in blockchain confidentiality, seepage may happen; despite users making dealings with their public key and private key [5]. Even a study also hinted on the fact that miscreants could trace real IP addresses of the users. However, Blockchain technology itself is non-disputable and has processed over the years in an impeccable manner and is being applied favorably to both financial and non-financial world applications [11].

Furthermore, it is being acclaimed as a technological advancement which is expected to transform how society trades and interacts. This status is in particular attributable to its properties of allowing jointly mistrusting entities to exchange pecuniary value and work together devoid of relying on a dependable third party. A blockchain besides provides reliability with protected data storage space and process lucidity [51].

This chapter discusses the mitigation of various wastages generated across the supply chain. The study is expected to act as a preliminary manual for practitioners in the field and the blockchain technology professionals to understand the various critical factors for increasing the efficiency and responsiveness by adopting a sound execution plan.

16.2   RELATED WORKS

Latest works on blockchain, which was coined by Nakamoto [53], has mostly engaged itself in monetary transactions and disseminated ledger systems. As a disruptive technology, the potential utilities of blockchain's technology might excel in the financial trade sector [28]. Because blockchain permits the protected swap over of data in a distributed manner, the knowledge could affect the configuration and control of supply chains as well as related maintenance and details of the transaction sharing between supply chain nodes. Blockchain, if incorporated with magnetic/infrared field sensing technology such as the Internet of Things (IoT), may generate enduring, benefiting to all and active accounts of products, creating readable digital footprints throughout the entire supply chain. The calibration may sound complicated, but such superior visibility will allow product tracking, originality, and legitimacy-every single one of these are vital to the sectors like food, pharmaceutical, and luxury-item supply chains. However, there have been many studies in the very recent past, involving authors emphasizing on varied aspects of the benefits and challenges which are discussed in Table 16.1.

TABLE 16.1   Literature Review

Author	Emphasis	Conclusion
Michelman [54]	Trust, safety, security	'Shared source of truth' is the most influential factor over integrating blockchain in the supply chain.
Collomb and Sok [55] Patel et al. [56]	One source of data	Trust related to data security has a massive implication on its applicability.
Bonino and Vergori [57; Wang et al. [58]; Xu et al. [59]	Geographical dispersion	Due to being geographically apart, it becomes complex for the business nodes to operate with efficiency.
Li et al. [60]	Entire visibility	Discussed the visibility with the other factors to be a significant affecting issue.
Nakasumi [61]	Symmetric information	Highlights the major challenge of information flow over the supply chain, which has been a setback in the conventional supply chain designs.
Abeyratne and Monfared [1]	Authenticity and legitimacy	Indicates, that as the purchasers are very cautious of what they purchase. Supply chain issues related to the two factors are prevalent.
Tian [62]	Sustainability	Suggests that sustainability-related issues can be addressed by deploying blockchain in the supply chain.
Engelenburg et al. [63]	Public security, curtail anti-social activity	Conclude with a view that due to the rigid and secured gateway, robbery, and terrorist threats can be truncated.
		They also proposed blockchain-based customs system.
Guo and Liang [64]	Legal issues	Argued that, blockchain can help in the arbitration of the legal issue efficiently with less wastage of time and effort.
Casey and Wong [65]; Lu and Xu [66]; Mansfield [67]	Transparency and visibility	To ensure the transactions are legitimate and authentic, transparency, and visibility over the blockchain can play a vital role.

TABLE 16.1 *(Continued)*

Author	Emphasis	Conclusion
Foerstl et al. [68]	Traceability of origin	Blockchain makes it easier to mark out the source and manufacturing process of food ingredient and textiles as well.
Mackey and Nayyar [69]	Tracking in the pipeline	Specifically advocates for integrating blockchain into the pharmaceuticals enabling tracking to deal with Counterfeits.
Polim et al. [70]	Partners in the pipeline	Allowing direct tenders to the third-party logistics (3PL) for shipments and removing fourth-party logistics from the network.
Casey and Wong [65]	Fast-paced transaction	Considering the expanse of the global trade and the volume transaction that take place in the global trade, blockchain allows the fast-paced transaction under its real-time attribute.
Bos et al. [71]	Authentication issues	The author has a view on the level of authentication as there are chances of theft of keys.
Swan [72]	Resource wastage	The amount spent on mining the data is way too higher. If the work put is not utilized wisely, it can lead to manifold loses.
Koteska, Karafiloski, and Mishev [22]	Size and bandwidth, data integrity and scalability	These are of the prime issues to be dealt with in the blockchain system itself before any integration.

E-Source: Google Scholar.

Regardless of the significance of blockchain technology mentioned above and various efforts in exercise, our understanding over the complete knowledge of the blockchain technology is remains till conscious level and implications on supply chains is a thing of the future to come.

16.3 BLOCKCHAIN TECHNOLOGY

In the finance sector bitcoins (BTC) were initially known as nerd money, or computer money [47]. It is a decentralized type of cryptocurrency that has developed much popularity recently [25]. It has highly accelerated the activities involved in enhancing security and related privacy measures [18]. A highly applied concept, blockchain technology has found its way in today's world in a very efficient way. However, before getting into the primary implementation phase of a blockchain, there are always some questions that are required to be answered. Some of them might be:

- What is blockchain?
- What is its way of work?
- What issues can be solved using a Blockchain? and most importantly
- How and where can it be implemented?

The concept of a blockchain originated in the year 1991, which was used to timestamp some digital documents such that it becomes easier to maintain their security. Application of blockchain made it tough to tamper with the documents. In 2009, "Satoshi Nakamoto" adopted this technology to develop a digital method of creating a cryptocurrency, popularly known as a "bitcoin."

Blockchain technology implements three significant theories that run as below:

1. Technology acceptance model proposed by Davis [73] says how users come forward to accept and use technology;
2. Technology readiness index introduced by Parasuraman [74] works on several items for a scale to find reliability and validity of the work; and
3. Theory of planned behavior proposed by Ajzen [75] says that an individual's attention towards his/her perception regarding certain items says a lot about his/her behavioral attentions.

Blockchain, in simple terms, is a chain of a certain number of blocks that contains information and can also be defined as a ledger that is distributed in nature. That means it can be open to anyone. One of its main properties is that once data has been inserted into a block of a blockchain, it becomes almost impossible to tamper or change the data. In order to clearly understand the basic structure of a block, let us have a look at its representation (Figure 16.1).

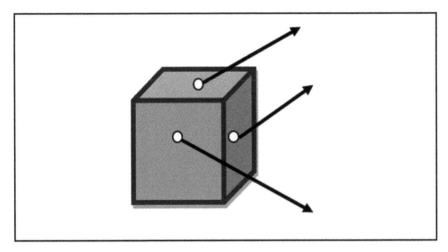

FIGURE 16.1 Sample representation of a block.

The data contains information that is required to be sent from the sender to the receiver. In the case of a BTC, the data includes three items mainly:

- The sender's information;
- The receiver's information; and
- The amount of information.

The hash can also be considered as a fingerprint that remains unique to the user. Just after the creation of the block, the hash is also calculated. If anything inside the block is changed, it automatically changes the hash. Furthermore, once it is changed, it is no longer the same block which makes it a very crucial part of the technology.

The database of blockchain has two types of records, namely: The transaction details which cannot be altered and the address of the previous block [9]. As explained by the name, the hash of the previous block contains the hash number of the previous block that ultimately leads to the formation of a chain. Now it may come to concern that if this is the case, then the first block of the chain will not do not have any previous block hash number. The first block of a blockchain which does not have any hash number of the previous block is called the 'Genesis Block.' Let us have a simple representation of a chain of three blocks in a blockchain (Figure 16.2).

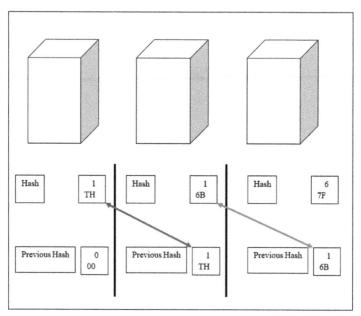

FIGURE 16.2   Sample representation of a blockchain.
Note: All the hash numbers are taken just for the sake of representation.

The above diagrammatic representation shows three sample blocks, namely Block 1, Block 2, and Block 3, which are connected through their unique hash numbers to form a blockchain. Now, for instance, if the hash number of Block 2 has tampered, it automatically will affect the next block in the chain, which will ultimately affect the entire blockchain.

It might be possible to track the hash numbers and can be tracked and tampered within the entire blockchain considering the technological developments of the modern-day world. In such situations, in order to mitigate such tampering, the concept of 'proof of work (PoW)' comes into play. The creation of new blocks is slowed down by the PoW mechanism. If we consider the BTC, it generally takes 10 minutes for the blockchain to create another block. Now if someone needs to tamper one block, it is highly necessary to tamper the entire blockchain.

Calculation of the entire PoW of the blockchain goes without saying, is a highly complex task to do. Thus, hashing and PoW mechanism makes it a secure chain.

Apart from this, blockchain follows a distributed network of peer-to-peer (P2P) nature. Here all members of the chain are given the authority to join, and thus they get a full copy of the blockchain. If a new block is required to be added, this block is then shared with all the participants of the P2P network. If all of the participants of the network find it authentic to be added to the chain, they send a green signal that it is good to go to the participants of the network. If anyone of the participants finds an anomaly, then the block is automatically rejected. So, in order to tamper, an intruder needs to tamper with all other blocks with the chain. He needs to do the PoW again and have the authorization of more than at least 50% of the P2P network. This makes the network connection to be a stringent one and pretty tough to tamper.

16.4 BLOCKCHAIN-DRIVEN SUPPLY CHAIN MANAGEMENT (SCM)

The methodology of blockchain technology can re-establish the way of procurement, manufacturing, logistics, and consumption. It can be useful in resolving various supply chain associated issues effectively. SCM consists of both integrated planning and execution of various processes within the supply chain. Efficient SCM helps in increased sales, decreased overhead costs, quality improvisation, smooth production, and distribution. However, maintaining a supply chain in practical terms is a much exhaustive task even for small and medium businesses [32].

Transparency and traceability or the ability to track products across the entire supply chain network have become a necessity these days with customers wanting to know from where all their items come. Lack of transparency and visibility may result in the creation of blind spots and expose to unnecessary risks in the supply chain. The traditional paper tracking methods and manual inspections introduce errors and delays in sharing information. Poor communication among the various stakeholders also causes inefficiencies, waste, and lead to mistrust as the various parties involved have little to no information about each other's actions [2].

Blockchain technology gives all the stakeholders in the supply chain a secured immutable digital ledger where all the parties can input data which is visible to everyone in the supply chain and allows viewing the origin of each order in real-time, thereby improving the visibility [6]. The traceability is achieved by connecting physical products with RFID, bar codes, or serial numbers and keeping them on the chain formed. Enhanced visibility also implies that inventory forecasting can be adjusted immediately, and thus inventory holding costs are lowered.

A supply chain, in simple terms, includes all the footfalls that are dropped in order to make a product or a service to be made available to the customer/

consumer. The main goal of an organization is to win the trust of the customer. Implementing blockchain technology with smart contracts (SCs) in SCM does the same.

Nick Szabo, a computer scientist and a law scholar and a cryptography expert, first used the word, SCs, in 1997. SCs are as same as a contract in the real world; the only difference being it to be a digitalized one. It is a tiny program code that is stored in a blockchain. Now, how does a smart contract help a supply chain to gain the trust of the customer? Let us understand it through the below example.

Let us take the example of Kickstarter. It is a funding platform that helps product teams to do a project, create a goal and initiate raising money from others who would like to invest in their plan. So, it is a third party that builds a bridge in between the product teams and the supporting teams. Thus, both the parties need to believe in Kickstarter to carry on with the process. If an investor is interested in an idea, he/she needs to make sure that the money reaches the correct receiver and on the other hand, the receiver needs to make sure that the money is received. If there is an anomaly, the money should get refunded. Now building the entire system is possible without the inclusion of a third party. Here comes the role of a smart contract. A smart contract holds all the received funds of the investors until the goal is reached. If the goal is reached, the investors can transfer the money through the contract. If the project fails to meet the goals, the funding goes back to the investors.

Moreover, since a smart contract is in a blockchain, it is non-duplicable and distributed. That signifies that once a contract has been made, it will remain unchanged so that no one can change the code or the digitalized contract between the investor and the product team. A distributed blockchain makes sure that all the members of the network validate the contents of the contract.

For any e-commercial website, if the money to be transferred is stored in a smart contract, then the customer gets to pay the amount only after they have received the item in the quantity and quality that was expected by them. In case the products do not turn out to be the same, the smart contract refunds the money to the customer.

16.5   MITIGATION OF SUPPLY CHAIN WASTAGES WITH THE HELP OF BLOCKCHAIN TECHNOLOGY

Businesses investigate all the areas of their supply chains to lower costs and wastages [27]. Blockchain cut down costs in the supply chain network of businesses by offering cost-saving techniques that former SCM techniques could not offer. However, it is essential to note that different businesses and industries may vary in the way they function, so not every business will have the same capabilities or success while implementing the strategies mentioned here. The objective is to establish a methodology that businesses can cite when improving their operations through blockchain technology to achieve key performance outcomes. Blockchain has entirely changed the modus operandi of inserting digitization into the supply chain processes by developing a system of record for data [7].

16.5.1 Mitigating Wastages in the Procurement Process of Supply Chain Management (SCM) by Blockchain

The consequences of blockchain technology for procurement and supplier management in the supply chain are highly significant. The procurement function of SCM has extensively changed and transformed over the last decade. The digitization of the procurement activities has resulted in this evolution, where the benefits provided by electronic procurement or e-procurement are to boost the efficiency of the various procurement activities [33].

One of the most significant issues in procurement or any SCM system is ensuring trust and transparency where multiple agents or even countries are involved in the various stages of the process. In the traditional supply chains, usually, companies take data of products in batches by adopting a combination of both paper-based trials and information generated from systems which create inefficiencies. However, the feature of transparency and visibility of blockchain is the most useful in the process of procurement of the SCM and can revolutionize the traditional procurement cycle. That is, there is a fluctuation from the traditional method of procurement, backed by a system which collects all activities and information in related to procurement and thus creates a direct "one-to-one" communication between the buyer and supplier [4].

Trust is the key to building a long-lasting buyer-supplier association. Companies may likely be unwilling to work with new suppliers because of diligence [30]. The contracts in the traditional supply chains always have the payment gap amidst the physical delivery of the product, creation of the invoice and the final full settlement of payment. Blockchain technology, with the help of SCs, can help with this process as it involves having access to the transaction records of all parties present on the network and provide advantages to ensure the digital identities of parties involved are well-established.

Also, the risk of any potential loss can be retaliated with the help of SCs, which ensure that the payment made by a company will be only discharged if the logistics service providers confirm that the raw materials are delivered in an appropriate way and according to the predefined conditions [17]. SCs can self-verify and self-execute by releasing the payment to the appropriate party. The terms are paid upon receipt of the order, and proof of delivery will be generated from the logistics carrier, which will immediately provoke automatic digital invoicing and payments between the banking systems. The result can have a direct effect on the bottom line by reducing working capital demands and simplifying the finance operations radically.

Order validation and authorization, the processing of the invoice, multi-way matching of the transactions and the entire order request-to-receipt procedure could be enhanced through the employment of blockchain in the procurement cycle. Blockchain provides immunity to all the necessary supplier credentials, qualification statuses and certificates from forgery and other compromises. Furthermore, blockchain can help track the closest and most cost-effective vendor within the supply chain network, resulting in decreased lead time and workload related to vendor searches [48]. Buyers can even rate the quality of the goods and services provided and the vendor's performance as well.

Blockchain helps to provide a secure, corruption less environment where the encrypted business transactions do not need third parties to intervene, and the buyers and sellers in the supply chain can directly communicate and make payments. Removing intermediaries from the process prevent extra costs, reduce the chances of duplicating the products, frauds, or counterfeits. With the help of blockchain, the costly central intermediaries can be removed as blockchain synchronizes data in a verifiable way across the network, and the complex negotiations can be executed in an absolute peer-to-peer way. Rather than depending on economic intermediaries like banks, payments can be handled directly between the parties of the supply chain with SCs. Blockchain technology enables the buyer with a sense of empowerment to ensure the traceability and authenticity of all goods throughout the procurement cycle.

16.5.2 MITIGATING WASTAGES IN THE MANUFACTURING PROCESS OF SUPPLY CHAIN MANAGEMENT (SCM) BY BLOCKCHAIN

Blockchain's most significant potential for delivering business value is in the manufacturing sector. Companies traditionally operate under a reactive model when it comes to manufacturing and replenishment. When the demand rises, they escalate the production and output, and when products go out of stock, they place the orders in fulfillment centers. This practice is highly inefficient as companies are always one step behind the market or customer demands. The demand is, however, continually fluctuating, and the number of SKUs (stock-keeping units) is increasing at an accelerating rate, and the lead times of manufacturing also vary automatically from one product to another [42]. The reactive approach of manufacturing leads to various demand-supply imbalances. As a result, companies either end up having excess inventory leading to massive carrying costs or end up having stock-outs leading to lost sales. Implementation of blockchain can significantly help reduce the imbalances and inabilities faced by the manufacturers [20]. The cornerstone to the blockchain technology is the fact that data flows in a seamless manner between all the parties involved in real-time. As a result, demand forecasting is done more accurately, leading to proactive planning for manufacturing and stock replenishment, instead of merely reacting to stock-outs. The companies can thus enhance revenue and profitability while eliminating the possibilities of carrying costs and lost sales.

In the manufacturing process, the vendors place their order request along with their product requirements in the distributed database of blockchain. After receiving the order from the vendor, the manufacturer can easily access the data stored on the blockchain with a hash address of the file. The manufacturer can then contact the suppliers to deliver the raw materials according to the product requirements. The manufacturer then adds the details like the type of the materials and quantity required to the blockchain, which immediately triggers the SCs and notify the suppliers about the new order. The vendors can continuously monitor how much and from where the raw materials have been ordered for manufacturing since the information stored on the blockchain is traceable and immutable. When the manufacturer starts manufacturing, IoT enabled manufacturing machines can allow vendors to track the production of their orders in real-time. If they discover issues or changes, they can report to the manufacturers directly in a more efficient manner. Since blockchain maintains the time-stamped records of all the transactions, it is possible to know which transaction occurred at what time and date [29].

In case of bidding, a blockchain integrated with SCs can create a system where the entire supply chain is automated and documented from both the directions at every step. A purchaser posts an RFQ (request for quotation) for a new order, and every manufacturer available in the marketplace is permitted to bid. Even the minute information such as capacity and costs can be part of the bid. As soon as the winning bid is chosen, and the funds have been committed, the manufacturing process starts. The manufacturer and the vendor are both able to track each stage of manufacturing, shipping, and the final delivery, with the minimum costs and maximum flexibility.

16.5.3   MITIGATING WASTAGES IN THE INVENTORY MANAGEMENT OF SUPPLY CHAIN MANAGEMENT (SCM) BY BLOCKCHAIN

One domain that is profoundly affected is inventory. Managing inventory requires an enormous amount of effort, time, and resources. If the inventory is not managed appropriately, visibility gaps occur, which result in higher costs, shortages, and unhappy customers who expect timely deliveries [52]. There exists a distinct trade-off between keeping inventory and wastages low and keeping customers happy.

The problems involved in managing inventories include dead-stock, spoilage, storage costs, stock-outs, and employee errors. The existing systems used by companies to manage their inventories are mostly outdated. Blockchain helps to eliminate these problems and simplifies the process of managing inventories. The fundamental goal of using blockchain in supply chains is to enhance the transparency and visibility of the supply chain, based on record-keeping functionalities [19]. The decentralized record-keeping system of blockchain makes everything in the network visible and transparent so that the various parties involved in managing the inventory can work with honesty and integrity.

By observing everything in real-time, it is much easier for the companies to manage excess inventory, stock-outs, clearing dead stocks and reducing employee errors. Companies can put up their SCs to exchange information. When inventory starts running low, the company's smart contract would spontaneously request for more inventories based on preset terms and conditions from the manufacturing contract. The blockchain, in fact, automatically sends payment to the manufacture once the terms are validated. Human intervention is very less due to which inaccuracies can be easily calculated and thus helps in reducing costs and maximizing revenue, providing a competitive advantage over the competitors.

A critical area of expense related to replenishment in inventory management is invoice match failures [40]. Companies must ensure that the invoices match with the original order and the received document. When they do not tally, the accounts payable department investigates to resolve the differences which consume much time and add to ordering costs. By incorporating a smart contract within the blockchain, the need to do reconciliation could be eradicated. As a result, the costs associated with invoice match failures as well as cycle stock could be reduced, thereby mitigating the overall ordering costs.

In a blockchain, every block records the transaction showing the current number of units sold in the supply network at every point, integrated with AI-based forecasting models. These blocks could provide permission to all the parties in the network who could improve planning based on the information on the blockchain. This aggregated information will reduce the uncertainty of demand significantly, mitigating the need for holding safety stocks, and ultimately lowering carrying costs.

16.5.4 MITIGATING WASTAGES IN THE LOGISTICS OR DISTRIBUTION PROCESS OF SUPPLY CHAIN MANAGEMENT (SCM) BY BLOCKCHAIN

For the logistics industry, blockchain is considered to be as promising as artificial intelligence (AI). It ensures transparency of all documents and transactions across the goods being shipped, ultimately improving the responsiveness, efficiency, flexibility of supply chain networks. It also guarantees that each product reaches its intended place at a minimal time, cost, and damage. The existing current practice of moving goods from the source to destination is very complex and has a deficit of a single source of truth (SSOT) to store and track all the products and transactions involved. It always has tons of documentation processes which add to more inefficiency and slow down the end-to-end transportation [41]. Blockchain mitigates these inefficiencies by tracking shipments, recording transactions, and constructing an efficient and transparent system for managing all the documents associated with the process of logistics [23]. Armed with such data, companies can implement faster routes and eradicate unnecessary steps in the delivery process. A record of a blockchain can consist of a unique tag or label for every product in a transaction. When the tagged product passes through the supply chain network, the ledger automatically creates an electronic record of the product's authenticity, origin, chain of custody and storage, transportation, or any other necessary information in a distinct verifiable ledger.

The logistics service provider adds details like type of product, the number of cartons, driver's name, vehicle number, and dispatch time on the blockchain. *Transportation of the goods through IoT-enabled vehicle could help the manufacturers and vendors to get the location of the vehicle in real-time through blockchain.* The transparency reduces the chance of any frauds or the involvement of any intermediaries, thus ensuring minimal costs.

16.6   CONCLUSION

Blockchain is a more transparent, better-automated, and less corruptible alternative to the centralized existing databases, whose primary purpose is to bring transparency in transactions and to build the trust of the various stakeholders involved in the supply chain while mitigating wastages in the form of unnecessary costs and removing the intermediaries who act as the central authorities between the parties and the system.

Even though the instances of various use cases of blockchain have augmented over the past couple of years, blockchain faces a range of obstacles in acceptance and functioning over the supply chain networks. Communication challenges are likely to be poor, where supply chain nodes are geographically scattered through different civilizations and cultures [36]. It is imperative to understand that the integration of blockchain into a supply chain would require extensive usage of electronics as well as improvised technological aids. Nevertheless, besides the scalable benefits, there are few significant concerns relating to the usage of resources and minimization of wastages. The wastages that might occur at the installation phase or while administering the processes are also crucial for not compromising blockchain system security. The technology is at its early stages of growth with a range of bottlenecks from various aspects [11], but hopefully, blockchain-driven SCM systems can be the criterion in the future.

KEYWORDS

- **blockchain**
- **costs**
- **distributed ledger technology**
- **inventory management**
- **logistics**
- **manufacturing**
- **procurement**
- **smart contracts**
- **supply chain**
- **wastages**

REFERENCES

1. Abeyratne, S. A., & Monfared, R. P., (2016). Blockchain ready manufacturing supply chain using distributed ledger. *International Journal of Research in Engineering and Technology, 5*(9), 1–10.

2. Asaad, J., (2018). *Fixing the Five Big Problems in the Food Supply Chain.* Retrieved from: https://supplychainbeyond.com/5-big-problems-in-the-food-supply-chain (accessed on 29th July 2021).

3. Atre, J., (2018). *How Blockchain Can Help Reduce Costs in Supply Chain Management?* Retrieved from: https://businessblockchainhq.com/business-blockchain-news/how-blockchain-can-help-reduce-costs-in-supply-chain-management (accessed on 29th July 2021).

4. Bienhaus, F., & Haddud, A., (2018). Procurement 4.0: Factors influencing the digitization of procurement and supply chains. *Business Process Management Journal, 24*(4), 965–984. doi: 10.1108/BPMJ-06-2017-0139.

5. Biryukov, A., Khovratovich, D., & Pustogarov, I., (2014). Deanonymization of clients in Bitcoin P2P network. *Proceedings of the 2014 ACM SIGSAC Conference on Computer and Communications Security*, 15–29.

6. BlockApps, (2018). *How Blockchain Cuts Costs in Supply Chain Management.* Retrieved from: https://blockapps.net/how-blockchain-cuts-costs-in-supply-chain-management (accessed on 29th July 2021).

7. Busch, J., Mitchell, P., Lamoureux, M., &Karpie, A. (2017). The Impact of Disruptive Technologies and Solutions on Strategic Procurement Technologies (Analytics, Sourcing, Supplier and Contract Management). *13*, 158–172.

8. Calvaresi, D., Dubovitskaya, A., Calbimonte, J. P., Taveter, K., & Schumacher, M., (2018). Multi-agent systems and blockchain: Results from a systematic literature review. *International Conference on Practical Applications of Agents and Multi-Agent Systems* (pp. 110–126). Springer, Cham.

9. Carlozo, L., (2017). What is blockchain? *Journal of Accountancy, 224*(1), 29.

10. Casino, F., Dasaklis, T. K., & Patsakis, C., (2019). A systematic literature review of blockchain-based applications: Current status, classification and open issues. *Telematics and Informatics, 36*, 55–81.

11. Crosby, M., Pattanayak, P., Verma, S., & Kalyanaraman, V., (2016). Blockchain technology: Beyond bitcoin. *Applied Innovation, 2*, 6–9.

12. Di Vaio, A., & Varriale, L., (2020). Blockchain technology in supply chain management for sustainable performance: Evidence from the airport industry. *International Journal of Information Management, 52*, 102014.

13. Eyal, I., & Sirer, E. G., (2014). Majority is not enough: Bitcoin mining is vulnerable. *International Conference on Financial Cryptography and Data Security* (pp. 436–464). Springer, Berlin, Heidelberg.

14. Fawcett, S. E., Ogden, J. A., Magnan, G. M., & Bixby, C. M., (2006). Organizational commitment and governance for supply chain success. *International Journal of Physical Distribution & Logistics Management, 36*(1), 22–35.

15. Francisco, K., & Swanson, D., (2018). The supply chain has no clothes: Technology adoption of blockchain for supply chain transparency. *Logistics, 2*(1), 2.

16. Govindan, K., & Hasanagic, M., (2018). A systematic review on drivers, barriers, and practices towards circular economy: A supply chain perspective. *International Journal of Production Research, 56*(1, 2), 278–311.

17. Hackius, N., & Petersen, M., (2017). Blockchain in logistics and supply chain: Trick or treat? *Digitalization in Supply Chain Management and Logistics*, 3–18.

18. Halpin, H., & Piekarska, M., (2017). Introduction to security and privacy on the blockchain. *IEEE European Symposium on Security and Privacy Workshops (EuroS&PW)*, 1–3.

19. Ivanov, D., Dolgui, A., & Sokolov, B., (2018). The impact of digital technology and industry 4.0 on the ripple effect and supply chain risk analytics. *International Journal of Production Research, 25*, 1–18.

20. Joshi, N., (2019). *How Manufacturers Can Use Blockchain for Inventory Management.* Retrieved from: https://www.allerin.com/blog/how-manufacturers-can-use-blockchain-for-inventory-management (accessed on 29th July 2021).

21. Konstantinidis, I., Siaminos, G., Timplalexis, C., Zervas, P., Peristeras, V., & Decker, S., (2018). Blockchain for business applications: A systematic literature review. *International Conference on Business Information Systems* (pp. 384–399). Springer, Cham.

22. Koteska, B., Karafiloski, E., & Mishev, A., (2017). Blockchain implementation quality challenges: A literature. *SQAMIA 2017: 6*th *Workshop of Software Quality, Analysis, Monitoring, Improvement, and Applications*, 11–13.

23. Lans, D., (2019). *7 Main Challenges of Supply Chain Management and How You Can Work Around it.* Retrieved from: https://yourstory.com/mystory/7-main-challenges-in-supply-chain-management-and-h-rdq2oy6mh9 (accessed on 29th July 2021).

24. Lemieux, V. L., & Lemieux, V. L., (2016). Trusting records: Is blockchain technology the answer? *Records Management Journal, 26*(2), 110–139.

25. Maesa, D. D. F., Marino, A., & Ricci, L., (2016). Uncovering the bitcoin blockchain: An analysis of the full users graph. *IEEE International Conference on Data Science and Advanced Analytics (DSAA)*, 537–546.

26. McDermott, B., Malladi, K., Abele, K., & Harkins, J. (2017). Tomorrow's Value Chain How blockchain drives visibility, trust and efficiency from Tomorrow's Value Chain: How Blockchain Drives Visibility, Trust and E (slideshare.net).

27. Murray, M., (2018). *Reducing Waste in the Supply Chain*. Retrieved from: https://www. thebalancesmb.com/reducing-waste-in-the-supply-chain-2221088 (accessed on 29th July 2021).

28. Nofer, M., Gomber, P., Hinz, O., & Schiereck, D., (2017). Blockchain. *Business & Information Systems Engineering, 59*(3), 183–187.

29. Perboli, G., Musso, S., & Rosano, M., (2018). Blockchain in logistics and supply chain: A lean approach for designing real-world use cases. *IEEE Access,* (99), 1–1. doi: 10.1109/ACCESS.2018.2875782.

30. Maltaverne, B. (2017, 17 July). Blockchain: What are the opportunities for procurement? Retrieved from https://medium.com/procurement-tidbits/blockchain-whatare-the-opportunities-for-procurement-d38cfd5446fa (accessed on 29th July 2021).

31. Pournader, M., Shi, Y., Seuring, S., & Koh, S. L., (2019). Blockchain applications in supply chains, transport and logistics: A systematic review of the literature. *International Journal of Production Research*, 1–19.

32. Pratap, M., (2018). *Blockchain Technology Explained: Introduction, Meaning and Applications.* Retrieved from: https://hackernoon.com/blockchain-technology-explained-introduction-meaning-and-applications-edbd6759a2b2 (accessed on 29th July 2021).

33. Rejeb, A., Süle, E., & Keogh, J. G., (2018). Exploring new technologies in procurement. *Transport & Logistics: The International Journal, 18*(46), 76–86.

34. Rodrigo, M. N. N., Perera, S., Senaratne, S., & Jin, X., (2018). Blockchain for construction supply chains: A literature synthesis. *Proceedings of ICEC-PAQS Conference 2018.*

35. Saberi, S., Kouhizadeh, M., Sarkis, J., & Shen, L., (2018). Blockchain technology and its relationships to sustainable supply chain management. *International Journal of Production Research, 57*(7), 2117–2135. doi: 10.1080/00207543.2018.1533261.

36. Sajjad, A., Eweje, G., & Tappin, D., (2015). Sustainable supply chain management: Motivators and barriers. *Business Strategy and the Environment, 24*(7), 643–655.

37. Schuetz, S., & Venkatesh, V., (2020). Blockchain, adoption, and financial inclusion in India: Research opportunities. *International Journal of Information Management, 52,* 101936.

38. Seebacher, S., & Schüritz, R., (2017). Blockchain technology as an enabler of service systems: A structured literature review. *International Conference on Exploring Services Science* (pp. 12–23). Springer, Cham.

39. Shen, C., & Pena-Mora, F., (2018). Blockchain for cities— A systematic literature review. *IEEE Access, 6,* 76787–76819.

40. Singh, N. (2020, February 20). Blockchain in Logistics: The Role of Blockchain 2020. Retrieved from https://101blockchains.com/blockchain-in-logistics (accessed on 29th July 2021).

41. Singh, N., (2020). *Blockchain in Logistics: The Role of Blockchain*. Retrieved from: https://101blockchains.com/blockchain-in-logistics (accessed on 29th July 2021).

42. Soni, P., (2018). *Reactive to Proactive: Blockchain for Inventory Planning, Replenishment.* Retrieved from: https://medium.com/@marcellvollmer/how-will-blockchain-impact-procurement-and-supply-chain-43d7bd511089 (accessed on 29th July 2021).

43. Steiner, J., & Baker, J., (2015). *Blockchain: The Solution for Transparency in Product Supply Chains.* Retrieved from: https://www.provenance.org/whitepaper (accessed on 29th July 2021).

44. Taylor, P. J., Dargahi, T., Dehghantanha, A., Parizi, R. M., & Choo, K. K. R., (2019). A systematic literature review of blockchain cybersecurity. *Digital Communications and Networks.*

45. Tribis, Y., El Bouchti, A., & Bouayad, H., (2018). Supply chain management based on blockchain: A systematic mapping study. *MATEC Web of Conferences* (Vol. 200). EDP Sciences, 00020.

46. Tse, D., Zhang, B., Yang, Y., Cheng, C., & Mu, H., (2017). Blockchain application in food supply information security. *International Conference on Industrial Engineering and Engineering Management (IEEM)*, 1357–1361. IEEE.

47. Vollmer, M. (2018, August 1). How will Blockchain impact Procurement and Supply Chain?. Retrieved from https://medium.com/@marcellvollmer/ how-will-blockchainimpact-procurement-and-supply-chain-43d7bd51.

48. Vollmer, M., (2018). *How will Blockchain impact Procurement and Supply Chain?* Retrieved from: https://medium.com/@marcellvollmer/how-will-blockchain-impact-procurement-and-supply-chain-43d7bd511089 (accessed on 29th July 2021).

49. Wang, Y., Han, J. H., & Beynon-Davies, P., (2019). Understanding blockchain technology for future supply chains: A systematic literature review and research agenda. *Supply Chain Management: An International Journal.*

50. Wang, Y., Singgih, M., Wang, J., & Rit, M., (2019). Making sense of blockchain technology: How will it transform supply chains? *International Journal of Production Economics, 211*, 221–236.

51. Wüst, K., & Gervais, A., (2018). Do you need a blockchain? *IEEE Crypto Valley Conference on Blockchain Technology (CVCBT)*, 46–54.

52. Yusuf, Z., Bhatia, A., Gill, U., Kranz, M., Fleury, M., & Nannra, A., (2018). *Pairing Blockchain with IoT to Cut Supply Chain Costs.* Retrieved from: https://www.bcg. com/publications/2018/pairing-blockchain-with-iot-to-cut-supply-chain-costs.aspx (accessed on 29th July 2021).

53. Nakamoto, S. (2008). Bitcoin: A peer-to-peer electronic cash system. *Decentralized Business Review*, 21260.

54. Michelman, P. (2017). Seeing beyond the blockchain hype. *MIT Sloan Management Review, 58*(4), 17.

55. Collomb, A., & Sok, K. (2016). Blockchain/distributed ledger technology (DLT): What impact on the financial sector?. *Digiworld Economic Journal, 103.*

56. Patel, D., Bothra, J., & Patel, V. (2017, January). Blockchain exhumed. In 2017 ISEA Asia Security and Privacy (ISEASP). *IEEE.* pp. 1–12.

57. Bonino, D., & Vergori, P. (2017, July). Agent marketplaces and deep learning in enterprises: The composition project. In 2017 IEEE 41st Annual Computer Software and Applications Conference (COMPSAC) (Vol. 1, pp. 749–754). *IEEE.*

58. Wang, J., Wu, P., Wang, X., & Shou, W. (2017). The outlook of blockchain technology for construction engineering management. *Frontiers of Engineering Management,* 67–75.

59. Lu, Q., & Xu, X. (2017). Adaptable blockchain-based systems: A case study for product traceability. *IEEE Software, 34*(6), 21–27.

60. Li, Y., You, S., Brown, M. S., & Tan, R. T. (2017). Haze visibility enhancement: A survey and quantitative benchmarking. *Computer Vision and Image Understanding, 165*, 1–16.
61. Nakasumi, M. (2017, July). Information sharing for supply chain management based on block chain technology. In 2017 IEEE 19th conference on business informatics (CBI) (Vol. 1, pp. 140–149). *IEEE.*
62. Tian, F. (2016, June). An agri-food supply chain traceability system for China based on RFID & blockchain technology. In 2016 13th international conference on service systems and service management (ICSSSM) (pp. 1ˆ6). *IEEE.*
63. Engelenburg, S. V., Janssen, M., & Klievink, B. (2017). Design of a software architecture supporting business-to-government information sharing to improve public safety and security. *Journal of Intelligent Information Systems, 52*(3), 595–618.
64. Guo, Y., & Liang, C. (2016). Blockchain application and outlook in the banking industry. *Financial Innovation, 2*(1), 1–12.
65. Casey, M. J., & Wong, P. (2017). Global supply chains are about to get better, thanks to blockchain. *Harvard Business Review, 13*, 1-6.
66. Lu, Q., & Xu, X. (2017). Adaptable blockchain-based systems: A case study for product traceability. *IEEE Software, 34*(6), 21–27.
67. Mansfield-Devine, S. (2017). Beyond Bitcoin: using blockchain technology to provide assurance in the commercial world. *Computer Fraud & Security, 2017*(5), 14–18.
68. Foerstl, K., Schleper, M. C., & Henke, M. (2017). Purchasing and supply management: From efficiency to effectiveness in an integrated supply chain. *Journal of Purchasing and Supply Management, 23*(4).
69. Mackey, T. K., & Nayyar, G. (2017). A review of existing and emerging digital technologies to combat the global trade in fake medicines. *Expert Opinion On Drug Safety, 16*(5), 587–602.
70. Polim, R., Hu, Q., & Kumara, S. (2017). Blockchain in megacity logistics. In IIE Annual Conference. Proceedings (pp. 1589–1594). *Institute of Industrial and Systems Engineers (IISE).*
71. Bos, J. W., Halderman, J. A., Heninger, N., Moore, J., Naehrig, M., &Wustrow, E. (2014, March). Elliptic curve cryptography in practice. In *International Conference on Financial Cryptography and Data Security* (pp. 157–175). Springer, Berlin, Heidelberg.
72. Swan, M. (2017). Anticipating the economic benefits of blockchain. *Technology Innovation Management Review, 7*(10), 6–13.
73. Davis, F. D. (1989). Perceived Usefulness, Perceived Ease of Use, and User Acceptance of Information Technology. *MIS Quarterly, 13*(3), 319. https://doi.org/10.2307/249008.
74. Parasuraman, A. (2000). Technology Readiness Index (TRI) a multiple-item scale to measure readiness to embrace new technologies. *Journal of Service Research, 2*(4), 307–320.
75. Ajzen, I. (1985). From intentions to actions: A theory of planned behavior. In *Action Control* (pp. 11–39). Springer, Berlin, Heidelberg.

CHAPTER 17

Traceable and Reliable Food Supply Chain Through Blockchain-Based Technology in Association with Marginalized Farmers

SANJUKTA GHOSH and SUJATA PUDALE

Srishti Manipal Institute of Art Design and Technology, Bangalore, India

ABSTRACT

A reliable and traceable food supply chain is one of the most critical and indispensable aspects of agri-food industry market. Agriculture and food production continues to be the fundamental tool for a nation's growth and development. It can significantly contribute to reduce poverty and enhance food security in emerging economic countries like India. Food traceability has now been at the core of recent food safety and quality discussions across industry and academia, particularly with new development in blockchain applications. The chapter started with a system-level thinking to Service Design, followed by Integrated Supply Chain Information and Secured blockchain Framework. The system map displayed the elements and their boundaries within the food supply chain framework considering macro- and micro-environmental factors. A secured blockchain framework can be implemented through a collaborative venture with the farmers working at grass root level and primarily contributing to this food production ecosystem. Therefore, it becomes important to understand the level of appreciation and adaptation of technology among the farmers. Considering the fact that attitude and capability towards technology adoption may vary among the farmers so it is important to cluster them into different categories based on their willingness to share data, understanding the need for technology, fear associated to technology adoption, reliance on technology, spend time on learning technology, collaborate with technocrats and facilitate them in experimentation and implementation. Four groups were evolved from this clustering process.

17.1 INTRODUCTION

A reliable and traceable food supply chain has become a crucial and essential factor in agri-food industry [1]. Agriculture and food production continues to be an essential tool for development at grassroots level, it can majorly contribute to reduce poverty and enhance food security in emerging economic countries like India [2, 3]. Information and communication can be considered as one of the resourceful domains which can significantly contribute to improvement in the agricultural sector and result to sustainable food production and security [4]. Food safety and quality control is currently of great importance among various stakeholders in the sector. Consumers, including a significant amount of stakeholders, are highly skeptical about the continuing unethical practices associated with food production and distribution [5]. Currently, supply chain mechanism do not encourage the retailers and consumers to review the data related to the origin of the food. The current practices are not so capable of taking care of food safety at every touchpoint. A collaborative initiative is expected from all the major drivers, including consumers and government entities to develop a safe and ethical food production and distribution network. There is also a need for innovative information technology intervention to improve the quality of the production along with transparency at all levels. India is world's largest source of human resources related to information technology. Moreover, mobile telecommunication service plays a significant role in business opportunity development in various sector [6, 7]. As per McKinsey Global Institute report, 2019, 40% of the Indian population has an Internet subscription and one of the largest and rapid growth of digital consumers. It is growing at a faster speed compared to many mature and emerging economies. However, the immergence of information technology integrated with a robust farming is still evolving in India [8, 9]. World Bank had also developed a blue print and highlighted the scope of primary improvement by applying advanced information and communication technologies (ICTs) in agri-food production system [10]. There has been an increasing usage of digital communication technology to increase the income and capabilities of farmers [11].

Food traceability has now been at the core of recent food safety and quality discussions across industry and academia, particularly with new development in Blockchain applications. The present distribution mechanism related to food with less shelf life is extremely susceptible to human errors, which significantly affect the quality of the food. When foodborne diseases affect public health. The primary step for a causal analysis is to track down the source of contamination and there is no tolerance for uncertainty [12]. These situations demand for a well-organized data capturing, storing, and processing mechanism to retrieve the relevant information related to the root cause analysis. Due to inefficiencies in the food supply chain and lack of information retrieval mechanism, the food industry faces major challenges in various crisis situations. However, the emerging blockchain technology can improve food safety through distributed network among multiple stakeholders like farmers, processors, retailers, and consumers.

The existing infrastructure related to food ecosystem causes traceability a lengthy and difficult task. As most of the involved entities record information through manual, chaos, and complex process. However, the structure of Blockchain ensures that each entity within the food value chain would generate and securely share information to develop a reliable and accountable system. Large amount of information or data with labels that clarify ownership can be promptly recorded without any deviation. This result to the development of an accurate food item journey mapping from farm to table and can be monitored in real-time.

The chapter will start with a system level thinking pertaining to the current food supply chain then it will elaborate on multiple steps associated with Service Design, followed by Integrated Supply Chain Information and Secured Blockchain Framework.

17.2 SYSTEM MAPPING

A system map displays the elements and their boundaries within a system framework considering macro and micro environmental factors at a particular time [13]. A system map is a graphical representation of a list of elements and is easier to assimilate. System map helps to arrange a system structure for effective explanation to others. The map facilitates clarification of thoughts and ideas at an initial stage of analysis. This lead to a detailed structural diagram of different type which facilitate to experiment with boundaries based on the interest level. In most of the cases, a system map is self-explanatory. System maps can be created from a system thinking approach which facilitate us to shift our worldview perspective. This provides an impetus to ideate and develop novel multidimensional framework. This requires a detailed mapping of the dynamics and inter-connectedness of the systems at play. Literature suggests eight achievable objectives through system thinking approach as mentioned in subsections [14].

17.2.1   TO IDENTIFY THE COMPONENTS AND PROCESSES WITHIN THE SYSTEM

Different components related to the agricultural production system starting from farm activity to final consumption and disposal need to be identified as shown in Figure 17.1. The major system components are soil health management; nurturing and maintaining the growth of crops, fruits, and vegetables (growing); irrigation system for efficient and optimal water supply in the farm; pest control is one of the tricky component in the system

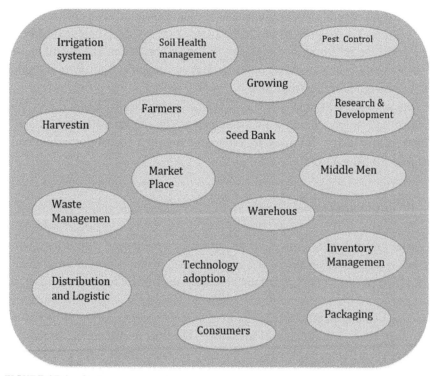

FIGURE 17.1   System components.

which balances between unwanted pests and pesticide usage; harvesting is one of the primary component of agricultural production system which involves complex decision making process around multiple factors like ripening time, storage facility, market accessibility and demand; packaging is an important component within this system for quality control and minimize food loss and also creates an esthetic appeal to the consumers; Warehouse plays an important role in managing inventory based on the shelf life of agricultural produces and also help to stabilize the price of agricultural commodities by checking the propensity towards post-harvest sales among the farmers; Distribution and logistics is one of primary component of this system which involves various resources like transport, human resource, storage facilities, etc. Another major component within this system is market place this is the phase where the final conviction and consumption decision takes place and can be virtual or physical space through fair trade practice. Waste management is another important and difficult process in this system. Minimizing food loss or wastage at every stage starting from growing to consumption should be of utmost priority in a country like India.

17.2.2   TO IDENTIFY RELATIONSHIPS AMONG THE SYSTEM'S COMPONENTS AND PROCESSES

Systems theory is an interdisciplinary study of systems [15]. System has a specific boundary governed by time, space, and environment. It should have a well-defined structure and objective which is expressed through its functioning. A system can be simple or complex, which is a unified assembly of interconnected and symbiotic components which can be natural or manmade. Change in one component can affect other component. Therefore, it is crucial to identify those direct and indirect relationships in order to map a complex system. The expression of this characteristic within agricultural production system is acknowledgment of the connection between different components. For example, Soil Health Management is directly connected to Pest Control. Minimal and judicial usage of pesticides improves soil fertility. Again, soil health is directly connected to the growing of crops, vegetables, and fruits. Growing of crops, vegetables, and fruits is also connected to the Irrigation system (Figure 17.2).

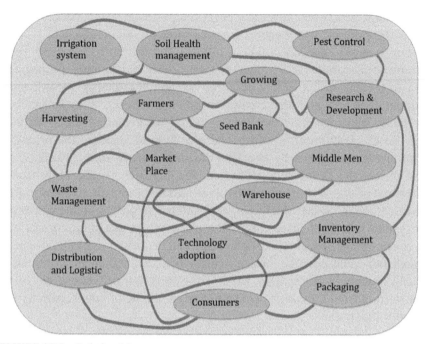

FIGURE 17.2   Relationship among system components.

17.2.3   TO ORGANIZE THE SYSTEM'S COMPONENTS AND PROCESSES WITHIN A FRAMEWORK OF RELATIONSHIPS

Once the components and the processes are identified and their relationships are mapped. Then the framework of the entire system can be sketched out. Here a simple framework has been established to organize the above-mentioned components in Figures 17.1 and 17.2. The broader framework will consist of three major phases production, distribution, and consumption. The components or processes mentioned can be organized as per these three phases.

17.2.4   TO GENERALIZE THE SYSTEM MAP OR FRAMEWORK

System is circumscribed by time and space and guided by its environment with a structure and objective. The outcome of a system can be more than the sum of its components if it expresses synergy among the part and revel an emergent behavior. This feature of a system is specific to a certain framework which is validated on the ground. With increased number of validations in multiple similar contexts lead to a generalized framework. Similar such framework can be developed for a traceable and reliable supply chain for food quality and safety through blockchain based technology.

17.2.5 TO IDENTIFY DYNAMIC RELATIONSHIPS WITHIN THE SYSTEM

The central concept of system dynamics is to identify and explore how the elements or processes in a system interact with one another. Understanding the dynamic nature within the relationships existing in the system facilitates the impact of change in one element on the other [16]. For example, within this agricultural production system waste management process is directly connected to Warehouse element. If some design intervention is initiated to improve warehouse efficiency and that intervention minimizes food loss or wastage, then this might improve farmers profit margin through a transparent fair-trade framework.

17.2.6 TO EXPLORE THE HIDDEN DIMENSIONS WITHIN THE SYSTEM

Acknowledging patterns and inter-relationships, which are not seen, on the surface is one of the important aspects of system mapping. Identifying those hidden dimensions within a system is very tricky and crucial. In this chapter, the hidden dimensions in the production phase can be farmers' motivation to adopt new knowledge to improve their production, lack of fair-trade practice among the intermediaries, lack of farmers' interest to connect to the market and improve their profit margin.

17.2.7 TO UNDERSTAND THE CYCLIC NATURE OF THE SYSTEM

The cyclic nature of an agricultural production system can be explained through the process of production perishable fruits and vegetables in the farm, then these produces through a supply chain process comes to a market place then the unsold rotten fruits and vegetables goes back to the farm and used for preparation of compost as shown in Figure 17.3.

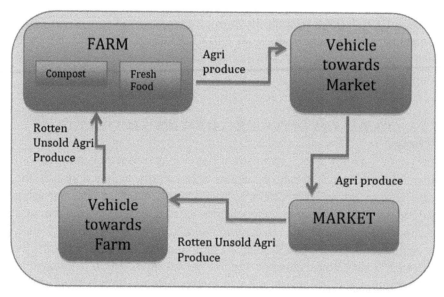

FIGURE 17.3   Cyclic nature of system.

17.2.8   TO THINK TEMPORALLY FOR RETROSPECTION AND PREDICTION

Perform causal analysis to explore and understand that some of the current outcome or interaction within the system can be the result of some past actions. Similarly, future prediction can be performed based on some present or past events and interactions. This notion can be explained within this system through an example like soil fertility is predicted to be affected by uncontrolled usage of fertilizers and pesticides for a long period of time. Multiple health disorders have been recorded due to the usage of genetically modified seeds for growing fruits and vegetables.

This system thinking approach facilitates to develop a robust system map. This system map further enables to design a detailed service blue print.

17.3   SERVICE DESIGN

Service design is enabled by complex service systems, which comprises of human resource, material sourcing and distribution, technologies, etc. Multiple service subsystem can interact with each other and may lead to optimizing single portion in isolation and not considering the inclusive performance of the system. However, optimizing one sub unit in isolation may improve the overall functioning of the service design. Therefore, it is suggested that service design is an integration of multiple service systems. However, service design approaches need to focus on one level at a time for a multi-level service. Creating optimal service experience for end-users is the primary focus of service design process. This requires detailed holistic observations; understanding of related actors or subsystems and their inter-actions; different supporting materials and infrastructures [17]. Designing a service can be well documented and explained through service blueprint.

17.4 SERVICE BLUEPRINT

Service blueprint was introduced by Lynn [39]. This allows an organization to explore all the elements and processes essential for developing and handling a service. Designing a blueprint involves the following steps:

1. **Identifying the Processes:** The first step in creating a blueprint is mapping the processes along with the elements that constitute the service. The difficulty level in identifying and defining the process increases with the complexity of the service. This diagram can be developed from the system mapping activity mentioned in Figures 17.1 and 17.2.

2. **Isolating Fail Points:** Once the process diagram is ready, the designers and managers then need to analyze where the system might go awry. And accordingly, they need to design a sub-process to mitigate the possible failure points. Through this step, the penalties of service failures can be minimized to a greater extent. Therefore, service quality execution increases.

3. **Establishing Time Frame:** Once the service profile diagram and process identification are completed. Then the vulnerabilities within the service design diagram are considered, to develop necessary fail-safe measures. The designers then need to plan accurately for the execution of the service. Since services are time-dependent and is one of the major costs determining factor. Therefore, mapping a standard execution time becomes extremely important for blueprint development. As the blueprint is the model, the design should consider the aberrations from standard time of execution under a certain condition

to make it workable. The difficulty level or complexity of the service design will determine the leeway required in the time frame.

4. **Analyzing Profitability:** A standardized service time must be established for structured execution by excluding unprofitable businesses or activities and maintains productivity. This standard can facilitate performance measurement and control uniformity and quality.

This powerful framework describes critical and analytical features of service system design and develops a rudimentary building framework of a service. It facilitates to comprehend the present service encounters and to stipulate the designed service. The main components of service blue print as per the literature are mentioned below [18]:

1. **Physical Evidence:** This is the main interface which customers or other people in the organization come in contact with.
2. **Frontstage or Visible Employee Actions:** This stage is generally visible to the customers, and they can see all the actions of different staff members executing the service.
3. **Backstage or Invisible Contact Employee Actions:** Various processes and the employee actions and their responsibilities which is not visible to the customers. However, they are important to make the service possible.
4. **Support Processes:** The internal and additional processes that support the employees to provide the service.
5. **Lines:** The lines are important features of a service blueprints. These lines clarify how the components in a service process intermingle with each other. This allows various human resources to understand their role and possible sources of customer disappointment within a service experience framework.
   i. **Line of Interaction:** This is the line which marks the customers point of interaction with the service. For example, the points of interactions are ordering food through a smooth interface, being able to track the order in real-time, time of delivery, etc.
   ii. **Line of Visibility:** The activities which are not visible to the customers are plotted below the line of visibility. For example, the backend technology assistance including human resource and technology requirement for smooth interface and development of distributed data base through blockchain.

   iii.   **Line of Internal Action:** These actions are extremely internal to the service design model and has no interaction with customers. For example, in a traceable food supply chain, the customer can trace the warehouse where the food is stored. However, this warehouse outsources the maintenance support from a third-party vendor, which is an internal action of this service blueprint.

Along with Lynn [40], Kingman et al. [19], and Bitner et al. [20] also had a significant contribution in developing the service blueprint [18]. They introduced the concept through 'X' and 'Y' axis. The 'X'-axis depicts the activities in sequential order that generally occur during a service until its accomplishment. And the 'Y'-axis depicts connections or exchanges between the customers and service providers through identification of various touch-points and other entities [18].

Information is generally placed as text boxes in a grid-like structure within the layers. The text boxes are linked to each other with directional arrows portraying the interaction flow and map the interdependency between various processes and steps involved in the service. The service blueprint contributes to the development of traceable and reliable food supply chain through blockchain technology to improve food quality and safety. This service can be achieved through a secured, transparent supply chain data-sharing platform.

17.5 INTEGRATED SUPPLY CHAIN INFORMATION

The supply chain information processing and sharing is not limited at the level of business processes but also incorporates a high volume of data from multiple sensors (IoT) and sometimes from user generated content. New and innovative exchange of information services is expected to have considerable impact in expansion of supply chain function and the business models associated with it. Value creation among customers and various other stakeholders within this supply chain service network can be achieved through effective information sharing. An integrated supply chain information system aspires to achieve the key capability of delivering the right information to the right people at the right time through a protected platform that will enable informed decision-making [21].

Substantial benefits and value can be achieved through the digitalization of information associated with food distribution and consumption. The primary motivation to integrate supply chain network is related to governance costs minimization and other trading costs with other ecosystem members of the service design framework [22]. Information technology-based cost savings enable accurate and timely processing of more information from various sources [41]. Proper automation of information flow reduces human error [23].

17.6   SECURED BLOCKCHAIN FRAMEWORK

A secured blockchain-based technology provides protection of data while sharing between untrusted entities. Literature provides inadequate information related to data source, auditing, and data trailing mechanism. A secured blockchain framework bank on smart contracts (SCs) as explained in section for effective monitoring of the data behavior out of the custodians care [24].

Blockchain is an evolving technology for transactional data sharing through decentralized and vast network of untrusted members. The technology aids novel forms of software architectures which is distributed in nature, where contract on distributed states can be ascertained without entrusting a central integration point [25]. This facilitates a distributed, unchallengeable, transparent, and reliable databases shared by a community [25]. This robust information network can lead to a sustainable and safe food supply chain. Blockchain technology focuses on tracking relevant socio-environmental information that might contribute to certain major concerns around health and safety [41]. A blockchain-based supply chain delivers enhanced ethical network with an emphasis on basic rights towards humanity and fair-trade practices. An authentic and detailed product information ensures customers about sourcing of materials and production process which were verified before and sounds ethical.

17.6.1   SMART CONTRACT DESIGN

Smart contract design can be developed from service blueprint. The technology associated to Blockchain, stipulates distributed agreement, and potentially increases the room for contracts through SCs [26]. Smart contract is a self-executing script that reside on blockchain [27]. This facilitates integration and proper workflow of a distributed and automated system. SCs can moderate informational distortions and improve fair trade.

SCs facilitate the tracking rules to control justifiable conditions and governing policy framework through suitable corrections [42].

17.6.2 DAPP ARCHITECTURE

The protected and unchangeable programs run on a decentralized network. The programs can be integrated with traditional frontend and backend technologies. This type of application is called decentralized applications or DApps [28]. Though some can be partially centralized but most of the activities in the truly DApp should occur out of a central control [28]. DApps, involve an exclusive system design to accomplish high security and dependability. DApps have user's "crypto identity" bound to the web/mobile reserve [28].

Blockchain technology uses a decentralized approach for trusted application development without trusted mediators [28]. DApp are unchangeable secure programs running on distributed networks of nodes with the use of traditional frontend and backend technologies (refer to Figure 17.4). DApps follows three-layer architectural approach with special system design to achieve high reliability. DApps are divided into two sub-systems, the frontend interface communicates with the backend SCs through reliable application programming interfaces very well known in the full-stack development practices. Ethereum (ETH) blockchain is an ideal environment for decentralized computers that has dedicated programing language called solidity for designing SCs of your own [2]. SCs running on Blockchain are being used in the field of supply chain management (SCM), where trust and sound certifications are required. DApps support user's crypto identity bound to web resources.

This chapter tries to follow a distributed application development in the field of the food SCM to define system goals, research, and identify stakeholder's requirements, design service blueprints, develop, test, and deploy blockchain applications. The process is grounded on several agile design practices, which make use of specific concepts found in blockchain development describing the design of the system [28].

We have experimented a development process for "DApp," based on SCs running on the ETH blockchain (Figure 17.4). The process covers software life-cycle: research and system definition, service design, process implementation, testing, and development [29].

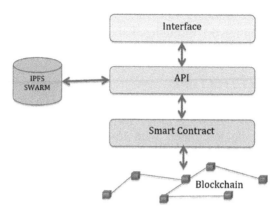

FIGURE 17.4 DApp architecture layers using smart contracts on blockchain.

The first step of the process demands the in-depth research to define the system goals that followed by the identification of the stakeholders who plays a significant role in the service design and leads to the business process definitions in terms of the service blueprint (Figure 17.5).

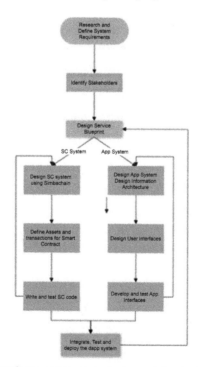

FIGURE 17.5 DApp development process to design and develop blockchain app.

The digitization of the service/business processes bifurcates to Smart Contract system (SC system) design and App system design. SC runs on all the nodes of the blockchain network. It is peculiar in terms of execution with strong constraint. SC execution makes sure that the all the outputs and state changes after execution will be the same all over the network nodes. SC has no direct access to the external world, external messages from the public interface are answered by SC through API.

SCs are designed using Solidity programming language. Solidity, the programming language of ETH, is presently the most used Blockchain programming language actually running SCs. SC realm comes with new concepts with respect to traditional programming. It refers SCs with "address," signs "transactions" to send messages from address, charge "GAS" to run SC, assigns, and transfers digital money between participants, provides external data through "oracle" with strong constraints.

17.6.3 IMPLEMENTATION OF DAPP USING SIMBA CHAIN AND VUE. JS

The application addresses the challenges like traceability, transparency, and reliability in food SCM.

The application makes use of critical GIS concepts to provide traceability in terms of the sources of the food produced in the food SCM. The application has two sub-systems:

- Smart contract solidity programs developed using SIMBA chain platform on Ethernet blockchain network (Solidity code is developed by SIMBA chain based on the asset-transaction graph) (Figure 17.6); and
- Web application developed on Vue.js software for full-stack developments of user interfaces.

These two sub-layers are communicating via SimbaAPI calls for transfer of messages between front end and backend processing of the applications.

The solidity code for the smart contract that gathers the information about the producer's geographical data and layers these details with produce quality data in the same map that makes the data accessible to the other stakeholders in the food supply chain.

Smart contract is designed following the asset-transaction graphs that depict the relationship among the business processes identified from the cyclic nature of the system, as shown in Figure 17.6.

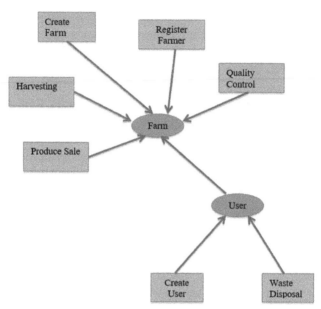

FIGURE 17.6 Assets (i.e., farm, user)-transactions graph used to code smart contract depicting the relationship among the business processes identified from the cyclic nature of the system (refer Figure 17.3).

Distributors, retailers, and consumers are equally participating to make the transactions transparent and also reliable by consensus of SCs. SCs are enforcing the mutual agreement of the transparency and reliability in overall business processes through the use of public ledgers on Ethernet Blockchain distributed network (Figure 17.7).

```
pragma solidity ^0.4.24;
contract Supplychain {
function Supplychain() public {}
enum Assets {
User,
Farm
}
Assets _create_user = Assets.User;
Assets _harvesting = Assets.Farm;
Assets _produce_sale = Assets.Farm;
Assets _qualitycheck = Assets.Farm;
Assets _farmer = Assets.Farm;
```

```
Assets _new_farm = Assets.Farm;
function create_user (
string assetId,
string username,
string password,
string first,
string last,
string role,
string email)
public {}
function harvesting (
string assetId,
string harvesting,
string produce,
string date,
int256 value_defined,
string weight_per_crate,
string no_of_crates,
string area,
string acres)
public {}
function produce_sale (
string assetId,
string distributor,
string retailer,
string consumer,
int256 price,
string date)
public {}
function qualitycheck (
string assetId,
string inspector,
int256 value,
string date,
string notes)
public {}
function farmer (
string assetId,
string address,
```

```
string notes,
string name,
string date,
string phone_no,
int256 estimated_cost)
public {}
function new_farm (
string assetId,
string addresss,
string date,
string owner,
int256 lat,
int256 lng,
int256 lat_neg,
int256 lng_neg,
string produce,
string rank,
string area,
string acres)
public {}
}
```

FIGURE 17.7   Solidity code for smart contract design using SIMBA chain.

The web interface enables the users to interact with SCs and access the information through web apps.

The blockchain-based DApp implementation has shown added advantages over the traditional web-based apps for food SCM systems as they offer trusted, scalable, and storage-free options for big player in the market (refer to Table 17.1).

TABLE 17.1   Comparative Analysis of Technology Performance for Blockchain-based Food Supply Chain Application versus Traditional Web-based Application

Performance Metrics	Blockchain-based Food Supply Chain Management Apps	Traditional Web-based Food Supply Chain Management Apps
Transaction finalization	Probabilistic	Immediate
Trust	Trusted by all parties	Semi-trusted
Performance speed	Medium	High

TABLE 17.1 *(Continued)*

Performance Metrics	Blockchain-based Food Supply Chain Management Apps	Traditional Web-based Food Supply Chain Management Apps
Cost of implementation	High	Medium
Scalability	High/infinite	Limited by server design
Storage consumption	Nil-decentralized over network nodes	High-centralized server space

The Trace Food app is implemented and deployed for testing by the key role players and stakeholders of the food-supply chain. The testing will continue for gaining the insights on the stability and accuracy of the app with the increasing complexity and functionalities in future (Figures 17.8 and 17.9).

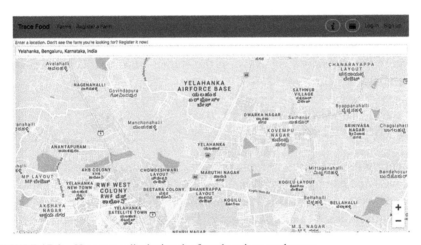

FIGURE 17.8 Homepage displaying the farm locations on the map.

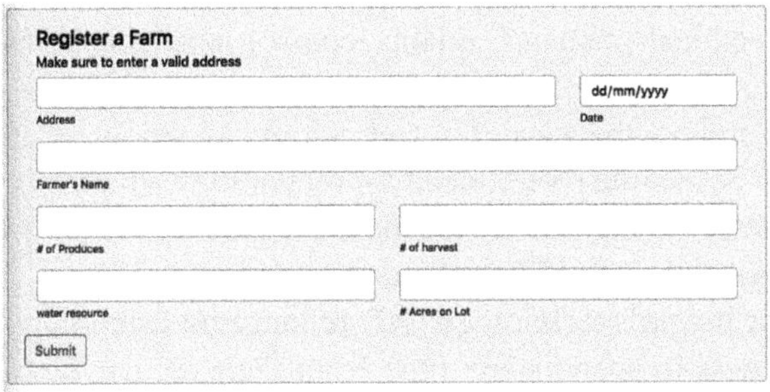

FIGURE 17.9 Screenshot of the register a farm page.

17.7 CLUSTERING FARMERS TO FACILITATE IMPLEMENTATION

This robust system can be implemented effectively through active participation and adaptation of the proposed service model. As the objective is to implement at grassroots level considering the marginalized farmers as one of the primary beneficiaries, it becomes important to analyze their behavior and attitude towards the service. Moreover, the journey for this entire service starts from the farmers. Therefore, no trial and testing for the proposed model can be possible without their participation. The variety of information lying within the farmer's knowledge system is highly contextual and personalized [30]. So, the information available from their personalized experience and practice is quite varied. Therefore, the actions and responses pertaining to digital and information technology adoption and acceptance will vary. Cluster analysis is performed to classify this heterogeneous community to a significantly homogeneous group.

Segmentation based on technology readiness and adoption among various populations has been conducted by various researchers in multiple domains to capture the diversity [31]. Most of the studies have conducted two-step cluster analysis technique [32, 33].

17.7.1 SURVEY TOOL

Considering the limitation of farmers' time, a very short and precise survey was conducted among 207 marginalized farmers from south 24 Parganas, a district of West Bengal. The survey was conducted in Bengali, which is their local language in a 5-point Likert scale. The content of the questionnaire is listed below:

- Willingness of the farmers to share all data related to production process-ethical practices, quality control, pesticide usage, etc. (DS: data sharing).
- Understanding the need for technology adoption to increase their revenue or income (NTA: need for technology adoption).
- Fear factor associated to technology adoption among the farmers (FTA: fear for technology adoption).
- Reliance towards technology (RT: reliance on technology).
- Willingness to spend some time from their regular farming or other activities and learn this technology from the young generation (STLT: spend time to learn technology).

- Willingness to collaborate with the technocrats to develop a prototype and allow them to experiment and implement on the field (CT: collaborate with technocrats).

17.7.2   RESULT AND ANALYSIS

Cluster Analysis was performed to distribute the entire set of farmers into multiple homogeneous groups. Clustering was conducted through R software and the output was analyzed, and the groups were explained. Hierarchical clustering was performed to identify the number of clusters, which informed K means clustering [32].

17.7.2.1   HIERARCHICAL CLUSTER ANALYSIS

Dendrogram derived from Hierarchical cluster analysis shows four distinguishable nodes as marked by arrow in Figure 17.10. Therefore, four clusters were considered for K-means Cluster analysis.

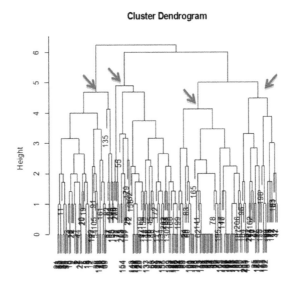

FIGURE 17.10   Output of hierarchical cluster analysis: Dendrogram.

R code for dendrogram

```
> mydata=read.csv("~/Documents/ANALYTICS R/Cluster/Trace.csv")
> d<-dist(as.matrix(mydata))
> d<-dist(as.matrix(mydata))
> hc<-hclust(d)
> hc<-hclust(d)
> plot(hc)
```

17.7.2.2   K MEANS CLUSTER ANALYSIS

The objective of K means clustering is to segregate n observations into k clusters in which each observation belongs to the cluster with the nearest mean, which are known as cluster centers or cluster centroid [34].

17.7.3   CLUSTER SUMMARY

Cluster summary for each group is developed from Final Cluster Center scores mentioned in Table 17.2. The scores are in five-point scale with 1 as lowest and 5 as highest:

- ➢ **Group 1:** The purple color oval set represents Group 1. The farmers belonging to this group are more or less willing to share data related to the production process. This group of farmers also agree to a certain extend that there is a need for technology adoption to increase their revenue or income. They are moderately interested to spend time to learn this technology.
- ➢ **Group 2:** This group of farmers represented by green set in Figure 17.11 more or less agrees towards the need for technology adoption and they moderately agree to the fact that technology is reliable. This group of farmers is slightly interested to spend some time and learn this technology from young generation. However, this group is comparatively smaller than the other groups.
- ➢ **Group 3:** This group of farmers belonging to this group are more or less willing to share data related to the production process and represented by blue set. This group of farmers is moderately scared about the activities related to technology and they fear to adopt it. The group is significantly big compared to other groups.

➢ **Group 4:** This group of farmers is represented by red oval set in Figure 17.11. The group is willing to collaborate with the technocrats and help them with all necessary support to develop a prototype and allow them to experiment on the field.

TABLE 17.2 K Means Cluster Analysis

R Codes for K Means Cluster Analysis

```
> fit<-kmeans(mydata,4)
> aggregate(mydata, by=list(fit$cluster),FUN=mean)
```

Result: Final Cluster Center

Group DS NTA FTA
1 3.428571 3.023810 1.500000
2 2.020000 2.800000 1.740000
3 2.600000 2.461538 3.400000
4 1.612245 2.755102 2.102041
Group RT STLT CT
1 1.42857 2.880952 1.619048
2 3.24000 2.560000 1.580000
3 1.507692 2.261538 1.646154
4 1.163265 3.755102 1.122449

Graphical representation of K-means clustering

R code
clusplot(mydata, fit$cluster, color=TRUE, shade=TRUE,labels=2, lines=0)

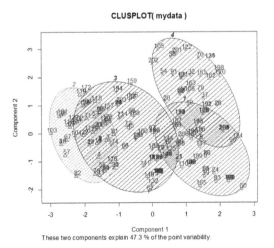

FIGURE 17.11 Graphical representation of groups through K means clustering.

17.8   CONCLUSION

Various food companies and retailers to create a traceable and reliable supply chain network can adopt blockchain technology. The technology can also provide an additional layer of security in the distribution process. Transparency is one of the major features of this service enabled by this technology. From production to sales, all the information can be recorded and accessible to all the stakeholders in the distribution process to mitigate fraudulent practices. Blockchain-enabled food supply chain will be capable enough to overcome the human error caused by traditional paper-based recording system. Any investigation related to quality issues can be carried out seamlessly as the technology provides end-to-end traceability. Through blockchain, the information transfer among the stakeholders in the supply chain network can be much quicker and safer. Therefore, food quality and safety can be enhanced through this distributed data storing and sharing platform. The system can be implemented through a collaborative venture with the farmers working at grass root level and primarily contributing to this food production ecosystem. Therefore, it becomes important to understand the level of appreciation and adaptation of technology among the farmers. Considering the fact that attitude and capability towards technology adoption may vary among the farmers so it is important to segment them into different categories based on their willingness to share data, understanding the need for technology, fear associated to technology adoption, reliance on technology, spend time on learning technology, collaborate with technocrats and facilitate them in experimentation and implementation. Four groups were evolved from this clustering process, and the group descriptions will facilitate to adopt multiple strategies to tackle these farmers' groups separately. Therefore, this segmentation strategy and mobilize farmers towards technology adoption and data sharing will facilitate the implementation process. And this can eventually lead towards the implementation of the proposed traceable and reliable food supply chain service system.

KEYWORDS

- **blockchain**
- **cluster analysis**
- **food traceability**
- **service blueprint**
- **service design**
- **supply chain**

REFERENCES

1. Sun, S., Wang, X., & Zhang, Y., (2017). Sustainable traceability in the food supply chain: The impact of consumer willingness to pay. *Sustainability, 9*(6), 999.
2. Dhahri, S., & Omri, A., (2020). Foreign capital towards SDGs 1 & 2-ending poverty and hunger: The role of agricultural production. *Structural Change and Economic Dynamics.*
3. Movilla-Pateiro, L., Mahou-Lago, X. M., Doval, M. I., & Simal-Gandara, J., (2020). Toward a sustainable metric and indicators for the goal of sustainability in agricultural and food production. *Critical Reviews in Food Science and Nutrition,* 1–22.
4. Pradhan, R., & Beriha, G. S., (2020). Impact of mobile phone on sustainable farming. *Our Heritage, 68*(30).
5. Beulens, A. J., Broens, D. F., Folstar, P., & Hofstede, G. J., (2005). Food Safety and transparency in food chains and networks relationships and challenges. *Food Control, 16*(6), 481–486.
6. Baumüller, H., (2012). Facilitating Agricultural Technology Adoption Among the Poor: The Role of Service Delivery Through Mobile Phones, Working Paper Series 93, Centre For Development Research, University of Bonn.
7. Kabbiri, R., Dora, M., Kumar, V., Elepu, G., & Gellynck, X., (2018). Mobile phone adoption in agri-food sector: Are farmers in Sub-Saharan Africa connected? *Technological Forecasting and Social Change, 131,* 253–261.
8. Seth, A. N. K. U. R., & Ganguly, K. A. V. E. R. Y., (2017). Digital technologies transforming Indian agriculture. *The Global Innovation Index,* 105–111.
9. Majchrzak, A., Markus, M. L., & Wareham, J., (2016). Designing for digital transformation: Lessons for information systems research from the study of ICT and societal challenges. *MIS Quarterly, 40*(2), 267–277.
10. World Bank, (2016). *World Development Report 2016: Digital Dividends.* World Bank: Washington, DC, USA.
11. Pick, J. B., Gollakota, K., & Singh, M., (2014). Technology for development: Understanding influences on use of rural telecenters in India. *Information Technology for Development, 20*(4), 296–323.
12. Obadina, A. O., & Olotu, I. O., (2017). Food safety in Africa-applications of risk analysis. *Agricultural Risk Management in Africa,* 169.
13. Alford, C., (2017). *How Systems Mapping Can Help You Build a Better Theory of Change.* Too Deep.
14. Assaraf, O. B. Z., & Orion, N., (2005). Development of system thinking skills in the context of earth system education. *Journal of Research in Science Teaching: The Official Journal of the National Association for Research in Science Teaching, 42*(5), 518–560.
15. Ford, D. H., & Richard, M. L., (1992). *Developmental Systems Theory: An Integrative Approach.* Sage Publications, Inc.
16. Allison, J. T., & Herber, D. R., (2014). Special section on multidisciplinary design optimization: Multidisciplinary design optimization of dynamic engineering systems. *AIAA Journal, 52*(4), 691–710.
17. Patrício, L., Fisk, R. P., Falcãoe, C. J., & Constantine, L., (2011). Multilevel service design: From customer value constellation to service experience blueprinting. *Journal of Service Research, 14*(2), 180–200.

18. Lobo, S., Sharma, S., Hirom, U., Mahamuni, R., & Khambete, P., (2019). Extending service blueprint for new-age services. In: *Research into Design for a Connected World* (pp. 809–821). Springer, Singapore.

19. Kingman-Brundage, J., George, W. R., & Bowen, D. E., (1995). Service logic: Achieving service system integration. *International Journal of Service Industry Management, 6*(4), 20–39.

20. Bitner, M., Ostrom, A., & Morgan, F., (2008). Service blueprinting: A practical technique for service innovation. *California Management Review, 50*(3), 66–94.

21. Dinter, B., (2013). Success factors for information logistics strategy — An empirical investigation, *Decis. Support Syst., 54*, 1207–1218, 02.

22. Santos, F. M., & Eisenhardt, K. M., (2005). Organizational boundaries and theories of organization. *Organization Science, 16*(5), 491–508.

23. La Londe, B. J., & Masters, J. M., (1994). Emerging logistics strategies: Blueprints for the next century. *International Journal of Physical Distribution & Logistics Management, 24*(7), 35–47.

24. Tian, F., (2016). An agri-food supply chain traceability system for China based on RFID & blockchain technology. In: *13th International Conference on Service Systems and Service Management (ICSSSM)* (pp. 1–6). IEEE.

25. Xu, X., Weber, I., Staples, M., Zhu, L., Bosch, J., Bass, L., & Rimba, P., (2017). A taxonomy of blockchain-based systems for architecture design. In: *2017 IEEE International Conference on Software Architecture (ICSA)* (pp. 243–252). IEEE.

26. Cong, L. W., & He, Z., (2019). Blockchain disruption and smart contracts. *The Review of Financial Studies, 32*(5), 1754–1797.

27. Christidis, K., & Devetsikiotis, M., (2016). Blockchains and smart contracts for the Internet of Things. *IEEE Access, 4*, 2292–2303.

28. Marchesi, L., Marchesi, M., & Tonelli, R., (2019). *ABCDE--Agile BlockChain DApp Engineering. arXiv preprint arXiv:1912.09074.*

29. Kamilaris, A., Fonts, A., & Prenafeta-Boldú, F. X. (2019). The rise of blockchain technology in agriculture and food supply chains. *Trends in Food Science & Technology, 91*, 640–652.

30. Starasts, A., (2015). Unearthing farmers' information seeking contexts and challenges in digital, local and industry environments. *Library & Information Science Research, 37*(2), 156–163.

31. Wiese, M., & Humbani, M., (2020). Exploring technology readiness for mobile payment app users. *The International Review of Retail, Distribution and Consumer Research, 30*(2), 123–142.

32. Ghosh, S., Datta, B., & Barai, P., (2016). Modeling and promoting organic food purchase. *Journal of Food Products Marketing, 22*(6), 623–642.

33. Stylianou, A., Sdrali, D., & Apostolopoulos, C. D., (2020). Capturing the diversity of Mediterranean farming systems prior to their sustainability assessment: The case of Cyprus. *Land Use Policy, 96*, 104722.

34. Everitt, B., (2011). *Cluster Analysis.* Chichester, West Sussex, U.K: Wiley.

35. Lim, C. H., & Kim, K. J., (2014). Information service blueprint: A service blueprinting framework for information-intensive services. *Service Science, 6*(4), 296–312.

36. Mckinsey.com, (2019). Digital India Technology to Transform a Connected Nation., Mckinsey Global Institute Report.

37. Ul Haq, S., & Boz, I., (2020). Measuring environmental, economic, and social sustainability index of tea farms in Rize Province, Turkey. *Environment, Development and Sustainability, 22*(3), 2545–2567.

38. Costa, L. B., Luciano, F. B., Miyada, V. S., & Gois, F. D. (2013). Herbal extracts and organic acids as natural feed additives in pig diets. *South African Journal of Animal Science, 43*(2), 181–193.

39. Shostack, L. (1984). Designing services that deliver. *Harvard Business Review, 62*(1), 133–139.

40. Korpela, Kari, Jukka Hallikas, & Tomi Dahlberg. (2017). "Digital supply chain transformation toward blockchain integration." In *Proceedings of the 50th Hawaii International Conference on System Sciences.*

41. Adams, R., Kewell, B., & Parry, G. (2018). Blockchain for good? Digital ledger technology and sustainable development goals. In *Handbook of Sustainability and Social Science Research* (pp. 127–140). Springer, Cham.

42. Saberi, S., Kouhizadeh, M., Sarkis, J., & Shen, L. (2019). Blockchain technology and its relationships to sustainable supply chain management. *International Journal of Production Research, 57*(7), 2117–2135.

PART V

Applications of Blockchain in Education and Agriculture

CHAPTER 18

Agro-Chain: Blockchain Powered Micro-Financial Assistance for Farmers

G. M. ROOPA, N. PRADEEP, and G. H. ARUN KUMAR

Department of Computer Science and Engineering, Bapuji Institute of Engineering and Technology, Davangere, Karnataka, India (Affiliated to Visvesvaraya Technological University, Belagavi, Karnataka, India), E-mail: roopa.rgm@gmail.com (G. M. Roopa)

ABSTRACT

Agriculture acts as the fixed sector for all economics over a millennium which constitutes a critical indeed major part of the developing country's GDP and extends as a massive part of rural household monetary income. As the majority of farmers mainly depend on the government schemes for financial support where agriculture banking plays a significant role in the modernization of agricultural activities, but the majority of the farmers/peasants are excluded from such banking systems and do not have access rights. It is explored with the farmers' concern that the traditional rural financial paradigm was supported by public authorities that desire to facilitate access to the financial service. However, such banking systems were costly and unsustainable due to poor repayment. They ultimately did not have a great effect on agriculture production and, in turn, led to the shutting down of state-owned banks and finally gave rise to the concept of microfinance. Over time, technological advancement from the plow to fertilizers and biotechnology has revolutionized the industry. Blockchain is the advancement of new technology aspiring to upgrade agricultural activities related to farmers' financing. This chapter adopts distributed ledger technology (DLT) that allows the recorded data in the system to fan-out amongst the farmers, consumers, and all the actors involved in the system. Blockchain-based micro-financial assistance makes the transactional data more transparent and immutable task. It reduces the intervention of the middleman in the transaction who expects some percentage of commission from farmers for selling their food grains in the market. The work extends to allow the farmers to establish a digital identity and supports them to access various monetary offers. The system makes a significant difference in helping to reduce the rate of interest and also ensures efficient repayment of loans by making it more accessible through smart contracts (SCs). Every farmer/consumer can keep track of the transactions carried out by them in selling/buying the products through the platform. Further, update the data collected from these transactions to the credit history of the farmers that give them the aid of availing the credits. The system makes a significant difference in helping to reduce the rate of interest and also ensures efficient repayment of loans by making it more accessible through SCs.

18.1 INTRODUCTION

Investigation in the recent past has explored that majority of the incidents about farmers' suicide have occurred due to the poor crop yield, lack of financial assistance from Government schemes, and low-price values for the food grains in the agro-market. Countries with a rise in population have led to a significant dependency on the land for raising building and industrial area is increasing at a high rate. Due to which, the fertile land for farming is vanishing, and with every country's demand, more crop yield supply arises. Due to the high demand of such issues, farmers usually are suffering from the problems related to banking services support from Government schemes and the low pricing values for the food grains in the market like: (i) to raise the initial investment to set up the field area with the high banking interest rates; (ii) to fetch suitable price values for food products with the intervention of middleman in agro-market; (iii) analyzing the direct connection between the farmers and customers for selling/buying the food products; (iv) inefficient food supply chain for storage and transportation which leads to the crop deterioration.

Likewise, the customers also face the problem of the high-price values of commodities and product quality and force the customers to purchase in the market with the price that the seller sets. Further, it leads to the practices of black marketing, adulteration, and hoardings carried out by the middleman and further increased the farm-product prices. With the high demand for organic products, there is a lack of techniques to trace the stages of organic cultivation and ensure reliability. Thus, the biggest challenge faced by farmers in the agro-market is the dis-connectivity between the farmers and consumers and the support for financial assistance. Therefore, to overcome the issues faced by the small farmers in rural areas related to financial aid and agro-market, blockchain-powered micro-financial assistance is proposed that supports a transparent marketplace. Here, the farmers and consumers can set up a cooperative farming procedure with further support for small loans made available for people and typically excluded from traditional banking services by designing the programs to satisfy their particular needs.

18.1.1 AGRI-CHAIN: ROLE OF AGRO-SUPPLY CHAIN IN FARM ACTIVITIES

Supply Chain involves a sequence of decision-making processes. It flows that usually aims at meeting the final customer requirements, which involves various stages from production to final consumption [39]. The supply chain process includes various actors like farmers, suppliers, distributors, warehouses, retailers, and consumers. From a broader perspective, it also provides product development, marketing, distribution, finance, and customer service [46]. In Figure 18.1, the Physical Flow illustrates the initial design of the food supply system, main stages, and the parties involved. Without the addition of the below two layers from Figure 18.1, the system becomes unreliable and inefficient where, exchange of goods process becomes complicated and still adopts the paper settlement, which is not transparent and gives rise to high-risks that exist between the sellers/buyers during the exchange process.

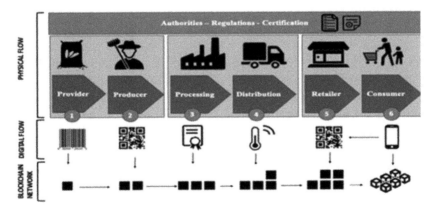

FIGURE 18.1 Simplified food supply chain system.
Source: Reprint from Ref. [3].

Due to the intervention of middleman majority of the transactions, becomes vulnerable to fraud, and they tend to increase the overall price values. Thus, there is a need for optimized supply chains, which effectively reduce the operating price values.

The main phases that characterize the standard agri-food supply chain:

1. **Production:** This phase represents the farm activities where the farmers use organic materials (seeds, fertilizers, and animal breeds) to grow food grains and livestock. Throughout the year, based on the production cycle and cultivation, farmers can expect one more yield or harvest.

2. **Processing:** Total/partial transformations of primary products to one or more other substitute products. After the packaging phase, a uniquely identified with batch code is assigned, including information on the production date and the raw materials used.

3. **Distribution:** After the products are packed-labeled and let out for distribution. Based on the food product type, the date-time set for delivery, or some of the products are stored.

4. **Retailing:** The food products are dispatched to retailers for product sale at the end of the distribution.

5. **Consumption:** Acts as the end-consumer of the supply chain process where she/he purchases the food products and demands for traceability details.

18.1.2 AGRI-DIGITAL: BLOCKCHAIN SUPPORT IN SUPPLY CHAIN AND AGRICULTURE

Today, blockchain practice has received massive success with its practical support in various crypto-currencies, organizations, agriculture, and micro-financing assistance that usually aims at exploiting its transparency and fault tolerance. The majority of blockchain applications are adopted to provide solutions for scenarios where several un-trusted actors are involved in the resource distribution, to give micro-financial assistance to farmers/poor [42] who are unbanked/under-banked and to solve the issues with farm activities.

Blockchain acts as a secure distributed public/private ledger platform which supports a decentralized system to record transactional details with processing procedures, authorize, and validate transactions to be stored into an immutable ledger [4]. It is also regarded as the "Internet of value" a secure means of storing and transacting details from contracts, currencies, and stocks from one party to another. It is an emerging technology powering the support for cryptocurrencies such as bitcoin (BTC) and Ethereum (ETH). In its simplest outline, Blockchain comprises linked chains which store auditable data in a small unit called block that contains the data, its hash-value, and pointer pointing to the previous block hash value. Here, parties take slots to record information on the system to generate, transact, and consume the products/services. This ledger is further inclusively supervised by the entire participating stakeholders along a peer-to-peer (P2P) network. Every time the system verifies the latest records before appending to the Blockchain.

Further, modifications to the recorded information must adhere to the consensus decision agreement, where the majority of the involved stakeholders should agree and approve. Here, any changes made to one record leads to the tempering of all its consequent records [9]. Due to which it is impossible to modify the data recorded in blockchain practice. Thus, Blockchain is an ICT transformative which can revolutionize the adoption of data for agriculture.

Figure 18.2 shows the complete data information move with the products passing from input to output along with the various value-added phases and the flow of the financial process from output to input. Here, we store the data/ information in terms of blocks that are generated by various stakeholders allround in the complete value-added stages through seed to sale of agri-products. Figure 18.2 shows the type of blockchain practices adopted (permissible/ non-permissible), applied platform (ETH/Hyperledger), type of consensus mechanisms (proof-of-work/proof-of-stake) and Byzantine techniques (faulttolerant) for accumulating the information at various phases in agricultural crop yield systems. The significant section of the framework is to provide an environment to build trust between multiple parties adopting the blockchain practices and also enhances the traceability of agriproducts [28].

18.1.3 DIGITIZATION OF MICRO-FINANCIAL SERVICES: BLOCKCHAIN IN FINANCIAL INCLUSION

As per the analysis drawn by Oradian Financial Inclusion Company-Africa and the estimation given by World-Bank Group around 2 billion rural people worldwide today do not have a basic bank account either in-state/ central banking system. Figure 18.3 shows the un-banked population in 25 countries, predominantly in Asia and Africa, as per statistics given from the world bank group [41]. Thus, excluding the rural population from such financial assistance makes their everyday life bitter. And difficult for them to overcome the risks they face, no support for them to invest in future/ save money, and finally lead a decent lifestyle which they usually do not deserve.

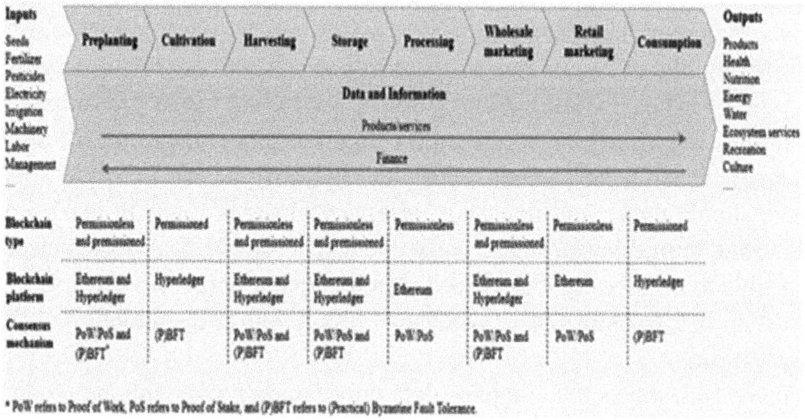

	Preplanting	Cultivation	Harvesting	Storage	Processing	Wholesale marketing	Retail marketing	Consumption	
Inputs Seeds Fertilizer Pesticides Electricity Irrigation Machinery Labor Management				Data and Information					**Outputs** Products Health Nutrition Energy Water Ecosystem services Recreation Culture
				Products/services					
				Finance					
Blockchain type	Permissionless and premissioned	Permissioned	Permissionless and premissioned	Permissionless and premissioned	Permissionless	Permissionless and premissioned	Permissionless	Permissioned	
Blockchain platform	Ethereum and Hyperledger	Hyperledger	Ethereum and Hyperledger	Ethereum and Hyperledger	Ethereum	Ethereum and Hyperledger	Ethereum	Hyperledger	
Consensus mechanism	PoW/PoS and (P)BFT	(P)BFT	PoW/PoS and (P)BFT	PoW/PoS and (P)BFT	PoW/PoS	PoW/PoS and (P)BFT	PoW/PoS	(P)BFT	

* PoW refers to Proof of Work, PoS refers to Proof of Stake, and (P)BFT refers to (Practical) Byzantine Fault Tolerance.

FIGURE 18.2 Complete data information pass from inputs to outputs along with the standard blockchain practices adopted.
Source: Reprint from: Hang, Tobias, Puqing, and Jiajin (2020).

Indeed, getting bank loans is the major problem for small farmers in the rural population in the world [44] and almost highly impossible in developing countries. Banks do not generate money with no credit history and minimal access to banks, and these small-scale farmers can only borrow money from local intermediaries who in turn takes away the majority of the farmers' profit. Hence, the concept of "microfinance" helps small farmers to get access to financial services [15]. However, the traditional micro-financial services are not banking and usually operate with the support of Private Sector banks. Due to which there was a substantial transactional cost that resulted in high interest rates and transaction rates for borrowers.

Further, these borrowers are based in remote locations and eliminate the business-relevant paperwork. Thus, the process of KYC becomes difficult and slow. Finally, the major problem with traditional microfinance practice is that the parties who put money into financial services are entirely unaware of the parties who get the financial services, and vice versa.

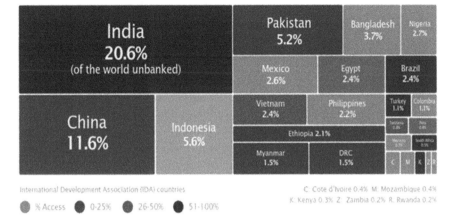

FIGURE 18.3   Un-banked population in 25 countries as per World Bank Group.
Source: Reprint from: https://oradian.com/wp-content/uploads/2018/07/Oradian-Blockchain-2018.pdf.

To summarize, macro-financial assistance in frontier markets plays an invaluable role in providing the financial-inclusion for unbanked and under-banked rural people even though traditional micro-financial services were sustainable and efficient in promoting services. But the conventional services faced many challenges like customer-identification, infrastructure support, fraud, and substantial operating cost.

Thus, the existing financial service extended by considering the potential role of Blockchain to address the challenges and its widespread adoption in agri-markets. Blockchain creates a direct connection between borrowers and lenders, thus cutting down the demand for local banks and significantly reduces the transactional costs.

18.1.4   *SMART CONTRACTS SUPPORT FOR SMALL FARMERS*

The idea of smart contracts had its existence over the past in which the work carried out was purely manually, and contracts were paper-based. Even with digital contracts, it was impossible to eliminate the involvement of trusted third-party that usually leads to security issues, fraud activities along with the increasing cost in transactions [16].

Thus, in the agricultural zone, self-performing smart contracts (SCs) with support for automatic payments would be a paradigm shift where the goal of intelligent contracts specifically in the area of the green bond, agri-insurance, and product traceability becomes very efficient. With the involvement of Blockchain in the field of digital technology space provides efficient solutions and allows all the parties in the network to interact with anyone in a distributed scheme to eliminate the requirement of third-party services.

Today, smart contracts with the digital-tokens bring transparency, efficiency, and fairness into farm contracts which allow the farmers to perform contract-farming among the sponsors from any place in the world in a non-predatory manner. By employing, SCs in everyday life can bring incredible change with its various benefits over conventional practices. ETH supports the most accepted blockchain environment for creating customizable intelligent contracts with the feature of "Turing Completeness."

The work presents a new technique that forefronts towards trusted cooperative schemes and financial assistance with the agro-chain between the small-framers and other parties involved in the chain. The system mainly aims at reducing the vast financial risks, increased transparency, and attains quick access to the global agricultural food markets. An example, of how blockchain practice is adopted to carry out an automatic transaction among the farmers (producers) and distributors/retailers with the use of smart contracts in Figure 18.4. It shows a hypothetical scenario of how smart contracts facilitate the sales of food products. The system executes automatic buyer access to the storage room. Blockchain also facilitates financial assistance for farmers' security against the unpredicted weather situations, which directly affects the crops and other risks (natural disasters).

18.1.5 MAIN MOTIVATION AND CONCERN

The main motivation for the adoption and assessing the potential of blockchain technology in the sphere of microfinance, food supply chain, and small contracts have been raised from two significant aspects predominantly for the providing the services related to the farming activities:

1. The asymmetric information among the lenders and the loan-holders exists in the regular commercial-banking system. Here, loan-holders usually lie in low-percentiles in terms of monetary literacy/banking services. Due to this majority of beneficiaries, it is left in a kind of vulnerable position when concepts like risks, contracts, interest rates, insurance, and liability are made open to them. Even though the loan process is fully made transparent, there always exists a room for a few agents gaining profit for them. The capital required

for per-transaction is considerably low, and on the other side, there are a considerable number of beneficiaries. The cost of gathering clients when estimated to the loan size is exceptionally high in traditional microfinance compared to regular commercial loans. Hence, the interest rates for micro-loans are high as the operational costs cannot be recovered from fees, and thus income gained from interest payments will be very high.

Thus, safe and efficient procedures to manage the transactions that drastically aids both lenders, high benefits, thereby eliminating / transforming the nature of intermediaries and loan-holders with less interest to pay and procedures to prevent over-leveraging.

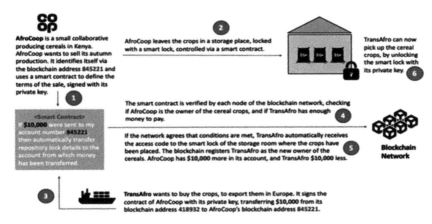

FIGURE 18.4   An example illustrating how smart-contract involves the following steps for enhancing and automating trusted transactions.
Source: Reprint from: Andreas, Agusti, and Francesc (2020).

2. Today, food supply chain persists to grow abruptly with high expectations of consumers for fresh exotic-foods products and leads to a massive expansion of the supply chain geographically with the involvement of many parties by making the system more complicated and confusing.

The significant challenges faced by various actors admitted in the supply chain are the complications in collecting and getting the food-products to the agro-market quickly, safely, and in the best possible condition [12].

The issues with food supply chain include:

- Lack of traceability: This issue further creates blind spots and exposes unnecessary risks which weaken the customer trust and leads to lower sales/profits due to the usage of outdated systems like traditional paper-tracking/manual work and delays in information sharing.
- Insufficient communication among the parties involved leads to mistrust among the suppliers and consumers, which becomes even worse when the transactions are global. High transactional costs involved in traditional practice are enormous.
- The food supply chain process becomes very challenging, either performed locally or globally, to ensure the highest level of product safety and quality at all times. Generally, with more supply chain visibility and good communication, the process can be effectively managed. With the focus on the adoption of core technologies, it is possible to deliver good results, invest in the best suppliers, reputed logistics partners, and more loyal customers.

18.2   RELATED STUDIES

With the concern of supply chain management (SCM), Halldórsson et al. [57] has identified SCM as a research discipline with its support for various theories from areas of non-logistics to demonstrate inter-organizational phenomena, Glazer [23] has drawn the gap that exists between the SCM practices and the theoretical procedures their work also presents many structures and methods to manage supply chains. Further, various factors like transaction cost, principal-agent theory, network perspective, and resource-based that mainly affects the characteristics of supply chain. Authors in Refs. [45, 52] through their work has found that it is not a good practice to rely upon a unified theory to explore the governance-structure and management-related decisions in the supply chain, and have introduced procedures applied with complementary ideas.

Blockchain is a revolutionary protocol that allows secure, simultaneous, and anonymous transactions by furnishing a tampered proof public/private ledger. They have suggested that technology can drive many digital-currencies and BTC. Authors in Refs. [32, 43] have found the potency of Blockchain and its support for distributed databases which maintains a continuously growing immutable data file list and revises even from the operators of the data-store nodes.

On launching the blockchain platform, it offers a considerable amount of resiliency, making them dependable and attractive in running decentralized services and software applications. The business blockchain [46] has provided a clear understanding of Blockchain for business potential and for people with business-minded to grasp various facets of blockchain practice. Authors in Refs. [47, 55] has ex-pounded the major principles defined for Blockchain and few of its cutting-edge applications. Their study has suggested the essential concepts, hybrid solutions, and the core features of the decentralized crypto ledger platforms with illustration on why Blockchain is disruptive and foundational-technologies.

The work presented by Porru et al.; and Underwood [27, 50] has highlighted on various applications of Blockchain has produced to support data security, trust, privacy-preserving, and digital identity. An anonymous-identity verification process allows everyone to read and verify the transactions within Blockchain besides anonymous verified identities will have transactions executed and further helps to create resiliency within the system to defend against attacks and preserve the data integrity in identity network.

The study carried by authors in Refs. [26, 51, 54] have suggested various mechanisms for blockchain security when applied to smart contracts by allowing digital-rights management. Such a tool support consensus (agreement) methods with credit score and build a hybrid blockchain using this new platform and proof-of-stake. Also, they have identified that the concept tries to prevent an attacker from monopolizing resources and to maintain secure blockchains.

Many researchers on exploring the utilization have conducted the study, integrating/developing an application with Blockchain and other emerging technologies (RFID, IoT, machine learning (ML), etc.) suggested by Brakeville and Perepa; and Xu et al. [25, 38]. The study further, demonstrates traceability with trusted data along the whole process of agri-supply chain, which efficiently guarantees product safety. Authors have presented the comparison/classification of blockchains and blockchain-related applications to support the design and assessment with its impact on system architectures. This ordering captures the blockchain-related architectural features with its implications for significant design decisions. Further, it intends to provide essential architecture considerations related to precise attributes of blockchain-based systems and its performance.

Authors in Ref. [41] have addressed the issues and challenges related to digital identity and access permissions within shared permissioned blockchains, they have studied that the Chain Anchor system can allow anonymous but verifiable identities for participation in supply chain on the blockchain. Caro et al. [21] have presented the use of cloud platform architecture with blockchain technology to support the supply chain-related application. On using the cloud storage/file scenario as a data entity with a fully secure and verifiable cloud data source by enclosing data provenance blockchain-related transactions. Performance evaluation model Prov-Chain has demonstrated that it contributes various security attributes including alteration proof data entity, privacy, and authenticity with less outlay on cloud-related applications [20].

Jha, Andre, and Jha [18] have presented the theory of blockchain application support in agro-chain networks and logistics. They have studied the challenges related to agro-chain, and further have proposed solutions to overcome the associated issues.

Authors in Refs. [10, 19] have shown that Blockchain practice can promise overpowering issues on trust and granting secure, faithless, and authorized system for logistics and agro-chain interchange in supply networks. The current implementation forms in the supply chain are transforming block-chain adoption to a broad perspective of distributed ledger approaches.

Maru et al. [14] has carried investigations on the implementation strategies of the Blockchain platform mainly targets on food grain quality affirmation by tracing the transactions in real-time schemes. Lucena et al.; and Lierow, Herzog, and Oest [12, 30] mainly concerned about creating standard practices to outline Blockchain adoption-related use cases, and applications relevant to financial assistance. Complete work has supported the actors involved in the decision-making process.

18.2.1 PROBLEM STATEMENT

Today around 40% of the world's population earn their income from agriculture where there is 50% lost in crop values during harvest to sale. The most common problem faced by farmers is impossible to solve, like crop-spoilage, poor-planning, un-reliable farm facilities, and lack of advanced equipment. From the survey, 2 billion of the worlds' un-banked are small farmers from the rural population who usually owns only one/two acres of land and earn money for their decent living. Such, small farmers rarely have access to advanced machinery/equipment to support the process of planting, growing, and crop-harvesting where such activities are carried out manually by farmers. The issues also extend with significant restrictions on market access for product sales and mainly depend on the local middleman. He offers a dis-proportionality low price below the market value for the food grains.

Thus, farmers usually have no access to any financial services such as macro-financial assistance (micro-loans), which can allow them to purchase better-quality farming products and even get advanced farming-machineries on rent. It is tough for them to access the information on the adoption of best-farming practices and how to get the best of their farmland.

However, farming cooperatives face significant challenges like poor record maintaining, lack of documentation, and corruption. Currently, these small farmers will continue to contribute to the majority of the extra 60% of food-grains required to feed the world population in the future years., farming cooperatives can play a significant task in helping farmers to increase their crop growth and earn more income. Finally, aiming at food quality assurance tracing in a real-time scenario should be initiated [1, 2].

18.2.2 CHALLENGES AND OPEN ISSUES IDENTIFIED

1. **Accessibility:** Blockchain needs to be easily/more accessible as the elementary digital platform can extensively become complicated with the integration of more components (IoT, sensors/actuators, RFID, robots, biometrics, etc.) [5, 7]. In order to provide precise functionality, Blockchain should depend on the remote systems to acquire exact details from the current world. These are "oracles" which associates the physical and digital worlds that generally arise from digitized sensors (hardware oracles), data retrieved from web-related applications (software oracles), and manual records (human). Blockchain practice must provide efficient procedures to tackle the oracle issues related to finance and smart-contracts-related applications.

   As Blockchain connects complex global chains, the required infrastructure information for operations, and maintaining the system will block the new users from accessing the market-services/food suppliers. Thus, blockchain practice can be a technical barrier to perform trading by reducing market access and competition.

2. **Lack of Skill Awareness:** Furthermore, there is an everyday insufficient skill awareness on the selection of blockchain practice. Meanwhile, training platforms are minimal. Other than, policy creators

empowering on the blockchain adoption remains elementary for the food price chain stakeholders. Currently, various startups are working on software development to make blockchain practice trouble-free usage for small farmers which aggregates essential data about the blockchain process related to farm activities and food products.

3. **Sustainability and Governance:** On contradictory, Blockchain acts as de-skilling practice for organizations and workers. Due to the increase in automation tasks/procedures within the supply chain and elimination of intermediates in the transactions might significantly reduce the human intervention. Blockchain practice leads to subsequent damage to skilled jobs.

The recent fall-off in the market segment and high volatility of financial price on desired crypto-currencies will diminish the overall public trust in the underlying blockchain practice. A minimum understanding between the policy creators and technical specialists remains on the usage concept.

18.2.3   TECHNICAL CHALLENGES WITH DESIGN DECISIONS

Numerous design decisions mainly influence the current Blockchain, which is under development. For instance, they are permissible (trusted participants), permission-less, open (anyone can join), closed-systems. Further, to decide on which party will own the Blockchain, delay in permission-less blockchain transactions which might run from few minutes to hours to complete before all the parties update their blocks when SCs are publicly accessed.

Such design choices will further disturb the operating process that, in turn, creates a lack of flexibility and makes blockchain-based solutions less efficient when compared to traditional centralized practices. Further, prioritizing blockchains can also improve time, yield, and threat in three areas: referencing data, trade payments, and consumer contribution [29].

Further, the current blockchain protocols have serious scalability problems as the existing transaction process is limited by applying the limited parameters like size and transaction-block interval. In general, Blockchain can provide opportunities for small farmers, can be a developmental aid for training/technology transfer to small farmers insight to bring precise solutions to particular situations to re-strain their socio-economic progression.

Finally, the details on a commonly defined ledger are updated, and their public keys recognize users. This form ensures to maintain high transparency, maintain/built trust, and also to retain stakeholder's privacy. Thus, maintaining privacy is especially very critical within the supply chain ecosystem, as huge parties are competitors in the agro-market.

18.2.4   PROPOSED SOLUTION

To overcome the problems with farming activities, an efficient digital technology-based system "Agro-chain: Blockchain-powered Micro-Financial Assistance for farmers" to support the transparent agro-market place where the farmers and consumers could design a cooperative farming method. The system allows the farmers to list the potential crops and the expected farm yield on the public distributed ledger. Further, the consumers can view the details and checks the farmer's credibility build on the past cultivation and supply and generates a clear and tampered proof digital-market platform for farm products. Thus, a consensus (agreement) between farmer and consumer where consumers can fund different crops or a field and further can purchase the yield/the profit based on the percentage of the market value.

Smart contracts provide a platform to design flexible and scalable businesses at low cost, and the overall manufacturing services could be enhanced. Microfinance mainly aims at granting micro-loans to small farmers/firms and entrepreneurs, outside the limit of conventional banking and group-based models. It works as an alternate source with small funds for underserved to balance the finance, has evolved as an efficient tool for up-lifting their needs and social standards, and to get work for them out of poverty.

Finally, a rating mechanism to build the farmer and consumer credibility based on past experiences in the agro-market.

The essential features supported by the proposed work are listed below:

- Small farmers need not wait for bank loans/other lending services to raise the initial amount for field setup. Consumers provide the funding service with zero/little interest.
- Consumers can expect quality products at cheap rates as they are funding the crops or fields right from the cultivation stage.
- Farmers need not have huge farmlands, even with small-scale farmers and small-household farmers can also sell their food products and yield with better profit.
- An efficient food supply chain can ensure a point-to-point update on the immutable chains, where customers could select specific farmers for specific products.

- The farmers can establish consumer loyalty based on the product quality and farming type that eventually yields them good profits.
- Even consumer groups with low income can support crop funds based on their requirements and can flee from the market fluctuations in product price values.
- Organic farming and quality measures guarantees with frequent quality-checking from the concerned parties where the immutable ledger will ensure transparency and reduces the chance of fraud.
- SCs can provide efficient agreement procedures among the farmers and consumers for any discrepancies that arise concerning natural-calamities, climate-variations, or situations with crop loss.

18.3 SYSTEM ARCHITECTURE OVERVIEW

Figure 18.5 shows the general processing of transactional request from the farmers/consumers. The designed scheme lets the small-farmers/consumers to enter an insured state for selling/buying their food products in the following stages:

1. **First Stage: Exploring:** Before joining the AGRO-CHAIN scheme, farmers provide essential details to make a transactional request. Based on this basic information and historic drought data, a short message service (SMS) suggests the farmer an amount that is set aside in their mobile wallet every year to overcome their risk. This amount of suggestion allows the farmers to set aside any amount freely. At the end of every season, based on satellite drought data and the defined standard index, all/part of the amount might be released and withdrawn by the farmers. Although, released funds primarily concern the list and also with the occurrence of an event which leads to crop-yield reduction.

2. **Second Stage: Gained-Trust:** With the usage of supply chain for selling their food products, farmers can first gain positive experiences with the financial scheme and continue to adhere/practice the rules and regulations defined with the services. Further, farmers can apply their mobile wallets under favorable situations based on the history of their savings which in turn involves increasing the fund amount over the years.

3. **Third Stage: Fully-Covered:** This stage will provide secure access to sufficient funds to deal bad agricultural season. It also gives

an option for farmers to leave the chain when they are not able to balance in their monetary issues, which can be processed only at the end of the season.

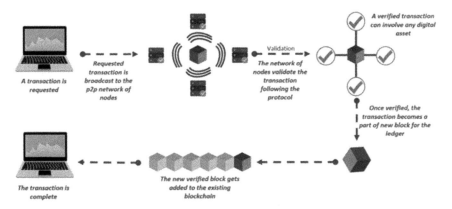

FIGURE 18.5   Processing of transactional request.
Source: Reprint from: https://www.postvilla.in/understanding-blockchain-technology-and-its-working/.

Aggregators (out-grower schemes, farmer's cooperatives, and NGOs) have strong trust and financial relationships with small farmers [30]. This assistance adds tremendous value in terms of capacity-building, marketing/relationship management, and a means to support the farmers in walking through the following stages of food supply chain.

Figure 18.6 depicts the overview of system architecture designed to demonstrate the complete transaction carried out at various stages to support supply chain services and financial assistance for farmers in the rural population.

The working procedure of blockchain-based food supply chain and smart-contracts:

1. **Storing the Details about the Food-Crops:** Farmers need to furnish the details of crops-grown (origination), crop type, sowing procedures, and storage information on the web portal or mobile-app or blockchain center location. Stored information by farmers is made accessible to all the stakeholders involved within the system. The regulations and policy rules framed in the SCs ensure to meet the acceptance by updating the information on the blockchain platform.

Then distribute the farm products for product processing refineries/ companies for the further testing process.

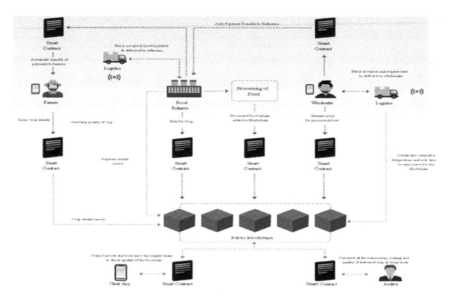

FIGURE 18.6 System architecture (food supply chain, smart contracts, and financial assistance).
Source: Reprint from: https://www.leewayhertz.com/supply chain-blockchain -reinventing-food-supply/.

2. **Adding the Processed Data on Blockchain:** After collecting the products from the farmers, food processing/refinery companies will tender for food products with contract platform. Further, the accepted bids are processed and the related information about the crop-refining is updated to the public Blockchain. This information later assists retailers/consumers to verify the appropriate processing of food with the defined regulations added to smart contracts, food-compliance to every processing step.
3. **Wholesalers Bid on the Processed Food-Products:** Wholesalers can request through the smart contract and as the bids are accepted, the processing parties will transport the food processed to the whole-salers and fill-in the transportation information on the blockchain [31]. Finally, wholesalers can hire local/remote logistic services to disseminate the food products to the re-seller. Adopting this practice

will aid in lowering the logistics charges and optimizes its routines [15].

4. **Food Quality Assurance for End Customers:** From start to the end of the process, the facts from farm-initiation, batch-unit, transportation-related factors, company processing data, storehouse temperature, expire details and other information to food products in the blockchain-based supply chain.

Every information bit gathered from the individual transaction is verified and certified by all the parties within the blockchain system to create an agreement (consensus). Update all the validated blocks transactions in the chain, creates a permanent and immutable record.

Later, consumers then assure food safety by backtracking the food products from supply blockchain. Finally, the consumer can decide on whether she or he should purchase a specific food product or not.

5. **Auditing:** Finally, the auditors (part of Government schemes support for small farmers) will execute audit checks on the price and quickly checks the crop quality from the stored records in the supply blockchain. The system can also support internal auditing if the internal audit does not meet the retailer's needs, then appropriate actions can be insisted upon/termination of the relationship.

6. **Financial-Assistance:** Small-farmers need not wait for bank loans/ other lending services to raise the initial amount for field setup. Consumers provide the funding service with zero/little interest. Several Government schemes can have a great impact on rising rural farmers' out-of-poverty and an essential tool for supporting financial inclusion and economic growth in small households.

In this context, Government schemes must take some initiatives to invest in educating and training programs for farmers for demonstrating the potency benefits of blockchain practice when applied to farm activities and also for financial assistance.

18.4 ALGORITHM IDENTIFIED AND APPLIED

18.4.1 SHA-256: SECURE HASHING ALGORITHM

SHA-256 is a deterministic cryptographic hash function with an input data (any type). It returns a fixed-length (256 bit or 64 characters long hexadecimal-code) which aids a unique fingerprint for the input data.

A block usually includes:

1. **Version:** Bitcoin software version used.
2. **Preceding Block Hash:** Refers to previous block hash value.
3. **Merkle-Root:** Set of Hash values of all transactions;
4. **Time-Stamp:** Block creation time.
5. **Target:** Block-PoW (Proof-of-work).
6. **Nonce:** Set Random number, so that the block-hash (input – data + nonce) contains a run of leading-zeros (like 0000e3st... or 0000003jg7...) mainly depends on the complexity of the network protocol (higher the complexity, large number of leading zeros).

The difficulty in finding a valid hash is less (starts with 0). The iteration begins with one and continues until the hash-value begins at 0 for 243-iteration.

Next, the difficulty is to readjust in finding the hash-value starting with '00' and is gained at 1389-iteration. This process continues in readjusting to find the next hash-value with '000' and so on, as shown in Figure 18.7.

Nonce	Difficulty (# of leading zeros)	Input Data	Nonce + Input String	Hash	Valid
1	1	hello world	1hello world	8c5b5db620b341dded0c285ddae3df91e8a761d3e5c637780bc92febf454bc83	No
2	1	hello world	2hello world	bc383cda4b61f65be27f3f94936a254605073bca147b5556ab33927a782538a0	No
...	1	hello world	<n>hello world	XXXX......	No
243	1	hello world	243hello world	0a996dfd9009d9c41b27097c2910f953c96157e7d49a0341d0fa7daaa2e8b0c3	Yes
244	2	hello world	244hello world	63b59556851cc13ebf2c3497c7422249e3965226b3de1fbd81409d2cb6d9c856	No
245	2	hello world	245hello world	ad47297a3acfca4c0507a8b8ff6a3ada73a99548c119ede8abc6ead3500de835	No
...	2	hello world	<n>hello world	XXXX......	No
1389	2	hello world	1389hello world	00e9fb46a9e282afcaf374f1fc66b5ce0a066b5d9600b6d009f4ce091ccf5357	Yes
1390	3	hello world	1390hello world	97b1ffdc9736a25f60edad554b145b81f6782e371000b631bcdf509a1ed11ac8	No
...	3	hello world	<n>hello world	XXXX......	No
8924	3	hello world	8924hello world	000d91554f0d81bc25bb7259cd01e0f89e3f0b553626fa439b0e57b10fd39d3c	Yes
8925	4	hello world	8924hello world	2dc076bf41d8137b48e8a2beaff4a3d7c2a32ac52662d37d0f6ac6cdb543f5a6	No
8926	4	hello world	8924hello world	a454448eba27eb2aca173b64149ad5a931413b7f6aad5a25206a0315f5a4a86b	No
...	4	hello world	<n>hello world	XXXX......	No
18014	4	hello world	8924hello world	0000d89cfa9bee16d7a2a67ab7c9b8795b22362dfa7c46e5bff514320a809beb	Yes
...	...	...	...	XXXX......	

FIGURE 18.7 Example for nonce concept for generating valid hash-values.
Source: Reprint from: https://prodmonk.com/trends-reviews/blockchain-cryptocurrency-simple-explanation/.

A mining concept adopted in Blockchain, where the hash-values generated are unique with the combination of input data and nonce.
SHA256 Hash-of-Blocks = SHA256 (Block # + Nonce + Data (Coin-base and Transaction Lists) + Previous-Block-Hash + Time-Stamp)

Every block in the chain points to the previous block using the field previous-hash value, and finally, blocks are links formed shown in Figure 18.8.

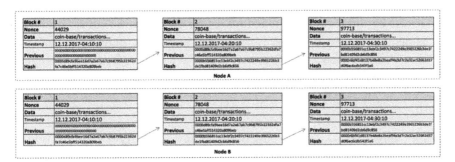

FIGURE 18.8 Chain of blocks created in blockchain.
Source: Reprint from: https://prodmonk.com/trends-reviews/blockchain-cryptocurrency -simple-explanation/.

Steps involved carrying out SHA-256-Hash algorithm:

1. Collect the entire valid transaction list from the transaction pool. Build a block that contains the transactional data and ensure the block size will not exceed 1 MB.

2. Calculate Hash value of the block by applying SHA256 Hash function *SHA256= (Block # + Nonce + Data (Coin-base* and *Transaction Lists) + Previous Block-Hash + Time-Stamp)*

   Previous Block Hash = SHA-256(SHA-256(Block Header))

3. Compare the computed hash value with the target hash value and check whether the calculated hash value is less than the target value.

   Consider, for example, if the target hash-value should have four leading-zeros, and the computed hash (say) 0000e3b is valid. Otherwise, increment nonce and repeat this step with the updated hash-value (until the calculated hash is less than the target hash).

4. Once, the miner process finds the winner block (i.e., hash-value less than target-hash), it then updates all the nodes about this new block. The node checks (long checklist) the winning block and if more than 50% of the nodes agree, then add this block to the existing Blockchain and the miner receives newly generated tokens as mining-reward.

5. They finally updated the ledger for every node.

6. However, set the creation time for a block to 10 minutes (as per BTC-protocol). The increased difficulty arises when the leading zeros number increases in target-hash. In such a case, high computational power to find the right hash s required.

Hence, by applying hash functions in the Blockchain acts as a critical component to ensure immutability because to modify transaction details, all the previous transactions and their respective copies should be changed simultaneously. Otherwise, the hash values will not match the existing record copies in all the Blocks.

18.4.2   THE SEQUENCE OF CONSENSUS ALGORITHM

The consensus algorithm acts as the vital technique in creating new blocks and then to append them on the Blockchain. All the algorithms exhibit proof-of-stake (PoS), proof-of-work (PoW), and proof-of-authority (PoA):

*   Then place the set of transactions to be executed in transactions pool.
*   Transaction genesis (main) block performs the execution by receiving the information of all the transactions, holds the average workload of these fragments in the "Work-Status-Model" and present workload and assigns each piece a transaction to process.
*   In every fragment, the transactions list is verified and ordered regarding the Check-List of 20-items.
*   Then add the verified transactions to the checklist, which is then updated to the block sequentially and distributed to the connected nodes.
*   During the block distribution process, then track the workload status of the present shared/completed, and then write the verified throughput in the workload status model.
*   Finally, delete all the processed transactions from the pool of transactions.

The steps involved in the blockchain network nodes:

*   Request to run a smart contract on the block node.
*   Computed results are verified and then passed to the consumer.
*   The consumer matches the received results to examine the transaction validity.

- Transactions set are ordered into blocks and distributed to the linking nodes.
- The linking nodes accept the verified block, confirm, and then save on the ledger.

The main attribute of blockchain: Merkle tree. The structure of Merkle tree supports for secure and efficient content verification in a large data block. Merkle trees generate a digital fingerprint of the complete transaction set by summarizing all transactions in the block and further enables the user to verify the transaction and check if the block is added or not.

Repeated hashing pairs of nodes create Merkle Trees unless only one hash is left (Root-Hash or Merkle-Root). Construct the tree using the bottom-up approach with an individual hash value for every transaction (Transaction-IDs).

The Merkle Root aggregates the related data to all the transactions and updates in the block header, as shown in Figure 18.9.

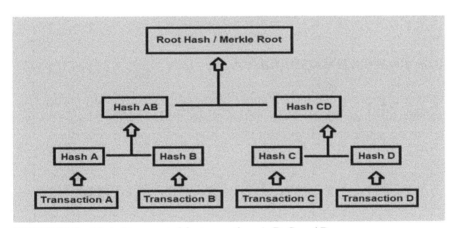

FIGURE 18.9   Merkel tree created for transactions A, B, C, and D.
Source: Reprint from: https://hackernoon.com/merkle-trees-181cb4bc30b4.

Working procedure of Merkle-tree:

- Instead of adding hash-values of all the transactions, just hash the leaves (transactions).
- Combine the leaves hash values and rerun the hash function (on combined hashed leaves) to create a first-level branch.
- Combine hash of the first-level branch and then run hash-function to get the second-level branch.

- Generate the single hash value referenced as Root-Hash with the repeated iterations.

Four major benefits of Merkle-Tree are:

- Easy monitoring of data validity and integrity.
- Merkle tree reduces the amount of memory space.
- The required proof/management needs only less amount of information transmitted over the network.
- Simplified payment verification (SPV): A path for verifying the transactions in a block without downloading an entire block.

Blockchain keeps its integrity and remains immutable due to its consensus mechanism; this mainly relies on two features: the "hashing" and the "proof-of-work" [14]. The proof-of-work (PoW) is a mathematical problem that guarantees that no user will be able to know up-front which node will be validating the transaction. Solving this problem is typically called "mining" and is also performed by members of the network.

18.5   PROPOSED METHODOLOGY

Figure 18.10 shows the methodology proposed to carry out the work. The fundamental phases are the Registration phase, the Detail Verification phase, the Testing phase, the Assignment phase, and the Farmer funding phase.

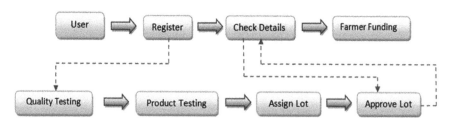

FIGURE 18.10   Micro financial assistance methodology (original).

Initially, to make a transaction in the Blockchain, the parties must get register to the system. On registration, individuals should furnish his or her identity by entering the details as per the birth certificate and other official documents that are released by Government Authorities rendering their name/address. Directly store this information onto the Blockchain.

The underlying technology uses Truffle for deploying the application and go-Ethereum (geth) for blockchain backend. Web3.js JavaScript provides a frontend API interface to interact and access the blockchain services.

In the quality testing stage, retrieved the farmer details from unique-ID assigned. Store these details as a structure using solidity-code onto the Blockchain. If the quality of the lot is up to the defined standard, then the lot will be approved, and update the information to the blockchain network. In product testing, customers can check the farmer's lot of details and status of the quality testing of the agricultural product. The customer enters the farmer-id and lot-number to see the details of the specific-lot. These farmers-profile and quality-report will be retrieved directly from the Blockchain.

Once the validity of the farmer is carried out, approval is given by the participating parties in the Blockchain for the farmer participation, the consumer interested in that particular lot, and the farmers can plan to fund them. The funding process, in the next supply chain process, can happen from the initial stage of cultivation. The direct connection between the farmer and consumer creates a strong bonding without any intervention of the middleman, thus reducing the processing cost. Pay the farmers they deserve and helps them to lead a better life. This funding process is kept private between the farmers and customers involved.

Figure 18.10 shows the complete flow of work to implement this work is shown in the methodology diagram.

18.6   IMPLEMENTATION DETAILS

18.6.1   ELEMENTS OF BLOCKCHAIN AND ITS IMPLICATIONS

In the presented work, we have used the ETH Blockchain to create a decentralized smart-contracts (SC_d) segment and the use of contract-compliance-checker (CCC) for deploying the decentralized smart-contract (SC_c) component:

1. **Contract-Compliance-Checker (CCC):** Adopt CCC, with its support for various features with the ease of integrating with a blockchain platform. It is an open-source tool intended for creating smart contracts and includes many RESTful service interfaces. At its core, it has permission to grants/removes rights, prohibitions, and obligations for the parties during the contract process execution.

To trigger CCC, from the following: (i) writes the contract rules and then is updated; (ii) load this file into the configuration/upload folder; (iii) deploy and instantiate CCC as a web-server, which waits for the events requesting for contract operation.

2. **Client Node:** The client node (business-partners) is responsible for hosting the request, like *BuyReq/SellReq* and funding. The SCc contract checks the request for validity and grants permission to access stores data. For instance, if the buyer wishes to access/retrieve the stores' data, the buyer should issue the appropriate operation request. Forwarded to the SCc and finally, SCc evaluates/verifies the operation concerning the business rules that are encoded in the contractual clauses and responds appropriately either accepted or rejected.

3. **Ethereum (ETH):** Design the prototype of the proposed work using ETH, which is an open-sourced Blockchain-based platform that supports SCs [36]. It supports the user ability to create applications by eliminating mediators, includes central servers to deposit information with less exposed to abuses from mediators and parties [37]. The work uses Solidity a contract-based, framework for implementing contracts, and Truffle a popular framework on the ETH platform. Geth (Go-Ethereum's) standalone CLI client used to run the ETH node on the network. A node running on the system allows farmers/consumers to execute transactions and communicate with SCs on the ETH blockchain. The digital identity of farmers/consumers is maintained using uPort's open identification system to register their identity on ETH, send/request credentials, signing transactions, and secure maintaining of keys and data. Few web services are accessed using RESTFUL-APIs.

18.6.2 SETTING UP SMART-CONTRACTS IN ETHEREUM (ETH) BLOCKCHAIN

Follow the seven steps to establish SCs in ETH blockchain:

➢ **Step 1: Creating/Running a Node in the Ethereum-Network:** To run full-ETH use the various command-line tools for programming languages (eth, geth, Pyeth, etc.). It provides several user interfaces like an interactive console, the command-line interface, an interactive console, and a JSON RPC server.

To run an ETH-node, we need to download its latest version with client-compatible and their specific OS and then install it on the system. On installation, it will start and connect to peers and the receiving blocks. In case to connect to "testnet," the chain opens the JSON configuration file and appropriately changes before triggering the client node through command-prompt. This point to the test network, which in turn makes the synchronous process faster. It is also possible to run anode by running the "Ethereum-Mist wallet-client," that helps to create addresses and send/receive transactions.

➢ **Step 2: Creating/Integrating-Wallets and Sending/Receiving-Ethers:** Post-running a node in the network within an ETH-client (with *geth*), one creates wallets and send/receive Ether and processed with *geth* through JavaScript-console or JSON-RPC using CURL.

Post network selection *"mainnet/testnet,"* prompts to choose or import an initially created wallet. On configuring the wallets send and received ethers from it.

➢ **Step 3: Creating Smart Contracts:** Write the SCs using Solidity similar to Serpent/JavaScript, which is the same as Python/LLL Lisp-based. Use browser-based IDEs to leverage the writing of smart contracts, like Cosmo, Ether.camp, etc. Ether-Camp IDE supports sandbox test-network with an auto-generated GUI for testing and transaction-explorer with *test.ether.camp*. Same explorer to carry live ETH-network with "frontier.ether.camp." Currently, there are readily available *"Dapp"* developing frameworks like Embark, Truffle, Populus, Dapple, etc., which provide IDE for writing Smart-Contracts.

➢ **Step 4: Compiling Smart-Contracts:** Smart-Contract, when written in Solidity, should use Solc (C++ libraries). Similarly, LLL, and Serpentine have their compilers. Dapp decentralized support frameworks (Embark/Truffle/Populus /Dapple) support a ready facility to compile Smart-Contracts. *eth_compileSolidity* supports to compile solidity-code via ETH-clients (geth)or with the JSON-RPC method. Post-compilation of smart-contract, store the generated.sol bytecode binary file in the bin folder.

➢ **Step 5: Deployment of Smart-Contracts on Ethereum:** For deploying Smart-Contract, we need to specify the amount of Ether that we wish to which s deposited in Smart-Contract, which is generated by compiler and source address in the hex bytecode format. The contract is signed using the node's wallet address or any other specified address. With post-deployment, we can get the contract's blockchain address with the ABI in JSON format of compiled contract's variables, methods, and events that are triggered.

Perform the transaction validation from the "*BlockExplorer*" via "*ethercan.io*" where the transactions are searched with the transaction id/address and are subsequently validate from (public key of nodes wallet) and to (contract's blockchain-address received during deployment) addresses, the Ether-value.

➢ **Step 6: Building the Dapps (Decentralized Applications):** Dapps has two essential parts: (a) the frontend; and (b) the smart contracts.

The front-end interface built with HTML/ JavaScript/CSS. The Smart-Contracts are created, compiled and deployed in the ETH network, which is initiated by Dapps using ETH JavaScript API (web3.js) and then performs some functionalities about the business rules defined.

➢ **Step 7: Running and Using the Dapps:** ETH supported Dapps owns community of involved parties. Identify the ownership by the crowd-sale mechanism which happens outset, wherein participating parties buy Dapp-tokens with the exchange of Ether. Two types of tokens are, usage-tokens (behaves like Dapp native-currency used to pay transaction fees and write to the Blockchain) and work-tokens (grants ownership-rights to the parties).

18.7 EXPERIMENTAL RESULTS AND DISCUSSION

The major participants involved in the presented system are farmer, quality assurance team, customer (buyer) and micro-financial investor.

Figure 18.11 depicts the registration form for farmers to participate in the supply chain process. Directly store the entered field details by the farmers on the Blockchain for further processing. The underlying application uses Truffle for deploying the presented application and go-Ethereum (geth) for blockchain backend. Web3.js (Java-script API) interface allows us to interact and access the blockchain services.

Figure 18.12 shows the MetaMask notification about the gas fee charged to carry out the transactions. MetaMask extension will detect and confirm the transaction to update on to the decentralized blockchain network.

FIGURE 18.11 Registration page (original).

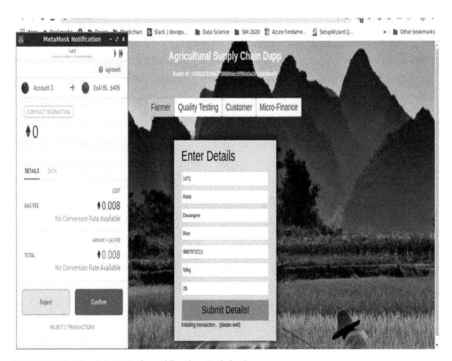

FIGURE 18.12 MetaMask notification (original).

Figure 18.13 reveals the Quality-Testing page with the block details generated as per the details stored on the Blockchain. After validating the details, it needs to be approved "Approve Details" for further processing and adding the farmers as the participant to the Blockchain.

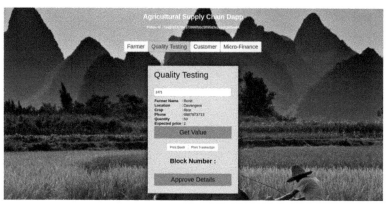

FIGURE 18.13   Quality-testing page (original).

Figure 18.14 shows the product details page that is part of the quality testing phase and Figure 18.15, retrieves the farmer details by farmers-Id and the time-stamp associated. Stored the particulars as a structure using a solidity code on the Blockchain. The product quality is tested based on the clauses defined in the Blockchain, and respond with an appropriate message by generating the MetaMask notification.

FIGURE 18.14   Product-details page (original).

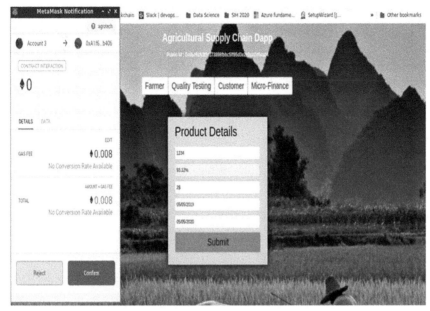

FIGURE 18.15 Notification page (original).

Figure 18.16 shows the customer page to check details and status of the quality-testing process of food products. The customer enters the farmer-ID and Lot-number to validate the details by directly retrieving from the Blockchain.

FIGURE 18.16 Customer page to check and validate the farmer/product information (original).

Figure 18.17 shows the farmer funding page used by customer to decide on the financial service support for specific farmer and further proceeds. Figure 18.18 shows the funding details and the amount the customer agrees to fund the farmer with the appropriate MetaMask notification.

The micro-finance form allows any customer to support financial assistance to farmers. Perform the funding process by providing the farmers public-id, product lot-number, and the funded amount.

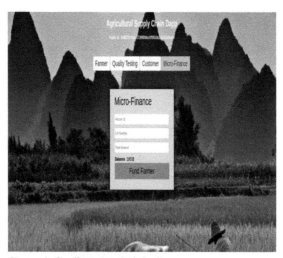

FIGURE 18.17   Farmer's funding page (original).

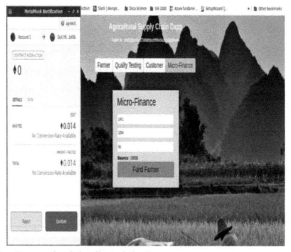

FIGURE 18.18   Farmer fund details (original).

18.8   CONCLUSION AND FUTURE SCOPE

The presented work demonstrates the current adoption of blockchain technology in various initiatives to establish a verified and trusted domain to implement more sustainable and transparent transactions, efficient distribution, integration of prime stakeholders in the food supply chain environment. However, various challenges still pertain, and issues need to be resolved, over the technical level.

To narrow the barriers on the usage of conventional farming practices, "Government must escort by an example" and encourage the digitization in the private/public domain. Government schemes must take initiatives to invest in educating and training programs to demonstrate the main benefits of blockchain adoption when applied to farm activities and also for financial assistance. From a policy viewpoint, various measures can support the widening of fascinating blockchain ecosystems in the area of farm activities and encouraging the general goal to optimize the competitiveness and to ensure the reliability of the Agri-food supply chain with smart contracts. The profitable sustainability of the current initiatives with Blockchain and the support system designed to help the farmers in the rural population still requires validation. Further, anticipating the outcome of the work presented to promote the demand of blockchain practice shortly when employed in the food supply chain practice and smart contracts with micro-financial services to bring a profound change in the lifestyle of the farmers by eliminating the middleman.

Summarizing, Identity blockchain as a rising technology regarding a transparent food supply chain, but various challenges and barriers exist, that hamper its wide recognition among consumers and farmers in supply chain systems. In the coming future, the Government should undertake such challenges to privatized sectors, to organize blockchain practice as a reliable, transparent, and secure and way to assure product safety and data integrity. Combining the blockchain practices with other emerging technologies can move towards a higher automation system to process the food supply chain with enhanced traceability and transparency.

The microfinance assistance for farmers has been a trademark in the economics sector over the past three decades. It has a great impact on rising rural farmers' out-of-poverty and is an essential tool for supporting financial inclusion and economic growth in small households.

They are leveraging the potential of the blockchain platform, to enhance security, transparency, and scalability and help businesses in a massive transformation of their supply chain network. Transparency and traceability are the major foundations in supply chain and logistics both at global and scalar business networks. Blockchain adds precise measures to supply chain in terms of in-efficiencies of the absence of inconsistent data and interoperability. It creates a tampered and transparent proof digital agro-market platform for farm activities/products.

In the future context, Blockchain will revolutionize the business practices in various organizations/industries, but its acceptance desires efforts and time. However, in the coming future, it is expected that many Government schemes will finally confirm on blockchain assistance and start to apply it for enhancing public/financial services.

KEYWORDS

- **agro-market**
- **blockchain**
- **Ethereum**
- **food supply chain**
- **micro-financial assistance**
- **smart-contracts**

REFERENCES

1. Hang, X., Tobias, D., Puqing, W., & Jiajin, H., (2020). Blockchain technology for agriculture: Applications and rationale. *Blockchain Technology for Agriculture Frontiers in Blockchain* (Vol. 3). Article 7. www.frontiersin.org (accessed on 29th July 2021).

2. Jorge, P. C. B., (2019). *Blockchain: Decentralization as the Future of Microfinance and Financial Inclusion.* https://www.researchgate.net/publication/331101506 (accessed on 29th July 2021).

3. Andreas, K., Agusti F., & Prenafeta-Boldu, F. X., (2019). The rise of blockchain technology in agriculture and food supply chains. *Trends in Food Science & Technology, 91*, 640–652.

4. Creydt, M., & Fischer, M., (2019). Blockchain and more-algorithm driven food traceability. *Food Control, 105*, 45–51.

5. Montecchi, M., Plangger, K., & Etter, M., (2019). It's real, trust me! Establishing supply chain provenance using blockchain. Bus. *Horiz., 62*, 283–293. doi: 10.1016/j. bushor.2019.01.008.

6. Sylvester, G., (2019). *E-Agriculture in Action: Blockchain for Agriculture (Opportunities and Challenges)*. Bangkok: International Telecommunication Union (ITU).

7. Vroege, W., Dalhaus, T., & Finger, R., (2019). Index insurances for grasslands – a review for Europe and North America. *Agric. Syst. 168*, 101–111. doi: 10.1016/j.agsy.2018.10.009.

8. Chod, J., Trichakis, N., Tsoukalas, G., Aspegren, H., & Weber, M., (2019). On the financing benefits of supply chain transparency and blockchain adoption. *Manag. Sci.* (forthcoming) (in press). doi: 10.2139/ssrn.3078945.

9. Dujak, D., & Domagoj, S., (2019). Blockchain applications in supply chain. *SMART Supply Network*, 21–46. Springer, Cham.

10. Hald, K. S., & Kinra, A., (2019). How the blockchain enables and constrains supply chain performance. *International Journal of Physical Distribution & Logistics Management.*

11. Harz, D., Gudgeon, L., Gervais, A., & Knottenbelt, W. J., (2019). Balance: Dynamic adjustment of cryptocurrency deposits. In: *Proceedings of the 2019 ACM SIGSAC Conference on Computer and Communications Security* (pp. 1485–1502). New York, NY: ACM.

12. Lucena, P., Binotto, A. P., Momo, F. D. S., & Kim, H., (2018). *A Case Study for Grain Quality Assurance Tracking Based on a Blockchain Business Network.* arXiv preprint arXiv:1803.07877.

13. Molina-Jimenez, C., Ioannis, S., Linmao, S., & Irene, N., (2018). Implementation of smart contracts using hybrid architectures with on- and off-blockchain components. *Conference: 2018 IEEE 8th International Symposium on Cloud and Service Computing (SC2).*

14. Maru, A., Berne, D., Beer, J. D., Ballantyne, P. G., Pesce, V., Kalyesubula, S., Fourie, N., et al., (2018). *Digital and Data-Driven Agriculture: Harnessing the Power of Data for Smallholders.* Global Forum on Agricultural Research and Innovation.

15. Perboli, G., Musso, S., & Rosano, M., (2018). Blockchain in logistics and supply chain: A lean approach for designing real-world use cases. *IEEE Access, 6*, 62018–62028.

16. Gatteschi, V., Lamberti, F., Demartini, C., Pranteda, C., & Santamaría, V., (2018). Blockchain and smart contracts for insurance: Is the technology mature enough? *Future Internet.* doi: 10.3390/fi10020020.

17. Hyperledger, (2018). *Bringing Traceability and Accountability to the Supply Chain Through the Power of Hyperledger Sawtooth's distributed Ledger Technology.* https://sawtooth.hyperledger.org/examples/seafood.html (accessed on 29th July 2021).

18. Jha, S., Andre, B., & Jha, O., (2018). ARBOL: *Smart Contract Weather Risk Protection for Agriculture.* https://www.semanticscholar.org/paper/ARBOL%3A-Smart-Contract-Weather-Risk-Protection-for-Jha-Andre/255c9377a89aa27c0b36d50b80628b6df4bb334d (accessed on 29th July 2021).

19. Kim, M., Hilton, B., Burks, Z., & Reyes, J., (2018). Integrating blockchain, smart contract tokens, and IoT to design a food traceability solution. In: *9th Annual Information Technology, Electronics and Mobile Communication Conference (IEMCON)* (pp. 335–340). IEEE.

20. Galvez, J. F., Mejuto, J. C., & Simal-Gandara, J., (2018). Future challenges on the use of blockchain for food traceability analysis. *TrAC Trends in Analytical Chemistry.*

21. Caro, M. P., Ali, M. S., Vecchio, M., & Giaffreda, R., (2018). Blockchain-based traceability in Agri-Food supply chain management: A practical implementation. *IoT*

Vertical and Topical Summit on Agriculture-Tuscany (IOT Tuscany) (pp. 1–4). Tuscany, Italy: IEEE.

22. Carrefour, (2018). *The Food Blockchain.* https://actforfood.carrefour.com/ Why-takeaction/the-food-blockchain (accessed on 29th July 2021).

23. Glazer, P., (2018). *An Overview of Cryptocurrency Consensus Algorithms.* Hackernoon. http://cs.brown.edu/courses/csci1800/sources/018_03_14_HackerNoon_ AnOverviewOfCryptocurrencyConsensusAlgorithms.pdf (accessed on 29th July 2021).

24. Braendgaard, P., (2018). *A Personal Look at the Early Days of Internet vs Blockchain Today.* Retrieved from: https://medium.com/@pelleb/personal-look-at-the-early-days-of-internet-vs-blockchain-today-590a98cb009f (accessed on 29th July 2021).

25. Brakeville, S., & Perepa, B., (2018). *Blockchain Basics: Introduction to Distributed Ledgers: Get to Know This Game-Changing Technology and How to Start Using it.* https://developer.ibm.com/tutorials/cl-blockchain-basics-intro-bluemix-trs/.

26. Infante, R., (2018). *Building Ethereum DApps: Decentralized Applications on the Ethereum Blockchain.* Shelter Island, NY. USA. Manning Publications.

27. Porru, S., Pinna, A., Marchesi, M., & Tonelli, R., (2017). Blockchain-oriented software engineering: Challenges and new directions. In: *Proceedings 39th International Conference on Software Engineering Companion* (pp. 169–171).

28. Ge, L., Brewster, C., Spek, J., Smeenk, A., Top, J., Van, D. F., Klaase, B., et al., (2017). *Blockchain for Agriculture and Food* (p. 112). Wageningen Economic Research.

29. IBM, (2017). *Leading the Pack in Blockchain Banking: Trailblazers Set the Pace.* IBM Institute for Business Value, The Economist Intelligence Unit.

30. Lierow, M., Herzog, C., & Oest, P., (2017). *Blockchain: The Backbone of Digital Supply Chains.* Oliver Wyman.

31. Liang, X., Shetty, S., Tosh, D., Kamhoua, C., Kwiat, K., & Njilla, L., (2017). Provchain: A blockchain based data provenance architecture in a cloud environment with enhanced privacy and availability. In: *International Symposium on Cluster, Cloud and Grid Computing.* IEEE/ACM Baltimore, MD.

32. Etherscan.io. (2017). *The Ethereum Block Explorer.* https://etherscan.io (accessed on 29th July 2021).

33. Jacobs, C., & Lange-Haustein, C., (2017). *Blockchain Und Smart Contracts. Zivil- Und Aufsichtsrechtliche Bedingungen.* ITRB.

34. Subramanian, H., (2017). *Decentralized Blockchain-Based Electronic Marketplaces.* Communications of the ACM, 61(1), 78–84. ACM.

35. Beall, G., (2017). *Blockchain Platforms to Keep Your Eye on in 2018.* Retrieved from: https://www.business2community.com/finance/5-blockchain-plat-forms-keep-eye-2018-01979716 (accessed on 29th July 2021).

36. Kasireddy, P., (2017). *The Synergies Gained from Building on Ethereum's Decentralized App Ecosystem.* Retrieved from: https://medium.com/swlh/the-synergies-gained-from-building-on-ethereums-decentralized-app-ecosystem-22a709a675d2 (accessed on 29th July 2021).

37. Macdonald, M., Liu-Thorrold, L., & Julien, R., (2017). *The Blockchain: A Comparison of Platforms and Their Uses Beyond Bitcoin.* The University of Queensland. https://doi.org/10.13140/RG.2.2.23274.52164.

38. Xu, X., Weber, I., Staples, M., Zhu, L., Bosch, J., Bass, L., Pautasso, C., & Rimba, P., (2017). A taxonomy of blockchain-based systems for architecture design. In:

Proceedings IEEE International Conference on Software Architecture (ICSA) (pp. 243–252). NY, USA. IEEE.

39. Tapscott, D., & Tapscott, A., (2016). *Blockchain Revolution: How the Technology Behind Bitcoin is Changing Money, Business, and the World.* New York: Penguin.

40. Hardjono, T., & Pentland, A. S., (2016). *Verifiable Anonymous Identities and Access Control in Permissioned Blockchains.* http://www.venturecanvas.com/wpontent/abs/1903.04584. (accessed on 29th July 2021).

41. Kleineberg, K. K., & Helbing, D., (2016). A 'social bitcoin' could sustain a democratic digital world. *The European Physical Journal Special Topics, 225*(17, 18), 3231–3241.

42. Chinaka, M., (2016). *Blockchain Technology – Applications in Improving Financial Inclusion in Developing Economies: Case Study for Small Scale Agriculture in Africa.* Doctoral dissertation, Massachusetts Institute of Technology.

43. Del, C. M., (2016). *Ethereum Executes Blockchain Hard Fork to Return DAO Funds.* Coindesk. https://www.coindesk.com/ethereum-executes-block-chain-hard-fork-return-dao-investorfunds (accessed on 29th July 2021).

44. Fanning, K., & Centers, D. P., (2016). Blockchain and its coming impact on financial services. *Journal of Corporate Accounting & Finance, 27*(5), 53–57.

45. Kaulartz, M., & Heckmann, J., (2016). Smart-contracts-anwendungen der blockchain-technologie. *Computer. & Recht, 32*(9), 618–624.

46. Mougayar, W., (2016). The *Business Blockchain: Promise, Practice, and Application of the Next Internet Technology.* Hoboken: Wiley.

47. Pilkington, M., (2016). Blockchain technology: Principles and applications. In: Olleros, F. X., & Zhegu, M., (eds.), *Research Handbook on Digital Transformations.* Cheltenham.

48. Shrier, D., Wu, W., & Pentland, A., (2016). *Blockchain & Infrastructure (Identity, Data Security)* (pp. 1–19). Massachusetts Institute of Technology – Connection Science.

49. Watanabe, H., Fujimura, S., & Nakadaira, A., (2016). Blockchain contract: Securing a blockchain applied to smart contracts. In: *2016 IEEE International Conference on Consumer Electronics (ICCE).* Las Vegas.

50. Underwood, S., (2016). Blockchain Beyond Bitcoin. *Communications of the ACM, 59*(11), 15–17. https://doi.org/10.1145/2994581. ACM.

51. Cachin, C., (2016). *Architecture of the Hyperledger Blockchain Fabric.* Workshop on distributed cryptocurrencies and consensus Ledgers.Chicago. https://www.zurich.ibm.com/dccl/papers/cachin_dccl.pdf (accessed on 29th July 2021).

52. Tian, F., (2016). An agri-food supply chain traceability system for China based on RFID & blockchain technology. In: *Proceedings 13th International Conference on Service Systems and Service Management* (pp. 1–6). NY, USA. IEEE.

53. Yuan, Y., & Wang, F. Y., (2016). Towards blockchain-based intelligent transportation systems. In: *Proceedings IEEE 19th International Conference on Intelligent Transportation Systems (ITSC)* (pp. 2663–2668). NY, USA. IEEE.

54. Swan, M., (2015). *Blockchain: Blueprint for a New Economy.* Sebastopol: O'Reilly and Associates.

55. Zyskind, G., Nathan, O., & Pentland, A. S., (2015). Decentralizing Privacy: Using blockchain to protect personal data. In: *Proceedings-2015 IEEE Security and Privacy Workshops, SPW* (pp. 180–184). San Jose. https://doi.org/10.1109/SPW.2015.27. IEEE.

56. Wood, G., (2014). *Ethereum: A Secure Decentralized Generalized Transaction Ledger* (p. 151). Ethereum project yellow paper. https://www.win.tue.nl/~mholende/seminar/references/ethereum_yellowpaper.pdf. (accessed on 29th July 2021).

57. Halldórsson, A., Kotzab, H., Mikkola, J. H., & Skjøtt-Larsen, T., (2007). Complementary theories to supply chain management. *Supply Chain Management: An International Journal, 12*(4), 284–296.

CHAPTER 19

Transformation of Higher Education System Using Blockchain Technology

PRADEEP TOMAR,[1] HARSHIT BHARDWAJ,[1] UTTAM SHARMA,[1]
ADITI SAKALLE,[1] and ARPIT BHARDWAJ[2]

[1]*Department of Computer Science and Engineering, University School of Information and Communication Technology, Gautam Buddha University, Greater Noida, Uttar Pradesh, India, E-mails: parry.tomar@gmail.com (P. Tomar), hb151191@gmail.com (H. Bhardwaj), uttamsharma.usc@gmail.com (U. Sharma), aditi.sakalle@gmail.com (A. Sakalle)*

[2]*Associate Professor, Department of Computer Science and Engineering, BML Munjal University, Haryana, India,*
Email Id: arpit.bhardwaj@bmu.edu.in

ABSTRACT

As cloud storage grows, companies are storing and handling vast volumes of data throughout this century. Large knowledge arrives from businesses, the Internet of Things (IoT), and unstructured channels. Nevertheless, the findings of an experiment are not relevant merely because large data and research methods are accessible. The idea that data offers valid information relies on the right data. Organizations who want to implement Blockchain in their organizations would then have to ensure that they have the appropriate big data and the highest standards because it can no longer be changed while on a Blockchain. Blockchain could be a catalyst for better data if done properly, leading to improved insights. The characteristics of blockchain extend to any sector and tackle questions regarding data quality and confidentiality, which hinder the usage and exchange of big data. Organization that wants to use big data and Blockchain will find new instruments that have been developed to support this, like BigchainDB built on MongoDB. The introduction of the features required for business growth, including scalability, database capability, and audit trails would allow Blockchain-based apps that follow organizational requirements simpler for companies to create. Finance is the first place to consider Blockchain applied to Big Data. All transactions in Bitcoin (BTC) are kept inside the Blockchain Network. The transaction data is not private though BTC provides a degree of anonymity. Blockchains have a lot of potential, especially when private data are involved. Blockchains are used in healthcare to help reduce the effects of cybercrime and data theft. Training of data attacks is sadly not that far behind financial and healthcare. In 2017, schooling witnessed the most data breach in all sectors, comprising 13% of all violations. Student data may seem worthless, even if it is a good commodity among cybercriminals since most small children and adolescents have limited financial knowledge. Because recording and processing information is simpler, Blockchain can be used to improve library and educational institution knowledge systems. Many institutions participate in the worldwide education market, and it becomes even more difficult to monitor the status and authenticity of the diplomas and use different approaches/ procedures. Today higher education agencies are also offering grants to campuses in various cities and nations. Significant security concerns are therefore considered in securing data transactions, such as student profiles and certifications. The blockchain technology is a potential answer to this issue, as in many other fields.

19.1 INTRODUCTION

Blockchain is termed as a public ledger that records and checks transactions automatically [1–4]. bitcoin (BTC) [5, 6], Ethereum (ETH) [7, 8] and other virtual currencies (which took a stroke this month) are driven by distributed ledger (DLT) technology [2, 9]. All the ways in which DLT can turn many industries are less well known. There are a variety of cases for a simple, verifiable transaction data registry, as DLT operates through a decentralized network to resist fraud.

The blockchain concept was first used by Satoshi Nakamoto in his thesis "Bitcoin: A Peer-to-Peer Electronic Cash System," which was published in the year 2008 [11]. In this chapter, BTC scheme was first proposed by Nakamoto, but the term blockchain was not used. The block and chain are instead described as the data structure for recording BTC transaction history. The "block" corresponds to the distributed data, while the "chain" is the cryptographic block string. A continuous transaction ledger is the block and the chain together. Overall, blockchain technology often applies to distributed blockchain-based accounting technologies [12, 13], including mutual consensus [14], anonymity, and smart contracts (SCs) [15, 16]. Technology Blockchain comprises three fundamental notions: transaction, block, and chain. The transaction involves the processing of the ledger, such as the entry or removing of the object, which often leads to adjustments in the ledger status; the block tracks all transaction data results throughout the time. The following method is used for Blockchain technology. First, a distributed ledger must be accessible on the network that enables only fresh data to be introduced. This assumes that no data from the representative can be lost, which guarantees that the data will not tap. The blocks are linked to a clock chain, with the hash value of the previous block retained by each block. When the entire network joins a new ledger contract, the contract data block is registered and linked to the chain by means of the digital signature elliptical curve algorithm (ECDSA) [17–19]. This is not possible to falsify or fake the records. In the meanwhile, transaction details are distributed throughout the network, and all network nodes are checked to make them unlockable. The blockchain platform thus has a decentralized [20], de-trusted [21], distributed data storage structure. This method means that transaction data cannot be abused and retraced and checked by cryptographic techniques. Distributed and shared data collection and mutual management throughout the network preserves the distributed existence:

1. **Decentralized:** A central node to verify and monitor transaction data based on a decentralized distributed P2P network [22] is not required for the blockchain technology. Network nodes could directly share data through a trust scheme that enhances data exchange efficiency. In the meantime, data from the entire network will not be impacted by node disruption.

2. **De-Trusted:** The cryptographic hash value-based connection block on the one side is generated by the Blockchain technique and guarantees the protection of transactions via the digital signature created by asymmetric cryptography [23].

3. **Reliable:** The blockchain ledger needs distributed storage, which ensures that any node may be supplied with a copy. The data integrity and durability are secured by this storage mode. Therefore, all transaction records are recorded on time stamps and can be tracked back to the source itself [24].

4. **Collectively Maintained:** All nodes in the network holds data of blockchain jointly [25]. Since node is not excluded from maintenance, a fault in the single node has little impact on the entire network data.

5. **Privacy Safe:** By exposing node identification, the data is transmitted using the public key and the private key because of the digital signature algorithm. Throughout the transmitting process, the recipient becomes invisible. Blockchain technology provides the strongest approach to online education problems with strong reputation and stability. Without third-party supervision, blockchain will have absolute, transparent online education learning records, guaranteeing a fair certificate of course credits. On the one side, intelligent contracts will increase the flexibility of online learning sharing, while cryptographic data processing safeguards consumer privacy. Therefore, the blockchain technology is a promising attempt at online training [26].

The certificate plays a critical role in education, with the certification being strong proof for the pupil who successfully completes and completes the course. Nonetheless, certain records grant certificates individually with all of the institute's information so the individual can delete them at the end of the day without discussing their successes with others. The blockchain offers various credential programs by providing block certificates and getting the trust of all the members (institute, learner, third party).

The authenticity of educational credentials is of special interest to clients and those who ought to validate such diplomas. Perhaps that there are no longer issuing bodies, or that they do not hold accurate documents. In all instances, it may be challenging to check the validity of diplomas. More institutions participate in a worldwide education market, and the various procedures for checking the status and authenticity of diplomas are becoming harder to keep track. As in several other fields, the blockchain is a potential approach to the question of legitimacy, so it may boost the actual path of education for diplomas.

19.2   RELATED WORK

Lizcano et al. [27] assesses the advantages of blockchain (or distributable leader) technologies and promotes a democratic paradigm of confidence in academic cryptocurrency transactions. Blockchain is used in this approach to manage content, teaching, and skill transactions, which students, teachers, and managers decided that the "distance" between academics and workers would once and for all be eliminated. The goal of this chapter is to address the complexities of higher education, which is progressively decentralized, free, and all-round. Any training organization should adopt the model proposed to tailor its teaching to the unique needs of employer-validated skill profiles in the field. This concept is checked by the test with above appropriate performance.

Shen et al. [28] introduced about the online student quiz, which is used to implement higher education computerization and modernization. A web questionnaire based on the dual-layer Blockchain alliance is introduced for the remedy of the question that the scoring method is transparent, inequality, and the final outcomes are quick to modify. The software guarantees that students' answers are reviewed openly, and the response documents cannot be violated but monitored. Furthermore, group signature is used in blockchain applications to resolve the problem of anonymous abuse. Furthermore, by adding the Prime Chain Index to the subchains, the storage pressure of the substring nodes can be reduced further.

Harthy et al. [29] discusses about campuses in various cities and foreign nations are being sponsored by higher education organizations. Significant protection issues are also taken into consideration when protecting electronic transfers, including student accounts and certifications. Therefore, this research focuses on study covering the possibility of Blockchain in schools of education. Based on higher education analysis, the finalized recommendation will be produced.

Mikroyannidis et al. [30] examines the applications in data science training of Smart Blockchain Badges. We explore how smart blockchain badges will provide teachers with customized feedback based on their research accomplishments to advance their careers in data science. This research aims to enhance accreditation of data science by implementing a comprehensive Blockchain-based framework. A comprehensive, accessible, and clear accreditation program would help graduates, and advice on employment can suit their abilities and that improve their careers. This work thus helps to bridge the gap in data science by connecting data science education with industry.

Liu et al. [31] discusses about substantial intelligence asymmetry between colleges and employees occurs in the existing partnership between education and business. For student credit schemes and the historical data chain, it is important to build and preserve urgently. The chapter utilizes the simplicity and non-tempering capabilities of the decentralized network to incorporate the Hyperledger blockchain platform as a shared education and business program. With its certificate authority and transactions in the Hyperledger Framework, the system simulates the roles of universities and companies on the system and enables universities and companies to distribute information transparently. This enables the student's information and knowledge information to be symmetrized, business recruitment requirements and recent trends of the market. This represents a major attempt to use blockchain technology as a pilot application of education and industry collaboration for technologies deployment.

Rahardja et al. [32] research is intended to study the technological impact of blockchain in Indonesia's higher education system. Among other technological areas, Blockchain technology is broadly deployed, coupled with widespread application and adoption of schooling. Few private and public schools and tertiary institutions have, however, been in Indonesia. Indonesian societies and legislatures are grappling with increasing education efficiency around the world and with the challenging transformation of the digital transition 4.0. The policy insists on investments in intellectual resources and social resources in conjunction with the RIRN (Rencana Induk Riset Nasional). A response to educational problems at the state level may be very likely to be sought with blockchain technologies. Further, the creation of a digital intelligence portal for Indonesia's tertiary education will be carried out utilizing an advanced technology architecture.

Vidal et al. [33] discusses about the validity of academic degrees is of considerable importance in confirming the genuineness of academic degrees by employers and other authorities. The authorizing authorities may or do not have archives locked. In such circumstances, it is challenging to verify university diplomas. Most organizations are concerned in a worldwide education sector, so it is also hard to track the status of diplomas and to follow specific approaches/procedures. As in many other fields, blockchain technology is a potential approach to the above problem. In this chapter, the author proposed a method utilizing this technology used in Fernando Pessoa University. They also address the current problems raised by the usage of this technology for defense and deployment.

19.2.1 WHAT IS THE BLOCKCHAIN REVOLUTION?

The Internet links thousands of people worldwide today and is a perfect place to chat and share useful information online. But because it is meant to transfer knowledge and store it, nothing has been achieved to improve our way of doing business. If teachers give their students material such as an e-mail, lesson notes, a PowerPoint presentation, or an audio recording of a lesson, they must submit them a replica of it, not the original. They cannot print, say, money or diplomas, but printing versions of their PowerPoint file is all right for them. Therefore, we will rely on strong intermediaries for the exchange of value across the Internet of knowledge.

The Governments, the Banks, digital platforms, and universities (for example, Amazon, eBay, and AirBnB) have worked to establish our identity, ensure our confidence, and assist us in the procurement and transfer of assets and transactions. In general, they are doing a very nice job, but limitations are present. They use unified servers that are hackable. They not only prevent us from using our data to our benefit but often undermine our privacy. They collect our data. Sometimes these mediators are untrustworthy and are often sluggish. Around 2 billion citizens are disadvantaged and do not have adequate funds to support a bank account, not to mention education. Perhaps difficultly, they asymmetrically harness the benefits of the modern period.

What if there were a powerful internet – a highly safe and open global network, a server, or a database through which we could archive, share, and have shared trust without strong intermediaries. The blockchain is this. Collective, hard-coded self-centering to render our online communication safe and comfortable through this modern interactive platform in this indigenous country. Trust in technology is coded, so they name the Protocol of faith blockchain. Maybe they are a refugee who charges a big fee for bringing their children money home in their native land so they can go into college. Or maybe they are fed up with the lack of openness and responsibility of the higher education officials and legislators of their community. Or maybe they are a social network user who thinks it is worth anything-for them-for all the data they generate and their privacy matters. And as they publish, innovators create blockchain apps for these purposes. And these applications are just the start.

It turns out to be profoundly beneficial for every company, institution, government, and person. What about the business, a new capitalist pillar; they are creating the 21st-century businesses to be more feel like networks rather to be vertically organized industrial age hierarchies. With the advent of a digital, pair-by-peer infrastructure for identification, loyalty, credibility, and transactions, they should be able to reengineer complex organizational frameworks for creativity and value generated. Too well are students trained for such a prospect at the present college or university.

Soon, in the real universe, trillions of smart objects would be adaptive, react, connect, exchange essential information, and produce, buy, and sell their own resources using the Internet of Things (IoT). It turns out that this Internet of all would require a database of everything.

One of the greatest possibilities of Blockchain is to liberate us from an annoying paradox of wealth. There is a growth in the economy, but fewer people profit. We will adjust how income — and opportunities — are predicted first, because citizens everywhere will share more of this technology than seek to fix the issue of through socioeconomic inequality by redistribution alone. This platform may be used to spread the resources we produce more entirely [34].

19.3 HOW BLOCKCHAIN ESTABLISH TRUST?

Not all are digitally saved — from capital, securities, shares, intellectual property through to music and media, loyalty points, and student records — with high rates of encryption and global credibility. If a trade is made, millions of machines are placed worldwide. A multinational community is named miners 10–100 times bigger with huge computing capacity than Google worldwide. All transactions since past 10 minutes, like a heartbeat network, are grouped into one block every 10 minutes. The miners then compete to solve a difficult issue; everybody who resolves the problem receives a digital currency and a validation of the block. When it comes to the BTC ledger, the winner gets money.

This block is connected to the last block to build a blockchain. Each block is stamped on time, like a digital waxed seal. If they decide to decrypt a block and return the same BTC, they need to decrypt the blockchain along with all the previous iterations, not only on one machine but on millions of machines, in complete daylight, and using top-level cryptography. It is difficult to do. This is much better than the operating programs that we are using today [35].

The BTC ledger is one of these [36]. For example, the ETH blockchain was developed by a 22-year-old Canadian named Vitalik Buterin. ETH has some impressive know-how and services. For example, it helps programmers to create intelligent contracts, agreements converted to computer code lines that deal with the execution, processing, execution, and payment contracts between individuals. The network of ETH offers proposals to establish an alternative stock exchange and a modern political paradigm in which governments have responsibility for the people.

19.3.1   BLOCKCHAIN, IDENTITY, AND STUDENT RECORDS

The procedure of building individual identity normally begins with a certificate of birth provided by a state-approved medical professional. From that day on, the child begins to gather personal information that includes similar academic accomplishments.

The first challenge is to preserve the privacy and security of information digitally stored by these universities [37]. In 2013, the education consultative board (EAB) issued a collection of 157 methods for the collection of data concerning students and graduates from universities and colleges, and institutions succeeded. However, universities are not less susceptible than other major organizations, when it comes to data security.

The blockchain can be designed to archive practically all that is of human interest and significance, from birth certificates to records on schooling, Social Security Numbers, student loans, to anything else that can be represented in the database [38]. The blockchain uses a decentralized network with a shared key infrastructure (PKI). The blockchain of BTC is now the biggest civil PKI in the United States and the second biggest in the world. Department of Defense General Control Program. Its system has been tailored to the so-called free data-sharing framework by Sony Global Education, by which two parties worldwide will freely exchange official academy.

Validity is a second challenge [39]. At a period when evidence is available, transient, and mutable, it is becoming extremely necessary for workers to validate statements from a workplace. According to CareerBuilder, 57% of the applicants for a job have upgraded their skills, and 33% have lied about their degree in education. However, universities frequently charge transaction fees when it comes to processing requests. For example, at MIT, the cost "base of a transcript is $8.00." For every transcript ordered online, a $2.00 charge is applied. Sony's solution may make it easier and comparatively cost-free to transfer this information. Imagine how a program such as this might support, say, refugees pursuing education or finding jobs in a foreign world.

Time is a third problem. In the US, only 25% of students go to school at the residential campuses in full time. The rest is work and home balancing. These semi-students spend twice their graduation time and only 25% of the students graduate. Initiatives like OpenBadges, Blockchain Certificates, and Learning is Earning 2026 explore ways of rewarding students for their experience regardless of their environment [40].

19.4 BLOCKCHAIN AND THE NEW PEDAGOGY

There is a need to rediscover the pedagogical model in education [41]. Too many public schools and major universities do follow the broadcast learning paradigm, where the instructor is the broadcaster and the student seem to be the potential eager receiver. The message is one-way. It goes in this manner: "I am a professor, and I have knowledge. Get ready; it is coming." Their aim is to take the whole details and grasp it into memory so that when one test them, they can recollect it and give correct answer back. That is true about colleges, universities, and enterprises — and many of their professors and instructors that they try to upgrade the broadcasting model by means of essays, practical labs, working experiences, and even discussions in seminars. The lecture, however, remains overall dominant. Professors who stay relevant, must give up their traditional professorship and begin to listen and talk to students. To begin by engaging with immersive and self-paced computer learning curricula outside a classroom, students can master the ability (everything with the right or the wrong answer) and allow students and faculty to spend time in classes on what matters: conversation, debate, and team projects. One must also be clear about the educational purpose. It is not about abilities, and it is not about knowledge, to some degree. What counts today is the ability to learn for the whole of life. It is important to research, analyze, synthesize, contextualize, critically evaluate data and to collaborate and communicate. Overall, there is a need for a good model for classroom collaboration that make things feasible for the professors and students to make education easier. Consensus Systems (ConsenSys) [42], one of ETH's first software development firms, is an outstanding platform for classroom collaboration. In management science, it breaks new ground in the line of holacracy, a mechanism that determines the work to be done collectively rather than hierarchically. Holacratic principles include "dynamic responsibilities and not conventional job descriptions; dispersed authority and not delegated authority; straightforward rules and not office politics; and quick reiterations instead of big reorganizations," that everything explain how technology blockchain works. The ConsenSys layout, the way it generates value and the way it is handled, is different from the classroom as well as the traditional online course. The majority of ConsenSys members choose between two or five projects. No top-down assignments, and no boss. Everyone owns directly or indirectly a part of each project, and the ETH platform offers tokens that members can exchange for others and then turn into others currency. The goal is to reconcile independence with interdependence. The watchwords for the classroom are maturity, transparency, and consensus: recognize what needs to be learned, assign students who are willing and able to do so; decide on responsibilities, assignments, and rewards. Teachers as well as students require preparation in order to take part.

19.5   APPLICATION OF BLOCKCHAIN TECHNOLOGY IN ONLINE EDUCATION

The technical characteristics of the blockchain can inspire several good solutions to online education problems. Across the following areas of online education, some areas seek to incorporate blockchain technologies are as in subsections.

19.5.1   COMPLETE LEARNING RECORDS

The blockchain stores data in a distributed ledger, with sequential timestamps recording of data blocks. Unable to uninstall the latest data boxes. To prevent data from being tampered with, the cryptographical algorithm is used, which raises the issue of fraud. Many online learning sites are decentralized, offering poor quality courses. Worse, due to a lack of a formal recognition program, the outcomes of the learning are not widely acknowledged. No wonder that online education does not yield fruitful results. The sequential analysis of blockchain data offers an excellent way to document online learning data.

The learning data of the student, including time of study, course files, and test grades, can be chronologically registered on the blockchain, and any time mark can be identified. Cryptographic storage is used to ensure data integrity, reduce threats such as manipulations and deletions. With sharing, de-centralizing, and collectively managing the blockchain, any educational platform or entity would be able, across regions and times, to monitor students' learning paths. This increases the performance of the network and decreases hardware costs. In addition to the complete recording of learning information from students, the blockchain learning record prevents manipulation and deletion, offering a strong guarantee of the integrity of learning information for students. At the same time, learning knowledge can be transmitted via the network and conveniently accessed from the employer with encryption technology's reliability. The employer will know more about the students' learning status and review their details based on blockchain-based data. Blockchain technologies can also effectively prevent document fraud, academic fake credentials and other abuse in higher education and set up a reliable network for students, education providers, and employers [43].

19.5.2 LEGITIMATE CERTIFICATION ON LEARNING

Although the online educational platforms are extremely successful, after learning a few courses, students are not enthusiastic because learning outcomes are either not publicly accepted or officially accredited. It is due to the gap in the evaluation of learning outcomes. Currently, third-party organizations are refusing to certify online education. This style is unlikely in the future to fulfill the needs of the online education boom. If a student hunts a job, his certificates are archived by the employer on the educational platform or school. If the applicant loses a certificate, he or she will have to undergo a difficult and lengthy procedure in order to get another copy of the credential. Nevertheless, blockchain technology offers a simple, reliable way of certifying learning outcomes, university certification. Even if they are misplaced, student certificates can be easily checked [44].

In order to ensure data security and legitimacy, the blockchain requires a cryptographic algorithm to authentication. A collection of qualification results can be established on this basis. The first is to record the students' learning information based on blockchain technology on the online education platform or the publishing organization, including basic information, course results, date of issuance, etc., and encrypt the data by the platform or organization's own private key. The authenticated digital certificates are then distributed to students and other network recipients. In this case, the employer can validate the digital certificates by using the site or organization's public key.

Due to the non-temperable and cryptographic nature of blockchain data, blockchain technology may provide a reliable framework for the certification of learning outcomes. This program does not allow students to worry about loss of certificate, it will simplify the preparation of certificates, and the employer can spend less on checking the results of the training. Overall, online training results can be applied more effectively to work.

19.5.3 SHARING OF ONLINE EDUCATIONAL RESOURCES IN A DECENTRALIZED MANNER

There are currently numerous online education sites that provide different course materials. However, due to constraints such as education, copyright, and other issues, courses are not shared across the platforms. The user experience is very bad for learners of various types of courses as they must log into different platforms. It is also very difficult to learn skills in another school or discipline for higher education students. Owing to the absence of organized and effective usage, many standard course resources are wasted. The growth of the sharing economy (e.g., shared bicycles) demands that resources be used better. Online Resource sharing represents the growth of the future in the field of education. Blockchain technology can make use of SCs that allows online education to share resources. Smart contract is a software framework based on an encryption protection mechanism, as a typical application of blockchain technology. Without human involvement, it can complete complex transaction operations. The program also allows automated execution and verification. Intelligent contract technology can simplify transaction processes, carry outsmart, automated transactions and enhance transaction protection. transaction protection.

SCs require the construction of a broad online training resource sharing network. The online education platform will buy, settle, and acceptance output based on a smart contract without paying any workload accurately. Blockchain's distributed saving and co-keeping enables students to leverage the resources of various networks by logging into only one node in the blockchain network. In addition, educational data resource will not be disabled if attacks harm individual nodes, which is a significant assurance of protection in the data. In addition, the blockchain network can be combined with global knowledge networks, such as Wikipedia, research institutions, academic journals, and other educational data, by blockchain technology providing an international base of information. Such information tools can be accessed by nodes in any blockchain network. It increases learning performance dramatically and strengthens methods of learning [45].

19.5.4 PROTECTION OF INTELLECTUAL PROPERTY

As part of their work, professors publish frequently articles, publish research papers and reports. In the conventional method, when a professor starts his research, there is no way to know whether a similar academic study is underway. Moreover, the work itself has a great deal of piracy. By using blockchain, these issues are addressed. Blockchain could allow educators, without restricting the source material, to publish content openly while keeping track of reuse. Such a program will encourage teachers, similarly to the compensation they earn based on the amount of citations to their research papers, to be compensated based on the actual use and reuse of their teaching materials.

Students and organizations will then agree on the parameters of teaching materials. Teachers should announce the publication and linking of their resources and the other resources used to produce the content. The extent of reuse of their respective resources may be accorded to teachers with crypto-coins. Coins will be used in an open-ended situation, rather than spending, to assess the popularity of the author. Coins would have intrinsic value in a closed situation which would give rise to monetary compensation.

More advanced implementation could search resources automatically to identify the percentage of other reused resources. For instance, a "intelligent" (or self-executive) contract might allocate payment to authors based on the frequency of citations or usage. Writers do not have to go through intermediaries like research newspapers, which also restrict their usage through high access charges [46].

19.5.5 MICRO-CREDENTIALING AND OWNERSHIP OF LEARNING CREDENTIALS

Blockchain may provide a more secure and versatile framework for storing student certificates as they pass across their career and secondary education, from course to course. Via blockchain, credentials cannot be modified, offering a more secure storage system for a learning life. Blockchain enables the learner to hold personal details. In a safe place open to anyone who wants to check it, students obtain control and custody of all their educational data, including accreditation and job portfolios. Public blockchains allow the autonomy of individuals by allowing them the sole arbiter to access and use their data and personal data. Within an educational context, this concept is becoming synonymous with the empowerment of individual students, who have no need to use the educational institution as a trusted mediator, to own, control, and share their qualifications. It will be more and more critical as the sector continues to shift towards education based on competences. Blockchain enables students to store formal or informal proof of learning themselves, share it with a chosen audience, and guarantee instant monitoring. This ensures that students can conveniently share an individual curriculum vitae with employers. In the meantime, employers may minimize their workload as they do not have to review CVs and can simply test whether applicants possess the requisite skills [47].

19.5.6   TRANSFER OF CREDIT

Transfer of credit was another persistent problem for universities, which several times put students at a disadvantage because, for example, they discovered that they needed to retake courses in order to meet the requirements of a new school. It is also difficult for students to move to another university, while also retaining and demonstrating the courses taken in a previous college. This issue is much more apparent when a student chooses to transfer to an institution in another country, where there are possible barriers to language and different processes. Furthermore, record storage requirements differ, making it impossible to share interinstitutional documents.

19.6   HOW BLOCKCHAIN COULD DISRUPT THE EDUCATION INDUSTRY?

Unquestionably, Blockchain is a brilliant innovation. This technology has begun to support virtual currency, but blockchain is fast becoming clearly more than BTC. BTC's encrypted ledger technology is specifically intended to influence many industries in the future. Whether it is healthcare, banking, the media, or government, the blockchain technology has changed the industry revolutionarily. All sectors like education are certainly affected by the technology. The fact that the educational system is far from where it will be is clear. A great deal can be achieved in the educational sector with this technology. Let us see the disruption of the education sector by this revolutionary technology.

19.6.1   DIGITAL CERTIFICATES

In 2017, MIT awarded its graduates virtual diplomas. On their smartphones, students got them. It varied in a variety of ways from a standard paper diploma. In comparison to a paper certificate, blockchain certificate is never lost [48]. Nor will it be counterfeited. Wherever they go, the certificates will always stay with them. The need for the conventional clearinghouse or the university to issue the transcripts can be removed with digital blockchain certificates. Today, it would be much easier for students to transfer between universities if they were to save all certificates and badges on a blockchain. Barriers such as credit transfer do not obstruct the educational journey. It is almost impossible to change the data once the data is applied. All their certificates are therefore secure and safe. And with a button tap, they can access it.

19.6.2 MAKING EDUCATION CHEAPER

The U.S. educational debt of students is $152 trillion, according to Forbes, and it is attributed to 44 million lenders. The debt crisis for students has become very severe and the budget years following graduation are devastating. Owing to the high cost of tuition. The blockchain technology will help make education available to all [49].

Finally, in 2017, there were 81 ICOs that had been blocked by education, but which had vanished into the small. Dr. Edward from EIU.AC notes that Blockchain has applications which change the industries and LOL Token can solve real issues such as transfers and affordability. Blockchain is economical and scalable. The documents are incredibly secure because they are decentralized. The inherent architectural versatility enables records, documentation, and digital assets to be protected and stored without additional infrastructure and security expenses.

Institutions often save money on the expenses and legal obligations associated with data processing. The platform also enables members to own shared data independently from the open-source, saving vast quantities of digital system costs.

19.6.3 EXPANSION OF MOOC'S

There have been massive Free Online courses. By offering freely accessible certificates and enabling transactions to make microtransactions simpler in a course-by-course way, Blockchain technology will give these courses more credibility.

Around 101 million Huge Open Online Courses users are registered and as of 2018, 500 MOOC credentials will be available. In 2017, Coursera was seeing a 70% rise in paying customers. MOOC provider Udacity has 50,000 paid users. This is ample evidence for the fact that MOOCs give the users great value in paying for these courses. Blockchain technology may also contribute and provide the MOOC services with greater credibility simply by enabling transactions in order to fund smaller courses [50].

19.6.4   OPEN-SOURCE UNIVERSITIES

With the rising costs of education, cheaper alternatives are required. One of the reasons why the open-source University (OSU) model is common is the high education cost. OSUs allow one is to store their educational qualifications by attaching an air of authenticity to more accessible and alternative educational methods that still carry employers' weight. OSUs are also very accessible and have their own course. The credentials acquired via the OSUs are conveniently verifiable and available by storing them on a single Blockchain database. This also removes the risk of fraud on the diploma market [51].

19.6.5   TALENT MANAGEMENT

Colleges spend a huge amount on educating their students, but record-keeping their results is a challenging job because records must be maintained and held for a lifetime. The technology used to manage talent currently is vulnerable and not very safe to cyber-attacks. For talent management, Blockchain provides a transparent and secure framework. Also, after graduation and entering staff, students would have access to it [52].

19.6.6   SMART CONTRACT

When a set of requirements or criteria must be fulfilled, Blockchain can also be used for agreement execution. Intelligent contracts will reduce educational paperwork. For example, after a set of requirements have been met, they may be used to confirm attendance or task completions. SCs will lead to the transformation of education in many other ways.

The distributed technology ensures that the education environment would be overhauled. Blockchain's still in its infancy, and it is a long way from technology. What this technology has in store for the educational sector would be fascinating to see [53].

19.7 CHALLENGES IN HIGHER EDUCATION WITH BLOCKCHAIN TECHNOLOGY

While blockchain has demonstrated its capacities in the field of education, many challenges in the use of blockchain technology in education must be considered. Such challenges have been outlined in several main categories.

19.7.1 SECURITY OF DATA

Even though protection is the main function of blockchain technologies, it does not eliminate the possibility of malicious attacks. Maintaining and protecting privacy concurrently is extremely complicated, and when the future of an individual is at stake, this aspect is even more relevant (online approval of educational certificates and diplomas). Many blockchain implementations use private and public keys to ensure secrecy [54]. However, since data from each cryptographic key is publicly visible, and so the user's transactions can be connected to the disclosure of information by the user, blockchain cannot guarantee transactional privacy. Adequate protection and storage safety of all members' private keys are protection issue and must be treated amicably. Another problem of security is a data leakage, which can arise because the information is frequently updated.

19.7.2 SLOW-SPEED BLOCKCHAIN TRANSACTIONS

One of the problems that could be experienced in education systems is the slow speed blockchain transactions called the blockchain's scalability problem [55]. This is about the slow transactions in blockchain technology in large-scale blocks as transactions proceed and the record are growing. As a result of a large rise in users, the block size is also increasing. There are also a lot of data available in the education systems to monitor the students who travel from school to school continuously. The size of the block can be increased. Studies show that this problem can be a hindrance in blockchain education growth. There are many studies to address this issue, and several researchers have proposed ideas like Zilliqa, a new blockchain framework focused on shared technologies to break up the massive blockchain network into multiple shares validating transactions.

19.7.3 HIGH COST FOR ADOPTION OF BLOCKCHAIN TECHNOLOGY

The Blockchain technology is evolving and needs to be incorporated with the infrastructure of legacy. However, it can be very expensive to introduce and enforce this. In addition to the expense of implementing this, many blockchain technologies often expense the transaction or computational costs. The costs will increase as the number of blocks increases as the number of users increases to handle and store these large student's data.

19.7.4 SHARING OF CREDENTIAL DATA

In order to validate certificates, all institutions must consent to share their data. The question is, however, how are all organizations willing to share their data. Whether DLT / Blockchain is an upgrade on a more conventional, centralized booklet in many cases cannot be clarified. However, there is little evidence for corporate benefits for educational organizations, making it more difficult for blockchain to implement them. The possibility of changing the current training methods impacting both existing programs and the economy can be risky. How can approved organizations take the risk of offering private information of their students. If approved organizations do not consent to supply such details, further problems will occur during the process of authorization. Policymakers or other leaders in higher education will think how long the blockchain is to be trusted.

19.7.5 POOR USABILITY OF DATA

The usability of blockchain products needs to be enhanced by promoting interfaces so that people who have no technological experience can understand and use the technology easily. More work will therefore be expected in the field of usability testing. Good designs and simple terminology will enable the education sector to change blockchain. Through implementing further data protection checks for sites that rely on the user data, the data become inaccessible. There is a compromise between confidentiality and data access. Thus, access to these data becomes more difficult as users handle their own data in blockchain technology.

19.8   FIVE WAYS BLOCKCHAIN IS REVOLUTIONIZING HIGHER EDUCATION

Most of these things are Blockchain, the distributed ledger technology (DLT). The agreement is encouraged as it is a forum for record-keeping. It is transparent, as chain participants can download individual ledgers and validate them. And it is continuous because the ledgers cannot be changed. Blockchain, like education, is designed not only to translate content but the inherent value of that content. It is also no wonder that blockchain is the next chapter on higher education technologies in the lengthy process of decentralization.

The blank paper that defined a trusted, peer-to-peer (P2P), distributed-ledger network model initially embodied in the BTC cryptocurrency has only been published eight years ago. There is already several exciting projects and creativity going on as Blockchain is still early for usage in education and science:

1. **Student Records and Credentialing:** Training transforms into a design pattern, abilities, skills, and qualifications of each individual. The truth is, there is a lot of learning outside and inside the classroom. Throughout their lives, people learn from various sources. This life-long learning process does not almost encapsulate their academic record. Blockchain is the model that enables them to securely collect and share all their skill indicators, including academic records, badges, certificates, citations, recommendation letters, and similar. Consider your e-portfolio of your learning-oriented life experience as unchangeable, updated, and verifiable. Blockchain will be decisive in preventing fraud for the same reasons, provided they have a trustworthy means of establishing who they are [56].

2. **Partnership Platform:** The New York City College is one of several universities that evaluate BTC as a payment method. But the future effect of blockchain on the financial side of education goes far deeper than cryptocurrencies. Taken into consideration California's three-tiered college or SUNY system in New York, higher education has

evolved into a distributed model for some time now. Colleges and universities have recently formed consortia to pool their capital. One such effort-the Internet 2 Net + Initiative-provides the universities that participate with a range of applications, computers, and other cloud services. This group initiative ties in with Blockchain's P2P platform focused on transactions. Blockchain-based "SCs," distributed by more than two parties in encrypted digital transactions, could be used to guarantee speed and transparency.

3. **Copyright and Digital Rights Protection:** It is perfect to support academics, instructor leaders, and other senior directors build and distribute intellectual properties and control the way Blockchain can handle, exchange, and secure digital information. Blockchain Professors may, for example, be compensated for the real usage and reuse, including if they are recognized depending on cites in academic papers and ratings of their teaching materials. The creation of "online information servers" would also be important for Blockchain — what is now considered as libraries. The State San Jose University is the pioneer of the campaign Library 2.0. Blockchain can be used, among other fields, to curate digital assets and preserve digital rights.

4. **Innovation Learning Platform:** The prism by which many students see their prospects for education in entrepreneurship. And many students perceive blockchain technology as preparedness for the future century.

Carolina FinTech Center has recently concluded its Blockchain Generation Competition (sponsored by Oracle) in a Regional Group of businesses covering North and South Carolina. Students undergo lessons in blockchain skills as well. Blockchain is a 501C3, non-profit company operated by students at Berkeley that provides local companies with blockchain based awareness, consultancy, and analysis. Otherwise, Cornell Blockchain's mission statement is to "create, advise, and endorse increasing blockchain ventures and to establish a group of innovators."

Around 30 years earlier I was in school technology. My dad worked at Apple and took Apple II home early. I worked with Steve Jobs and saw him speak about the improvement of technology through Apple II. Today, other innovations and social trends — influential between them — pledge to revise schooling. For starters, the Train to Win program at the Center for the future presents a revolutionary vision of the national learning environment.

Computing into the cloud. If Blockchain is a blockchain infrastructure and is convenient for educational organizations and their broad constituent communities to use, it displays findings on a scale which can contribute to radical reform.

19.9 CONCLUSION

In 2020, the global revolution of the new decade became one of the leading developments in technology. In higher education, blockchain has tremendous potential-improved protection, accessibility, performance, and easier exposure to certificates, documents, and workflows on campuses shows that this platform is intended to dismantle a slow to evolve market, which is historically focused paper, such as higher education. On the other side, we recognize the worries of the boards and workers of the colleges, which show huge challenges in their execution.

The blockchain introduction in higher education will result in SCs, paperless transactions, and substantial long-term resource savings in the automation of processes and services. Universities would have to change IT departments and infrastructures to understand blockchain's maximum potential and enjoy the benefits of blockchain on campus. This will in turn improve student experience and overall campus efficiency, thus enabling universities in an increasingly volatile market to remain competitive.

19.10 FUTURE AND ROLE OF BLOCKCHAIN TECHNOLOGY FOR HIGHER EDUCATION

Collaboration among educational institutions can be a major area in which blockchain is of major benefit. As mentioned above, various educational institutions are exploring blockchain as a secure and efficient means for documenting the academic achievements of their students. This requires the issuance of certificates for candidates, as well as the various grades and abilities they have obtained. A future study in this direction will explore how blockchain can help educational institutions cooperate [10].

The SCs usage allows education institutions for storing and exchanging academic information with their students, such as academic transcripts, details of programs, major/minor criteria, and academic's probation. Students are then eligible to take courses in any of the participating institutions. Joint training programs can also be available to educational institutions. This form of technology would increase the versatility of students by allowing them access to all other institutions' academic programs. Educational institutions must also contribute to the expense of operations through common facilities, programs, and training programs.

Job-driven education would be another area for the use of blockchain technology. The emphasis of job-oriented education is on the delivery of training programs that relate to current and future work requirements and can lead to jobs for the students involved. In promoting this form of education, Blockchain may play an important role. Businesses may use blockchain to exchange the necessary knowledge and skills. This information can be reviewed periodically by educational institutions and used to develop training programs that respond to business needs. Students can also use blockchain as a booklet to store their knowledge and skills. Recruitment agencies should review and assess candidates based on their qualifications and recommend such training programs.

The application of blockchain technology in the accreditation and enhancement of the quality of online education is also relevant for further study. Although there are many benefits of online education, such as reduced costs, accessibility, and versatility, there are also disadvantages. Accreditation and poor efficiency are at the root of these disadvantages. Many educational institutions are accredited and deliver high-quality online courses. This problem can be resolved by Blockchain. It is a shared forum for the safe and accurate exchange of knowledge between students, educational institutions, and accreditation agencies. Educational institutions can store online course records, online curricula, instructors, and accreditations. After completing a course, students can exchange ratings both with their instructors and with them.

KEYWORDS

- **blockchain technology**
- **cryptographic data**
- **digital signature elliptical curve algorithm**
- **distributed ledger**
- **education consultative board**
- **open-source university**

REFERENCES

1. Rella, L. (2020), *Blockchain*, 351–358.
2. Natarajan, H., Krause, S., & Gradstein, H., (2017). *Distributed Ledger Technology and Blockchain*. World Bank.
3. Nofer, M., Gomber, P., Hinz, O., & Schiereck, D., (2017). Blockchain. *Business & Information Systems Engineering, 59*(3), 183–187.
4. Gupta, S. S., (2017). *Blockchain*. John Wiley & Sons, Inc.
5. Antonopoulos, A. M., (2014). *Mastering Bitcoin: Unlocking Digital Cryptocurrencies*. O'Reilly Media, Inc.
6. Underwood, S. (2016). Blockchain beyond bitcoin. *Communications of the ACM, 59*(11), 15–17.
7. Dannen, C., (2017). *Introducing Ethereum and Solidity* (Vol. 1). Berkeley: A press.
8. Vujičić, D., Jagodić, D., & Ranđić, S., (2018). Blockchain technology, bitcoin, and Ethereum: A brief overview. In: *2018 17th international symposium infoteh-jahorina (infoteh)* (pp. 1–6). IEEE.
9. Seijas, P. L., Thompson, S. J., & McAdams, D., (2016). Scripting smart contracts for distributed ledger technology. *IACR Cryptology E-Print Archive*, 1156.
10. Castañeda, L., & Selwyn, N. (2018). More than tools? Making sense of the ongoing digitizations of higher education, 1–10.
11. Nakamoto, S., & Bitcoin, A., (2008). *A Peer-to-Peer Electronic Cash System*. Bitcoin-URL: https://bitcoin.org/bitcoin. pdf (accessed on 29th July 2021).
12. Appelbaum, D., & Nehmer, R., (2017). *Designing and Auditing Accounting Systems Based on Blockchain and Distributed Ledger Principles*. Feliciano School of Business.
13. Vijai, C., Suriyalakshmi, S. M., & Joyce, D., (2019). The blockchain technology and modern ledgers through blockchain accounting. *Adalya Journal, 8*(12).
14. Rückeshäuser, N., (2017). Do we really want blockchain-based accounting? *Decentralized Consensus as Enabler of Management Override of Internal Controls*.
15. Sillaber, C., & Waltl, B., (2017). Life cycle of smart contracts in blockchain ecosystems. *Datenschutz und Datensicherheit-DuD, 41*(8), 497–500.
16. Rozario, A. M., & Vasarhelyi, M. A., (2018). Auditing with smart contracts. *International Journal of Digital Accounting Research, 18*.
17. Johnson, D., Menezes, A., & Vanstone, S., (2001). The elliptic curve digital signature algorithm (ECDSA). *International Journal of Information Security, 1*(1), 36–63.
18. ShenTu, Q., & Yu, J., (2015). *A Blind-Mixing Scheme for Bitcoin Based on an Elliptic Curve Cryptography Blind Digital Signature Algorithm*. arXiv preprint arXiv:1510.05833.
19. Bi, W., Jia, X., & Zheng, M., (2018). *A Secure Multiple Elliptic Curves Digital Signature Algorithm for Blockchain*. arXiv preprint arXiv:1808.02988.
20. Ali, S., Wang, G., White, B., & Cottrell, R. L., (2018). A blockchain-based decentralized data storage and access framework for Pinger. In: *2018 17th IEEE International Conference on Trust, Security and Privacy in Computing and Communications/12th IEEE International Conference on Big Data Science and Engineering (TrustCom/ BigDataSE)* (pp. 1303–1308). IEEE.
21. Lu, Y., (2019). The blockchain: State-of-the-art and research challenges. *Journal of Industrial Information Integration*.

22. Waterhouse, S., Doolin, D. M., Kan, G., & Faybishenko, Y., (2002). Distributed search in P2P networks. *IEEE Internet Computing*, (1), 68–72.
23. Ren, D., Wang, J., & Cheng, G., (2018). A Summary research on the security of IoT based on blockchain technology. In: *2018 Joint International Advanced Engineering and Technology Research Conference (JIAET 2018)*. Atlantis Press.
24. Lemieux, V. L., (2017). Blockchain and distributed ledgers as trusted recordkeeping systems. In: *Future Technologies Conference (FTC)* (Vol. 2017).
25. Sun, H., Wang, X., & Wang, X., (2018). Application of blockchain technology in online education. *International Journal of Emerging Technologies in Learning (iJET)*, *13*(10), 252–259.
26. Feng, Q., He, D., Zeadally, S., Khan, M. K., & Kumar, N., (2019). A survey on privacy protection in blockchain system. *Journal of Network and Computer Applications, 126*, 45–58.
27. Lizcano, D., Lara, J. A., White, B., & Aljawarneh, S., (2020). Blockchain-based approach to create a model of trust in open and ubiquitous higher education. *Journal of Computing in Higher Education, 32*(1), 109–134.
28. Shen, H., & Xiao, Y., (2018). Research on online quiz scheme based on double-layer consortium blockchain. In: *2018 9th International Conference on Information Technology in Medicine and Education (ITME)* (pp. 956–960). IEEE.
29. Al Harthy, K., Al Shuhaimi, F., & Al Ismaily, K. K. J., (2019). The upcoming blockchain adoption in higher education: Requirements and process. In: *2019 4th MEC International Conference on Big Data and Smart City (ICBDSC)* (pp. 1–5). IEEE.
30. Mikroyannidis, A., Domingue, J., Bachler, M., & Quick, K., (2018). Smart blockchain badges for data science education. In: *2018 IEEE Frontiers in Education Conference (FIE)* (pp. 1–5). IEEE.
31. Liu, Q., Guan, Q., Yang, X., Zhu, H., Green, G., & Yin, S., (2018). Education-industry cooperative system based on blockchain. In *2018 1st IEEE International Conference on Hot Information-Centric Networking (HotICN)* (pp. 207–211). IEEE.
32. Rahardja, U., Hidayanto, A. N., Hariguna, T., & Aini, Q., (2019). Design framework on tertiary education system in Indonesia using blockchain technology. In: *2019 7th International Conference on Cyber and IT Service Management (CITSM)* (Vol. 7, pp. 1–4). IEEE.
33. Vidal, F., Gouveia, F., & Soares, C., (2019). Analysis of blockchain technology for higher education. In: *2019 International Conference on Cyber-Enabled Distributed Computing and Knowledge Discovery (CyberC)* (pp. 28–33). IEEE.
34. Halaburda, H., (2018). Blockchain revolution without the blockchain? *Communications of the ACM, 61*(7), 27–29.
35. Tapscott, D., & Tapscott, A., (2017). How blockchain will change organizations. *MIT Sloan Management Review, 58*(2), 10.
36. Badertscher, C., Maurer, U., Tschudi, D., & Zikas, V., (2017). Bitcoin as a transaction ledger: A composable treatment. In: *Annual International Cryptology Conference* (pp. 324–356). Springer, Cham.
37. Zyskind, G., & Nathan, O., (2015). Decentralizing Privacy: Using blockchain to protect personal data. In: *2015 IEEE Security and Privacy Workshops* (pp. 180–184). IEEE.
38. Jacobovitz, O., (2016). *Blockchain for Identity Management*. The Lynne and William Frankel center for computer science department of computer science. Ben-Gurion University, Beer Sheva.

39. Funk, E., Riddell, J., Ankel, F., & Cabrera, D., (2018). Blockchain technology: A data framework to improve validity, trust, and accountability of information exchange in health professions education. *Academic Medicine, 93*(12), 1791–1794.

40. Albeanu, G., (2017). Blockchain technology and education. In: *The 12ᵗʰ International Conference on Virtual Learning ICVL* (pp. 271–275).

41. Gräther, W., Kolvenbach, S., Ruland, R., Schütte, J., Torres, C., & Wendland, F., (2018). Blockchain for education: Lifelong learning passport. In: *Proceedings of 1ˢᵗ ERCIM Blockchain Workshop 2018 European Society for Socially Embedded Technologies (EUSSET).*

42. Kokina, J., Mancha, R., & Pachamanova, D., (2017). Blockchain: Emergent industry adoption and implications for accounting. *Journal of Emerging Technologies in Accounting, 14*(2), 91–100.

43. Ocheja, P., Flanagan, B., & Ogata, H., (2018). Connecting decentralized learning records: A blockchain based learning analytics platform. In: *Proceedings of the 8ᵗʰ International Conference on Learning Analytics and Knowledge* (pp. 265–269).

44. Bond, F., Amati, F., & Blousson, G. (2015). Blockchain, academic verification use case. *Buenos Aires.*

45. Zheng, Z., Xie, S., Dai, H., Chen, X., & Wang, H., (2017). An overview of blockchain technology: Architecture, consensus, and future trends. In: *2017 IEEE International Congress on Big Data (BigData Congress)* (pp. 557–564). IEEE.

46. Sharples, M., & Domingue, J., (2016). The blockchain and kudos: A distributed system for educational record, reputation and reward. In: *European Conference on Technology Enhanced Learning* (pp. 490–496). Springer, Cham.

47. Lord, M., (2019). *Show What They Know, 28*(7), 34–37. ASEE prism.

48. Holotescu, C., (2018). Understanding blockchain opportunities and challenges. In: *Conference Proceedings of E-Learning and Software for Education (eLSE)* (Vol. 4, No. 14, pp. 275–283). Carol I" National Defense University Publishing House.

49. Grech, A., & Camilleri, A. F., (2017). *Blockchain in Education.*

50. Dai, Y., Li, G., & Xu, B., (2019). Study on learning resource authentication in MOOCs based on blockchain. *International Journal of Computational Science and Engineering, 18*(3), 314–320.

51. Scott, B., Loonam, J., & Kumar, V., (2017). Exploring the rise of blockchain technology: Towards distributed collaborative organizations. *Strategic Change, 26*(5), 423–428.

52. Chen, Y. C., Wu, H. J., Wang, C. P., Yeh, C. H., Lew, L. H., & Tsai, I. C., (2019). Applying blockchain technology to develop cross-domain digital talent. In: *2019 IEEE 11ᵗʰ International Conference on Engineering Education (ICEED)* (pp. 113–117). IEEE.

53. Cheng, J. C., Lee, N. Y., Chi, C., & Chen, Y. H., (2018). Blockchain and smart contract for digital certificate. In: *2018 IEEE International Conference on Applied System Invention (ICASI)* (pp. 1046–1051). IEEE.

54. Han, M., Li, Z., He, J., Wu, D., Xie, Y., & Baba, A., (2018). A novel blockchain-based education records verification solution. In: *Proceedings of the 19ᵗʰ Annual SIG Conference on Information Technology Education* (pp. 178–183).

55. Lemieux, V. L., (2016). Trusting records: Is blockchain technology the answer? *Records Management Journal.*

56. Jirgensons, M., & Kapenieks, J., (2018). Blockchain and the future of digital learning credential assessment and management. *Journal of Teacher Education for Sustainability, 20*(1), 145–156.

Index

Milton Keynes UK
Ingram Content Group UK Ltd.
UKHW050306161024
449569UK00037B/27